TECHNIQUES OF AUTORADIOGRAPHY

TECHNIQUES OF AUTORADIOGRAPHY

by

ANDREW W. ROGERS

School of Medicine, The Flinders University of South Australia, Bedford Park, South Australia 5042

Third completely revised edition

1979

ELSEVIER/NORTH-HOLLAND BIOMEDICAL PRESS
AMSTERDAM/NEW YORK/OXFORD

Published by:

ELSEVIER/NORTH-HOLLAND BIOMEDICAL PRESS
335 Jan van Galenstraat, P.O. Box 211
Amsterdam, The Netherlands

Sole Distributors for the U.S.A. and Canada:

ELSEVIER NORTH-HOLLAND INC.
52 Vanderbilt Avenue
New York, N.Y. 10017, U.S.A.

THIRD EDITION 1979

Library of Congress Cataloging in Publication Data

Rogers, Andrew W
    Techniques of autoradiography.

    Includes bibliographies and index.
    1. Autoradiography--Technique. 2. Radiobiology
--Technique. I. Title.
QH324.9.A9R63 1978   574'.028   78-16861
ISBN 0-444-80063-8

ISBN 0-444-80063-8

With 98 illustrations

Printed in The Netherlands

Preface to the Third Edition

In the five years that have passed since the preparation of the Second Edition, autoradiography has continued to make significant contributions to the biological sciences. Although these years have seen no major breakthrough in autoradiographic techniques, substantial advances in a number of areas have made it necessary to prepare a new edition rather than remain satisfied with minor alterations and a reprinting of the old edition. Many attempts have been made to improve the efficiency of autoradiography at all levels of work from the macroscopic to the electron microscopic: to provide a firm basis for assessing these proposed methods, the chapter on the photographic process has been substantially rewritten. The analysis of electron microscope autoradiographs has progressed steadily, and the advances in this area have had an influence on analytical techniques for light microscopy. Several emulsions have ceased to be available, notably those produced by Agfa-Gervaert, and new emulsions have appeared, such as the ^{3}H-Film of CEA-Verken and Product 129-01 from Eastman-Kodak. In many other places, new information or dissatisfaction with the clarity of the existing text have prompted me to rewrite.

The general format of the book remains the same. The first section deals with the principles underlying autoradiography. The second section is concerned with the collection and interpretation of data from autoradiographic experiments. The third section contains detailed descriptions of the major techniques available.

The text was prepared during a period of study leave spent in Britain, which provided an opportunity to meet many users of the technique. I am particularly grateful to Dr. M.A. Williams and Miss Dilys Parry for long discussions on electron microscope autoradiography; to Dr. P. Dörmer and Dr. W. Sawicki for their help with aspects of light microscope autoradiography; and to Professor S. Ullberg, Dr. S.A.M. Cross and Mr. R. McCullogh for bringing me up to date on the autoradiography of macroscopic specimens. The continued collaboration of **Dr. John M. England,** who is responsible for the section on the statistical analysis of autoradiographs, has been invaluable. To them and to the many others who have contributed to this book, I wish to express my thanks.

I am grateful to the following authors for permission to reproduce material which also appeared in the Second Edition: to Dr. H. Levi for Figs. 13–16 from work we carried out together; to Dr. M.M. Salpeter for either these figures themselves or the data from which they were derived,

Figs. 2, 5, 20–22, 24, 26–28, 67, 68, 72, 79–81 and 94–98; to Dr. L.G. Caro for Figs. 4 and 25; to Dr. Z. Darzynkiewicz for Fig. 43; to Dr. R. Ross for Fig. 45; to the Ilford Research Laboratories for Fig. 1; to Dr. L. Schwartz for Fig. 36; to Dr. W. Kinter for Figs. 46 and 47; to E. Leitz, GmbH, now a part of Wild Microinstruments, for Fig. 55; and to Dr. Falk for Fig. 33. The range-energy data in the Appendix is based on material published by Dr. P. Demers.

My thanks are due to the Editors of the following journals for permission to reproduce material already published by them, and which has already appeared in earlier editions of this book: *Proceedings of the Royal Danish Academy of Sciences*, Figs. 13–16; *Journal of Cell Biology*, Figs. 4, 25, 27, 79, 80 and 96; *Laboratory Investigation*, Figs. 26, 28, 95 and 97; *Journal of Anatomy*, Figs. 48 and 58; *Radiation Research*, Fig. 33; *Journal of Histochemistry and Cytochemistry*, Fig. 45; *Leitz Mitteilungen*, Figs. 54, 56 and 57; *Philosophical Transactions of the Royal Society*, Figs. 20–22; *Journal of Microscopy*, Figs. 60–62, 64 and 65; *Journal of Clinical Investigation*, Figs. 46 and 47; *Journal of Endocrinology*, Fig. 49; Academic Press, Fig. 34.

Of the material that is new to this Edition, the following figures have very kindly been provided by others. I am grateful to the authors and editors for permission to use their material: for Fig. 6, Dr. R. Rechenmann and *Journal de Microscopie et de Biologie Cellulaire*; for Fig. 23, Dr. S.A.M. Cross; for Fig. 88, Dr. P. Dörmer, and for Figs. 32, 63 and 69, Dr. P. Dörmer and Springer-Verlag, Berlin; for Fig. 35, Professor M.M. Salpeter and Van Nostrand Reinhold; for Fig. 51, Dr. D. Clarkson; for Fig. 52, Dr. P.M. Frederik and *Journal of Cell Science*; for Figs. 73, 74 and 90, Miss Dilys Parry; for Fig. 83, Professor S. Ullberg and *Science Tools*; and for Fig. 85, Mr. R. McCullogh.

I am deeply indebted to Mrs. Betty Hammond for organising the typing of the manuscript, and typing much of it herself under very difficult conditions.

The Flinders University,   ANDREW W. ROGERS
South Australia,
April, 1978

BIBLIOGRAPHY OF AUTORADIOGRAPHY

1 F. Passalacqua, *Biol. Latina*, 8, Suppl. 4 (1955) 7. Covering publications from 1924 to 1954.
2 G.A. Boyd, *Autoradiography in Biology and Medicine*, Academic Press New York 1955. Covering period up to 1954.
3 M.E. Johnston, *Univ. Calif. Radiat. Lab.*, *8400*, July, 1958. Covering period 1954 to 1957.
4 M.E. Johnston, *Univ. Calif. Radiat. Lab.*, *8901*, August, 1959. Covering period 1958 to 1959.

Professor Rogers studied at the London Hospital, where he obtained a B.Sc. in Anatomy, qualifying in Medicine in 1956. After seven years as a lecturer in Anatomy in Birmingham, during which he obtained a Ph.D. for a thesis on "Radio-isotope techniques in biological research", he spent two years as a Research Associate at the State University of New York, Buffalo. He then moved to the Neuroendocrinology Research Unit of the M.R.C. at Oxford, where he was awarded a D.Sc. He is at present Foundation Professor of Human Morphology at the Flinders University of South Australia.

CONTENTS

PART 1: THE THEORETICAL BASES OF AUTORADIOGRAPHY

CHAPTER 1

The Uses of Autoradiography

DEFINITION

Radiography is the visualisation of the patterns of distribution of radiation. In general, the radiation consists of X-rays or gamma rays, and the recording medium is a photographic film. The specimen to be examined is placed between a source of radiation and the film, and the absorption and scattering of radiation by the specimen produces its image on the film.

In autoradiography, the specimen itself is the source of the radiation, which originates from radioactive material incorporated in it. The specimen will absorb and scatter its own radiation, rather like similar radiation from an external source, so that the final image reflects in varying degree the pattern of distribution of radioactivity in the specimen and the modification of this pattern by the structure of the specimen. The recording medium which makes visible the resultant image is usually, though not always, a photographic emulsion.

THE HISTORY OF AUTORADIOGRAPHY

The first undoubted autoradiograph was obtained over a century ago. In 1867, Niepce de St. Victor[1] published an account of the blackening produced on emulsions of silver chloride and iodide by uranium nitrate and tartrate. It is curious that the blackening of photographic emulsions by uranium should have been seen so long before radioactivity was discovered. Niepce found this blackening to occur, even when the uranium salt was separated from the emulsion by sheets of glass of different colours. He interpreted his results in terms of luminescence.

In 1896, Henri Becquerel[2] repeated and extended Niepce's observations,

3

again in the belief that he was investigating mechanisms of fluorescence. He used crystals of uranyl sulphate, and showed that, after exposing them to sunlight, they were able to blacken a photographic plate through two layers of black paper. On one occasion, it seems that the sun did not shine for several days, and the uranyl sulphate remained in a closed drawer together with the photographic plate. This plate was also found to be blackened. Through this experiment, and the work of the Curies in 1898, radioactivity was first demonstrated. So autoradiography is in fact older than the knowledge of radioactivity itself, and contributed directly to its discovery.

After these first, almost accidental, autoradiographs of crystals of uranium salts, autoradiography remained a curious observation rather than a scientific technique for a quarter of a century. Not until 1924 did Lacassagne and his collaborators begin to use this response of photographic emulsions to ionising radiations in order to study the distribution of polonium in biological specimens[3,4]. Their work, which followed sporadic experiments by other investigators, was the first systematic and successful attempt to exploit the phenomenon observed by Becquerel as a means of observing the sites of localisation of radioactivity within biological specimens.

The development of autoradiography as a biological technique progressed very little from Lacassagne's work until after the 1939–45 war. Physicists were using photographic methods of recording and studying radioactive phenomena, but the application of similar techniques to biological material was limited by two factors. The first, and most important, was that the few naturally occurring radioactive substances were of very little biological interest. In the second place, autoradiography was dependent on emulsions prepared for photographic purposes: the few autoradiographs that were made involved pressing the specimen against a photographic plate. The first fifty years of autoradiography saw very little accomplished, apart from the study on a macroscopic scale of the distribution of various salts of radium, thorium, or uranium in a few plants and animals[5].

The revolutionary advances in physics during and after the Second World War brought a new impetus to autoradiography. The study of cosmic rays and of the particles which could increasingly be generated in the laboratory created the demand which led to the production of nuclear emulsions — photographic emulsions with specialised characteristics, which recorded the tracks of charged particles with greater precision and sensitivity. From the work of such men as C.F. Powell[6,7], a wealth of new information became available, both on the techniques of handling this new recording medium, and on the interpretation and analysis of the observed particle tracks. Several fundamental particles were first described on the basis of their tracks in nuclear emulsions.

Controlled nuclear fission brought a further impetus to autoradiography. The advent of the atomic bomb made it vitally important to know the distribution in plants and animals of the fission products of radioactive fallout. At the same time, new radioactive isotopes became available, opening up new possibilities in the investigation of biological systems. It is not surprising that the physicists and biologists working in these new fields should have adopted the techniques and emulsions of the particle physicists.

In 1940, Hamilton, Soley and Eichorn[8] demonstrated the uptake of radioactive iodine by the thyroid gland, and Leblond[9] soon afterwards prepared autoradiographs showing its distribution in the gland. These were still made with the old technique of placing the sectioned specimen in direct contact with a lantern plate. By 1946, Bélanger and Leblond[10] had evolved a technique with liquid emulsion that gave considerably better resolution. The molten emulsion was removed from the lantern slides, and painted on the specimens with a fine paintbrush. It was not long before Arnold[11], who was studying the retention of long-lived isotopes in the body, adapted this technique for use with nuclear emulsions. The following year (1955), Joftes and Warren[12] described dipping slides in molten nuclear emulsion, a technique which has been widely used, and is the basis for present-day liquid emulsion methods.

During the same decade, a parallel group of techniques was emerging. The lantern slide provided the starting point once again, and several authors[13,14] attempted to improve the contact between emulsion and specimen by stripping the emulsion off its glass support and applying it directly to the specimen. As was the case with liquid emulsions, a new technique employing a nuclear emulsion soon made its appearance. At the suggestion of S.R. Pelc, Kodak Ltd. began the manufacture of a special autoradiographic stripping film[15,16]. Since the publication of these two papers the stripping film technique, which brought great advances in resolution and in reproducibility over any of the techniques that had been tried up to that time, has been very widely used, and the stripping film, Kodak AR-10, is still available.

Both the liquid emulsion and the stripping film techniques produce emulsion layers a few microns thick over the surface of the specimen. Charged particles coming from the specimen only leave one or two silver grains to show their passage in this type of preparation. A few autoradiographers, however, saw possibilities in the more direct application to biology of the physicists' techniques of recording particle tracks. This approach is direct and simple in the case of α particles, which leave a very characteristic track that is easy to record and recognise. β Particles are not so amenable to track methods but, thanks to the pioneering work of physicists such as Hilde Levi[17-19] and C. Levinthal[20], β-track autoradiography has developed into a technique of great quantitative precision.

Cellular biology has been transformed by the development of the electron microscope. It was inevitable that attempts should be made to link the techniques of autoradiography to this new method of observing biological material. The first, and rather unpromising, autoradiographs viewed in the electron microscope were published in 1956 by Liquier-Milward[21]. Since then, new techniques have been proposed and new nuclear emulsions produced, to meet the requirements of this approach for extremely high resolution. It is now possible to resolve the site of incorporation of radioactive material to within 500–700 Å in favourable circumstances, and further improvements are certain to come.

Radioactivity is no longer the property of a few rare elements of only minor biological interest. An increasingly wide range of compounds is now available labelled with a radioactive isotope, opening up new possibilities in the study of living systems. In consequence, the blackening of an emulsion of silver halides by uranium salts observed by Niepce a century ago has evolved into a wide spectrum of techniques for recording and measuring radioactivity in biological material.

Although this book is primarily intended for biologists, autoradiography is not limited to biological specimens. Metallurgists, materials scientists and geologists have become autoradiographers, and problems ranging from the distribution of flour additives in biscuits to the detection of fingerprints on paper have been successfully tackled by autoradiography.

RADIOACTIVE ISOTOPES

What place do radioactive isotopes have in the study of living systems?

The majority of techniques available to the biologist are basically analytical. In other words, by their application a mixture of individuals (which may be molecules or cells or animals) can be separated into groups on the basis of some common similarity between the members of each group. The techniques of biochemical analysis, such as chromatography, for instance, can give detailed and quantitative information on the molecules out of which cells and cell products are made. The techniques of histology and histochemistry provide an analysis of the cells and tissues of the body on the basis of their appearance and chemical constitution.

In living systems molecules and cells, and even whole organisms, undergo rapid and often surprising transformations. An aminoacid may be synthesised into a protein, which is subsequently degraded, yielding the original aminoacid again. The large, multinucleate megakaryocyte forms the small blood platelets. By their very nature, analytical procedures are cumbersome and unreliable for the study of these transformations. The relative sizes of the aminoacid and protein compartments of a cell are a poor measure of the rate of transformation of the one into the other.

If, however, aminoacid molecules labelled with a radioactive isotope can be introduced into such a system, and their recognition combined with subsequent analysis, the synthetic pathways by which they are incorporated into specific proteins may be studied, and the rates of these transformations measured with considerable precision.

This is the basic pattern of the tracer experiment. Whatever the material under examination, the pattern is the same. A population that is heterogeneous is separated into homogeneous groups by an analytical technique after the addition to it of labelled members of one group. The possible transformations that may occur between that group and the others are then determined by looking for the distribution of radioactivity in the analysed population.

The chief value of radioactive isotopes in biological research has been to provide precisely this dynamic information to supplement the analytical techniques as they have been applied at every level from the molecular upwards. In every field of biology, the combination of radioisotope techniques with the analytical methods available has added another dimension to the observations that can be made. It is difficult to see how the work of the past thirty years on oxidative respiration, photosynthesis, or the control of protein synthesis by the nucleic acids, to quote only these examples, could have been carried out without the advances in nuclear physics that made radioactive isotopes so freely available.

In addition to this use of isotopes in the tracer experiment, radioactivity has also become the basis of a number of analytical techniques. The precision with which relatively small numbers of labelled atoms may be detected and measured has led to methods of analysis more sensitive than those otherwise available. In radioactivation analysis, for example, a method has developed for measuring the yield of certain elements in biological specimens at a sensitivity which is often far higher than that available with any other existing technique. The principle involved is neutron irradiation of the specimen in order to induce radioactivity in the element under study. The characteristic radiation from this activity is then detected and measured[22].

Another example of the use of radioactive isotopes as the basis for an analytical technique comes from histochemistry. In 1961, Ostrowski and Barnard[23] suggested the use of isotopically labelled enzyme inhibitors as histochemical reagents. Following their application to the tissue under study, the distribution of radioactivity could be observed by autoradiography. From this pattern, the distribution of the enzyme to which the inhibitor was bound could be inferred, and measurements of the radioactivity present in a particular cell or structure could be used to estimate the number of molecules of enzyme present there. With the development of specific antibodies and inhibitors against a very wide range of biological receptors, radioisotope cytochemistry has become a powerful technique.

The third main group of experiments involving radioactive isotopes is the study of the distribution and retention of ingested radioactive material in plants, animals and the human body. This is where autoradiography began, with the work of Lacassagne on polonium distribution, followed after World War Two by the study of ingestion and retention of fission products.

These are the three principal ways in which radioactive isotopes are used in studying living systems. The techniques available for recording and measuring radioisotopes will next be considered, to try and pinpoint the characteristics of nuclear emulsions which make them suitable for particular experiments, and to relate these features to the other methods of detecting radiation.

AUTORADIOGRAPHY IN RELATION TO OTHER TECHNIQUES OF DETECTING RADIOISOTOPES

The methods available for the detection and measurement of radioactivity can be classified under three headings.

The first of these is the group of electrical methods that depends on the production of ion pairs by the emitted radiation. The Geiger tube, the ionisation chamber, and the gas-flow counter are all examples of this approach, in which the ionisation caused by the passage of a particle or γ ray through the sensitive volume of the counter is recorded as an electrical pulse, which can then be amplified and registered.

The second group relies on the property, possessed by a number of materials, of absorbing energy from the incident radiation, and re-emitting this in the form of visible light. In a scintillation counter these minute flashes of light are detected and converted into electrical pulses by a photomultiplier tube, and may then be amplified and registered in the same way as in the ionisation detectors.

These two groups of techniques have much in common. A β particle entering the sensitive volume of the counter produces a transient effect which is converted into an electrical pulse. These pulses can be handled by data processing systems rapidly and reliably. The pulse counting techniques, whether based on ionisation or scintillation, can provide accurate measurements of the radioactivity in a source, but each measurement is a sum of the radiation entering the sensitive volume of the counter. Variations in radioactivity from one part of the sample to another are not detected.

Autoradiography differs from the pulse counting techniques in several important respects. Each crystal of silver halide in the photographic emulsion is an independent detector, insulated from the rest of the emulsion by its capsule of gelatin. Each crystal can respond to the passage through it of a charged particle, with the formation of a latent image that persists throughout the counting or exposure period, and is made permanent by the process of development. The record provided by the nuclear emulsion is cumulative, and spatially accurate.

By responding in this strictly localised fashion to incident charged particles, a nuclear emulsion is ideally suited to studies of the distribution of radioactivity within a sample, a function that the pulse counters cannot perform. But while the emulsion can and does respond in a quantitative fashion to radiation, it is often a slow and difficult process measuring the overall activity of a sample in this way, in comparison to the speed and simplicity of the pulse counters.

There is thus little point in autoradiographing a specimen that is homogeneous. But where the specimen is made up of different components the measurement of the radioactivity present in bulk samples by pulse counting techniques only gives a mean value for the whole specimen. An extreme case of heterogeneity within the specimen is provided by animal or plant tissues. Pulse counting from a gram of homogenised liver gives a rapid and accurate assessment of the total radioactivity present, but no evidence on whether it is intra- or extra-cellular, in parenchymal cells or other cell types, nuclear or cytoplasmic, and so on. The earliest experiments in autoradiography were concerned solely with the localisation of radioactivity within a specimen, and this probably remains the most frequent goal of biologists using nuclear emulsions.

The strict localisation of the response of a nuclear emulsion to those grains through which an incident particle passes, means that it is possible to study sources of very small size within a larger specimen. It is possible to observe the nucleus of a single cell, and determine whether or not it is labelled, or an individual chromosome in a squash preparation of a dividing cell.

It may be impossible to isolate sources as small as these from the tissue to present them to a pulse counting system. Even if microdissection is possible, the levels of radioactivity in such minute specimens are usually too low for detection against the background of the pulse counter. In such cases, there is no alternative to using the nuclear emulsion itself as a measuring instrument.

Nuclear emulsions have a very high efficiency for β particles, particularly those with low energies. Fortunately, many of the elements of interest to the biologist have suitable isotopes — tritium, carbon-14, sulphur-35 and iodine-125 for example. If the volume of emulsion to be examined is restricted to the immediate vicinity of the source, the effective volume of the detector may be as little as 100 cubic microns. Reducing the detector volume also reduces the probability of observing a background event, due to cosmic rays, for instance. It may be weeks or months before background in such small volumes of emulsion builds up to restrictive levels. It is possible, therefore, to combine a high efficiency for low energy β particles with very long counting times. With suitable techniques, radioactivity in sources the size of a single cell or smaller can be accurately measured at decay rates of 1 disintegration per day or less. By contrast, most commercially available pulse counters have backgrounds of 10–20 counts per minute.

In summary, then, autoradiography supplements the data provided by pulse counting techniques when the specimen is relatively large, indicating the distribution of radioactivity between the various parts of the specimen. With sources of cellular dimensions, pulse counting is often impossible, and measurements of radioactivity may have to be made by autoradiography. Quantitative methods of autoradiography have made great strides in recent years, and it is now often possible not only to compare the levels of radioactivity in microscopic sources, but even to measure them in absolute terms[24]. Absolute measurements of radioactivity have even been made on subcellular structures by electron microscope autoradiography[25].

The essential feature of autoradiography is that it allows the distribution of radioactivity to be related to the detailed structure of the specimen. For this reason, it often provides the biologist with information that no other technique can give. At the same time, the complex and varying geometry between specimen and emulsion introduces many difficulties into the interpretation of autoradiographs. In the next decade or so, interest is likely to swing more and more from purely cellular events to the problems of understanding the organisation of cells in complex tissues. Techniques such as autoradiography, which permit sites of particular molecules and rates of activity to be visualised while retaining tissue structure, are certain to continue developing and diversifying.

REFERENCES

1 N. Niepce de St. Victor, *Comp. Rend.*, 65 (1867) 505.
2 H. Becquerel, *Compt. Rend.*, 122 (1896) 420, 501, 689, 1086.
3 A. Lacassagne and J.S. Lattes, *Bull. Histol. Appl. et Tech. Microscop.*, 1 (1924) 279.
4 A. Lacassagne, J.S. Lattes and J. Lavedan, *J. Radiol. Electrol.*, 9 (1925) 1.
5 F. Passalacqua, *Biol. Latina*, 8 (1955) 7, Suppl. IV.
6 C.F. Powell and A.P.S. Occhialini, *Nuclear Physics in Photographs*, Clarendon, Oxford, 1947.
7 C.F. Powell, P.H. Fowler and D.H. Perkins, *The Study of Elementary Particles by the Photographic Method*, Pergamon, London, 1959.
8 J.G. Hamilton, M.H. Soley and K.B. Eichorn, *Univ. Calif. (Berkeley) Publ. Pharmacol.*, 1, No. 28 (1940) 339.
9 C.P. Leblond, *J. Anat.*, 77 (1943) 149.
10 L.F. Bélanger and C.P. Leblond, *Endocrinology*, 39 (1946) 8.
11 J.S. Arnold, *Proc. Soc. Exptl. Biol. Med.*, 85 (1954) 113.
12 D.L. Joftes and S. Warren, *J. Biol. Phot. Assoc.*, 23 (1955) 145.
13 A.M. McDonald, J. Cobb and A.K. Solomon, *Science*, 107 (1948) 550.
14 G.A. Boyd and A.I. Williams, *Proc. Soc. Exptl. Biol. Med.*, 69 (1948) 225.
15 I. Doniach and S.R. Pelc, *Brit. J. Radiol.*, 23 (1950) 184.
16 R.W. Berriman, R.H. Herz and G.W.W. Stevens, *Brit. J. Radiol.*, 23 (1950) 472.
17 H. Levi, *Exptl. Cell Res.*, 7 (1954) 44.
18 H. Levi, *Exptl. Cell Res.*, Suppl. 4 (1957) 207.
19 H. Levi, A.W. Rogers, M.W. Bentzon and A. Nielson, *Kgl. Danske Videnskab. Selskab, Mat.-Fys. Medd.*, 33, No 11 (1963).
20 C. Levinthal and C.A. Thomas, *Biochim. Biophys. Acta*, 23 (1957) 453.
21 J. Liquier-Milward, *Nature*, 177 (1956) 619.
22 J.M.A. Lenihan and S.J. Thomson (Eds.), *Advances in Activation Analysis*, Academic Press, London and New York, 1969.
23 K. Ostrowski and E.A. Barnard, *Exptl. Cell Res.*, 25 (1961) 465.
24 A.W. Rogers, Z. Darzynkiewicz, K. Ostrowski, E.A. Barnard and M.M. Salpeter, *J. Cell Biol.*, 41 (1969) 665.
25 M.M. Salpeter, *J. Cell Biol.*, 32 (1967) 379.

CHAPTER 2

Nuclear Emulsions and the Photographic Process

The responses of photographic emulsions to light and to the passage of charged particles are still the subject of research. It is relatively easy to construct a simple model of the photographic process which will allow the production of autoradiographs. Inevitably, however, as autoradiographers have tried to improve the sensitivity and precision of their techniques, they have pushed the photographic emulsions further towards their limits of performance, and have explored the theory of the photographic process in more and more detail. The working hypothesis needed by the scientist who wishes to understand the range of photographic techniques in current use in autoradiography is considerably more sophisticated now than it was ten years ago.

Many salts can be activated by light. The nuclear emulsions used in autoradiography are suspensions of crystals of silver bromide in gelatin. This type of emulsion began to be used around 1850. When light falls on such an emulsion a change is produced in the bromide crystals. This change is not directly visible but, on treatment of the emulsion with a developing agent, those crystals affected by light become converted into grains of metallic silver, whereas crystals which have not been illuminated remain unchanged. In photographic terms, the film that has been exposed possesses a "latent image", which can be converted into a true image by development.

We now believe that the latent image is due to the presence of metallic silver, situated at sensitivity specks inside the exposed crystals. In the presence of a developing agent this nucleus of metallic silver catalyses the conversion of the entire crystal into metallic silver. The bromide crystals which have not been reduced to silver are finally dissolved out of the emulsion by the photographic fixative, leaving a pattern of silver grains reproducing the pattern of light falling on the emulsion.

The tracks of charged particles are recorded in a nuclear emulsion by fundamentally the same process, except that the electrons initially liber-

13

ated in the crystal derive their energy from the passage of the particle through the crystal, and not from photons of visible light.

Each stage in this process will now be considered in rather more detail. Those readers who wish to explore the fundamental mechanisms of photography are referred to the short list of books at the end of this chapter[1,2].

SILVER BROMIDE CRYSTALS

Silver bromide crystallises in a cubic pattern, with the ions of silver and bromide regularly spaced, each silver ion surrounded by six bromide ions, and vice versa. Such crystals can be grown in controlled conditions to a considerable size. But light does not produce a latent image in populations of perfect crystals. Defects in the regular lattice of the crystal are essential for photosensitivity. They are introduced during manufacture by, for instance, inserting atoms of other elements such as sulphur in the crystal. The resulting dislocations in the regular lattice are known as sensitivity specks. They are often at the surface of the final crystal, particularly if the crystals are very small, but also occur more deeply.

The crystalline lattice of silver bromide has the property that electrons shared between adjacent ions can, on receiving very little added energy (of the order of 5 eV), be moved out of their orbit to enter a conductivity band, allowing them to travel large distances through the crystal from one ion to the next. This property is basic to latent image formation.

Probably the chief difference between the emulsions produced for light photography and those for nuclear work is the much higher ratio of silver bromide to gelatin in the latter. The higher this ratio becomes, the higher the density of the emulsion, and hence the greater its stopping power for charged particles — in other words, particle tracks will become relatively shorter.

For a given bromide-to-gelatin ratio, the silver bromide may form relatively few, large crystals, or be divided into a large number of smaller crystals. It is clear that the latter alternative will give more information per unit volume of emulsion: there are more possibilities for crystals to be activated or not, and the events occurring within the emulsion will be reproduced with greater accuracy, or resolution. Nuclear emulsions, designed to record the tracks of sub-atomic particles, tend to have smaller crystals than the emulsions designed for photography for this reason. Their crystals range in size from 0.02 to 0.5 μm diameter, and with equally wide variations in the sensitivity of the individual crystals. In general, the more precise the resolution of events in the emulsion that the experiment demands, the smaller the crystal diameter that will be appropriate. In X-ray films the crystal diameters are, in general, far larger, ranging from 0.2 to 3.0 μm. The significance of these differences between the two types of emulsion will be discussed further in relation to the definition of efficiency (see p. 88 and p. 314).

In general terms, the smaller the mean diameter of the silver halide crystals in an emulsion, the more difficult it becomes to achieve a high sensitivity. It is clear that a very small crystal can only contain a correspondingly small part of the trajectory of a charged particle, so that the total energy liberated within the crystal by the particle is relatively small. There must be a limiting crystal diameter below which the passage of a β particle is unlikely to be recorded. This stage may not yet have been reached but, as well be seen (p. 295), the emulsions with the smallest crystal diameters at present available are less sensitive than those with larger crystals.

Fig. 1 shows crystals of three Ilford emulsions, as seen under the electron microscope. The variation in crystal diameter from one to another is clearly seen. One other major difference between nuclear emulsions and those used for photography is the uniformity of their crystals. In conventional photography a certain amount of variation in size and in sensitivity from one crystal to another can be tolerated, and may even be a positive advantage, giving greater contrast rather than an all-or-none, soot-and-whitewash type of picture. In nuclear emulsions, however, uniformity in size and sensitivity are important characteristics of the silver bromide crystals.

GELATIN

The functions of gelatin are more important and more varied than might at first be thought. This complex protein is derived from the skins of cattle, and the emulsion manufacturers go to great lengths to ensure that their supplies are uniform in many different respects.

In the first place, gelatin forms a supporting medium for the silver bromide crystals, isolating them from each other so that one crystal may not catalyse the development of its neighbours. If an emulsion is centrifuged while molten, each crystal will apparently move within its own shell of gelatin, so that the concentrated silver bromide crystals will still remain isolated from each other. As a supporting matrix gelatin must permit the access of reagents to the crystals and, in the completed autoradiograph, allow for clear visibility of the developed grains. At the stage of formation of the latent image, and even more so later when

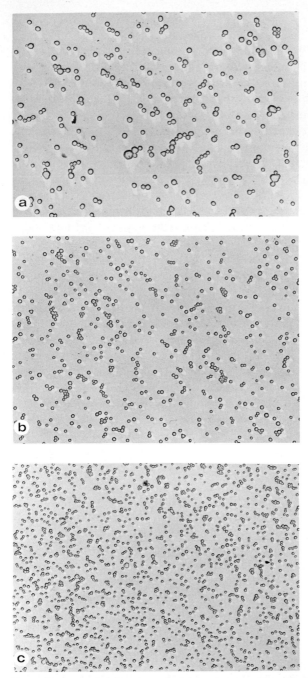

Fig. 1. Electron micrographs of silver halide crystals from Ilford nuclear research emulsions. The crystals have been shadowed so that the length of the shadow is twice the height of the crystal. 1a, G5 emulsion; 1b, K5 emulsion; and 1c, L4 emulsion. (× 7500). (Micrographs kindly made available by the Research Laboratory, Ilford Ltd.)

development is in progress, the deposition of metallic silver within the crystal is accompanied by the migration of bromine atoms out of the crystal. Gelatin acts as an acceptor of bromine, permitting these processes to take place without too high a probability of recombination between the bromine and the photolytic electron in the one case, and bromine and the silver atoms in the other.

In short, gelatin matters, and attempts to replace it with other materials have so far failed.

Two properties of gelatin deserve special mention. It undergoes sol–gel transformations reversibly at little above room temperature. It can thus be melted and poured conveniently. It is also capable of considerable changes in volume on hydration or drying. If it is spread on a plane support, these changes will usually appear as variations in thickness of the emulsion layer. Rapid drying may cause severe deformation of gelatin, and can even crack a glass support. The gelatin may also apply pressure to the silver bromide crystals, causing the appearance of latent images through mechanical stresses alone. All drying procedures should be gentle and slow, to achieve the best results.

Ilford add a plasticising agent to their emulsions to reduce the shrinkage that would otherwise occur on drying: this may be washed out of the emulsion in any procedure that involves soaking or diluting the emulsion before exposure. If this happens, the stress background will be considerably higher. Waller[3] has suggested adding 1% of glycerol to any aqueous bath in which the emulsion is soaked. This certainly reduces the stress background in nuclear emulsions, and this step is frequently included in the detailed techniques described later.

THE LATENT IMAGE

At a fairly simple level, the effect of light or radiation on a photographic emulsion is to cause a latent image, which is converted into the final image by development. The latent image can only be demonstrated by development. The most useful definition of the term "latent image" is a statistical one[1]. "The latent image is any exposure-induced change which increases the development probability from < 0.5 to $\geqslant 0.5$, under a specified set of development conditions." It would be much easier all round if we could define the latent image in more concrete terms, but exposure of an emulsion to light or charged particles causes a whole spectrum of changes in individual silver halide crystals: the larger the changes, the higher the probability that the crystals concerned will be developed by any given process. Changing the development process will change the population of exposed crystals that will be converted to developed silver grains, and hence, by definition, the number that contained a latent image.

Let us look in this section at the changes that can occur in exposed crystals, leaving to the next section consideration of how such changes can affect developability.

The smallest change occurring in an exposed crystal is that one electron within the lattice is raised to a higher energy level, entering a conductivity band and travelling through the lattice. Its journey usually ends at a dislocation in the lattice. Here, it is thought to be captured by a silver ion, which thus becomes an atom of metallic silver, withdrawn from the lattice. Such single electrons may be liberated by a single photon of light falling on a crystal. The single atoms of silver produced at sensitivity specks are very unstable. The atom of silver tends to lose its electron and to resume its place in the crystal, in a time which is dependent on the temperature. At room temperature this reaction has a half-time of about 1 second. At −70°C or lower, this return of the silver atom to the lattice is virtually halted.

If a second electron is raised to the energy level of a conductivity band before the first silver atom has rejoined the lattice, a second silver atom will result, with profound effects on the probability of either atom rejoining the lattice. In other words, a silver deposit of two or more atoms at a sensitivity speck is relatively stable, even at room temperature.

Many crystals contain more than one dislocation, or sensitivity speck. It appears to be quite a random process by which the first electron "selects" a sensitivity speck. The presence of one silver atom at a sensitivity speck seems to increase slightly the probability of the second electron ending its journey here; this probability becomes much higher once stable two-atom deposits are present. So such two-atom specks become much more efficient at collecting further electrons and silver ions.

For each silver ion that is withdrawn from the lattice to form metallic silver at a sensitivity speck, one ion of bromide is converted to an atom of bromine. Bromine is liberated from the surface of the crystal and migrates to acceptor sites in the gelatin.

If large numbers of electrons are liberated in a crystal almost simultaneously, many silver atoms will be formed at every sensitivity speck in the crystal. The amount of silver formed in this way may be so great as to become visible, if continuous exposure to bright light occurs, for instance.

Even the stable deposits of two or more atoms can be lost, with the silver rejoining the lattice. Certain conditions are known to favour this process of latent image fading. It is temperature dependent, occurring more readily in hot conditions: it is favoured by the presence of moisture. Oxidising agents will greatly accelerate its rate. It occurs as a random process, but affects more severely the silver deposits at sensitivity specks on the surfaces of the crystals, while specks inside the crystals survive better. It tends to be more severe, then, with emulsions with small crystals, in which all the sensitivity specks are likely to be surface ones. Latent image fading is a process that gradually reduces the size of silver deposits, reducing the number of grains that can be developed under constant conditions. It is clearly likely to be more of a problem the longer the time elapsing between exposure of the crystal and development.

Unfortunately, the changes described above within crystals exposed to light or radiation are not specific for those two agents, but can be caused by a number of others. Pressure is one, and beautiful fingerprints may be recorded on nuclear emulsions. The pressure exerted by gelatin during the process of drying has already been discussed (see p. 16). Heat may provide sufficient energy for latent image formation. Many chemicals, particularly reducing agents, can act directly on an emulsion to produce silver deposits in the crystals.

It should be remembered that the silver deposits in crystals that are hit by a charged particle do not lie exactly on the track of the particle through the crystal, but at dislocations in the lattice of silver bromide that was formed during manufacture. The effect of this on the resolution of autoradiographs at the electron microscope level is discussed on p. 76.

In most nuclear emulsions under reasonable conditions of development, the number of silver atoms per developable crystal probably lies between 10 and 1000. Crystals containing 10 to 2 atoms of silver have a progressively lower chance of development. The silver may be confined to one sensitivity speck or distributed between several of them; the probability of development is higher if all the silver is at one speck. Crystals that contain too few silver atoms, or in which the silver is distributed in too many small packets, fail to achieve a probability of development of 0.5: these are known as the latent sub-image.

CHEMICAL DEVELOPMENT OF THE LATENT IMAGE

In chemical development the emulsion is placed in a solution which reduces the silver bromide to metallic silver and hydrogen bromide. Most developing agents have the appropriate reducing potential at alkaline pH.

In the formation of silver deposits at sensitivity specks during exposure, the first step is the formation of a single silver atom, which is rather unstable. Once a two-atom nucleus is formed, the deposit becomes more stable, and more successful at capturing subsequent electrons. The same sequence of events occurs in chemical development: the larger the aggregation of silver atoms at a sensitivity speck, the higher the probability of further silver being added, while crystals without any silver atoms at sensitivity specks have to pass through the stage of establishing a stable two-atom nucleus before development can proceed. In short, the presence of metallic silver in a crystal catalyses the reduction of silver bromide.

These events are best considered as statistical probabilities occurring in large populations of crystals. Under constant conditions of development, crystals with large silver deposits have a high probability of development in a given time. Those with only 2 or 3 atoms of silver have a much lower probability, and those without any at all have an even lower probability. Even in the absence of metallic silver, the probability of development will be finite, however, and, since the number of such crystals is likely to be very high in an autoradiograph, some crystals in this group will become developed grains.

So the emulsion enters developer with a distribution of numbers of silver atoms per crystal from zero to perhaps 1000. The grains that will ultimately be developed will be selected from the whole distribution of crystals, with probabilities ranging from 1 in the case of very large silver deposits to nearly zero in the case of crystals without a single atom of metallic silver. If the crystals that have been hit by charged particles form a very low fraction of the total, crystals without silver deposits may form the majority of the grains that finally develop in spite of their low individual probability of development.

With increasing time in chemical developer, more and more silver is added to the silver deposit at a sensitivity speck, producing a ribbon of silver growing out from the speck. The process ends when all the silver available in the crystal has been converted to metallic silver; at this point, the developed grain looks like a coiled filament, and occupies a volume up to 3 times that of the original crystal (Fig. 2). The centre of this coil does not necessarily coincide with the centre of the original crystal. Electron microscopic studies of development show that several ribbons of silver may start to grow from within the same crystal, presumably representing silver deposits at several sensitivity specks. These ribbons coalesce during growth[4].

Some developers are only capable of reducing sensitivity specks at the surface of crystals: others can act on deeper specks also. The latter usually contain some solvent for silver bromide, etching the crystal lattice during development to expose the deeper sensitivity specks while developing all exposed specks. These will be further discussed later, in connection with the selection of a suitable developer (p. 32). Note that since the fading of latent images chiefly affects silver deposits at surface specks, developers that only attack such specks will be particularly sensitive to latent image fading[5].

21

Fig. 2. Electron micrograph of silver grains of Ilford L4 emulsion developed with Microdol X. These irregular, coiled ribbons of silver are characteristic of chemical development. (×36,000). (Material provided by Dr. M.M. Salpeter)

When is a grain considered "developed"? In the photographic literature, development is often defined as the complete conversion of the silver bromide crystal to a grain of metallic silver. In autoradiography, however, the production of a visible mass of metallic silver is the endpoint, and this can occur well short of complete development. Obviously, this endpoint will vary with the conditions of viewing. A grain will be visible in the electron microscope long before it is big enough to be seen in the light microscope. In light microscopy, dark field illumination will allow one to see grains which are too small for conventional transmitted light optics at the same magnification (Fig. 3). The optical background provided by a stained section may make it difficult to recognise grains which would be indisputably "developed" in an unstained preparation.

Development, then, is a process of amplification, increasing the size of the deposit of metallic silver in a crystal until it reaches a threshold at which it can be recognised, and this threshold is determined entirely by the conditions of viewing. At this threshold, the crystals become silver grains. Silver grains can grow above this critical size until they become fully developed, and all the silver bromide present in the original crystal is reduced.

A study of a series of autoradiographs, prepared, developed and viewed under identical conditions, varying only the time of development, produces a graph such as that in Fig. 4. Developed grains over the labelled source increase in number until they reach a plateau. The background grains in the emulsion away from the source increase much more slowly with time,

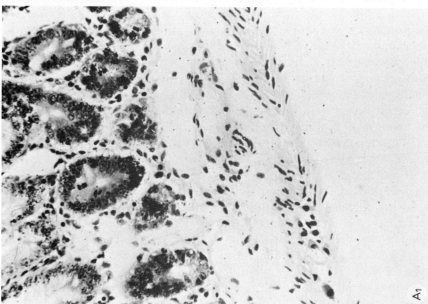

Fig. 3. Photomicrographs of the intestine of the galago after the administration of [³H]thymidine. Identical sections covered with Ilford G5 emulsion have been developed in Amidol for increasing times. The same field from each slide is viewed with transmitted light (1), and incident dark-field lighting (2). (A), 3-minute development. The grains are small and

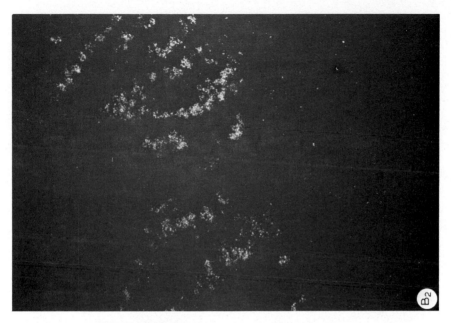

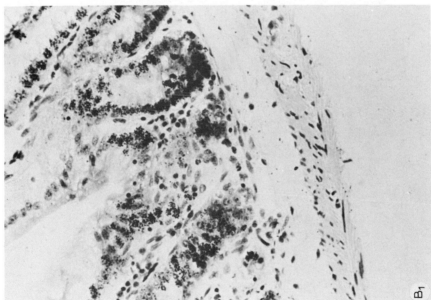

hard to see by transmitted light, clear and distinct by dark-field methods. (B), 7-minute development. The grains can now be seen by transmitted light, and are large by dark-field methods. (C), 11-minute development. The grains are clearly seen by transmitted light and the background remains low. By dark-field lighting, however, many tiny background grains are visible. Stain, Harris' haematoxylin. ($\times$ 170)

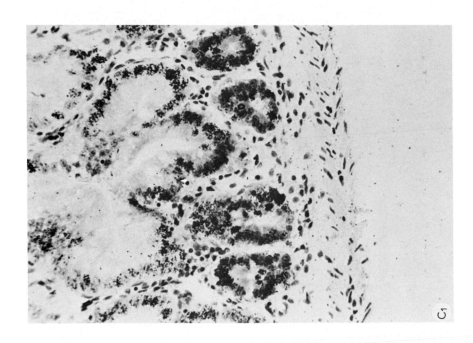

but finally climb exponentially. Ultimately, every crystal in the emulsion will be developed. The aim of development should be to produce the optimal ratio of grains over the source to background grains. With most emulsion–developer combinations, there is an appreciable range of times which gives fairly similar signal–noise ratios.

On this plateau, the number of grains over the source may not increase significantly, but the sizes of the grains in general will increase[6]. This is particularly important if the grains are to be measured by some photo-metric procedure rather than being counted. Each increment in developing time adds grains to the background at the threshold of visibility. At any given time on the plateau, the mean size of grains over the source will tend to be larger than that of background grains.

If we return for a minute to the emulsion just before the start of development, some of the crystals will have silver deposits in them sufficient to give a probability of 0.5 or more that they will be developed: these constitute the latent image. Others will have silver deposits so small or so disperse that their probability of development is less than 0.5: these are the latent sub-image. Others, the great majority of crystals in most cases, will have no silver deposit and only a very low probability of development. With conventional chemical development it is impossible to develop every crystal with a silver deposit without an unacceptably high background. As development is increased to raise the probability for two-atom deposits, the probability for unhit crystals also rises. Specialised techniques have evolved to try to increase the gap between these two sets of probabilities, and these will be discussed later.

Conditions affecting chemical development

The curves in the graph (Fig. 4) of the time course of development are valid in general terms for a great many combinations of emulsion, developer and conditions of viewing. The absolute values of the time base may, of course, be altered by a number of parameters.

First, some developers are more powerful reducing agents than others, and the rate of reaction can often be considerably varied for the same developer by altering the pH. For the same developer and pH, diluting the solution will tend to slow down the process.

Temperature is another factor which controls the speed of development: the higher the temperature, the faster the process[7]. One other variable is often not appreciated. This is the effect of agitating the developing solution. In the absence of agitation, the layer of solution next to the emulsion rapidly becomes depleted of developer, and the arrival of fresh molecules is brought about by the rather slow processes of diffusion and

26

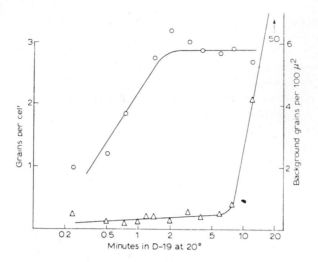

Fig. 4. The kinetics of development of Ilford L4 emulsion in D-19 developer. The grain count per labelled bacterium (□) remains constant after about 2 minutes development. The background grain density (Δ) increases slowly up to 6 minutes, and very rapidly thereafter. (From Caro and van Tubergen, 1962)

convection. Agitation maintains the concentration in the layers next to the emulsion[1].

How important are these variables quantitatively? Diluting the developer by its own volume of distilled water may lengthen the time to optimal development by 50–100%. Continuous gentle agitation by bubbling nitrogen through the developer can halve the time. Temperature can have a very great effect on the rate of development. At or below 5°C, development is so slow that it can be considered arrested, in conditions that would give development in 10–20 minutes at 22°C. Above about 23°C, it is often difficult with established developing routines to produce a plateau of developed grains, since the whole process represented by Fig. 4 is so compressed.

Development can continue after the emulsions are removed from the developer, since traces of the solution will be carried in the emulsion layer. Development is usually stopped by lowering the pH, using a stopbath of 1% acetic acid. Alternatively, with thin emulsion layers it may be sufficient to dilute out the developer by rinsing in distilled water. The undeveloped silver bromide is then solubilised in a fixing solution and washed out of the gelatin in running water, leaving the clear gelatin with the developed silver grains in it.

PHYSICAL DEVELOPMENT

The process of chemical development uses the silver bromide of each crystal as a source for the silver of the developed grain. The fully developed grains are large, irregular coils of silver, larger than the original crystal by a factor of up to 3 times (Fig. 2). This grain morphology may be troublesome in electron microscope autoradiography, obscuring underlying detail in the micrograph. It also contributes to the difficulty of analysing the autoradiograph, since the grain does not necessarily coincide in position with the original crystal.

There has thus been considerable interest in the possibility of producing small, dense, round grains for electron microscope work[4]. This has usually been done by physical development, since attempts to arrest chemical development at a stage prior to the appearance of a coiled ribbon of silver has produced such slender, short filaments that they have not been very stable in the microscope.

In physical development the silver is deposited from solution on the silver deposits at sensitivity specks. This produces a concentric growth of the deposit, giving a round grain the size of which can be fairly readily controlled (Fig. 5). There are two main groups of procedures in this category. The first is carried out before photographic fixation, by a developer such as amidol at acid pH, with the addition of silver nitrate or silver sodium sulphite to provide a source of silver. After development is stopped, the silver bromide of the crystal is removed by fixation in the usual way.

The second and more common group of techniques for physical development start with the removal of silver bromide from the emulsion by fixation and washing. Development then takes place with p-phenylenediamine or phenidone as the developing agent, and sodium sulphate and silver nitrate also present.

The p-phenylenediamine recipes have tended to give a rather low efficiency of development and to be not very reproducible. The Agfa-Gevaert formula for phenidone is probably the best one in routine use[4]. The shape of grain produced is illustrated in Fig. 5.

Note that the definition of a developed grain is still dependent on the conditions of viewing, and that the rate of development is still critically dependent on conditions such as dilution, temperature and agitation of the developer.

FIXATION AND WASHING OF THE EMULSION

The use of a 1% solution of acetic acid to stop development has already been referred to (p. 26), as well as the alternative of simply rinsing the

28

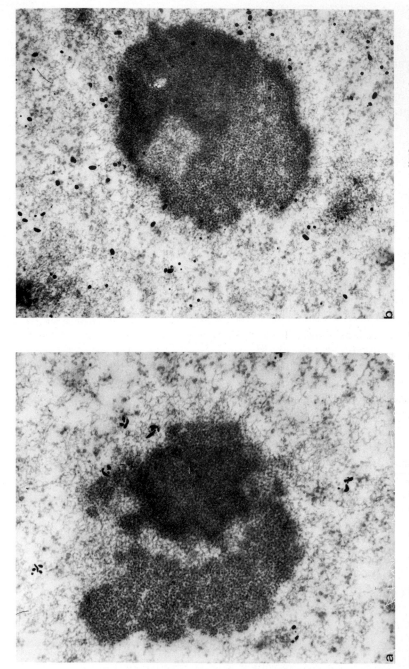

Fig. 5. Autoradiography of nucleoli from mesemchymatous cells of the newt after incorporation of [³H]thymidine. (a). Prepared with a monolayer of NTE emulsion developed in Dektol. (b). Prepared with a monolayer of L4 emulsion developed with *p*-phenylenediamine. (×28,000). (Photographs provided by Dr. M.M. Salpeter)

emulsion layer in distilled water. One bonus of the acetic acid stopbath is that it hardens the gelatin slightly, reducing its tendency to swell greatly in fixative and washes.

After chemical development, fixation dissolves away the silver bromide crystals that remain in the emulsion. This is usually carried out in a solution of sodium thiosulphate (hypo). The thiosulphate ion forms a series of soluble complexes with ionic silver, without affecting the developed grains. It is possible to increase the rate of fixation by agitating the solution, as with development, and by increasing its temperature. The end-point of fixation is generally taken as twice the period required for the emulsion to become transparent.

The concentration of sodium thiosulphate is not critical, in the range of 25–35%. It is usual to work with a 30% solution, but it is quite unnecessary to weigh out the hypo to the third decimal place.

It is important to remember that nuclear emulsions contain a higher percentage of silver bromide than do photographic emulsions, and that the speed of fixation drops off rapidly as the products of fixation accumulate in the solution. As a rough guide, no more than 24 autoradiographs with thin emulsion layers on 3 in. × 1 in. slides should be fixed in 250 ml of hypo. With thicker emulsion layers, larger volumes and frequent changes of hypo are necessary. The rate-limiting factor appears to be the diffusion of the silver–thiosulphate complexes out of the emulsion.

Rapid fixers of various sorts are often used in photography instead of plain hypo. They are based on the use of ammonium thiosulphate, and they can reduce the fixation time by as much as 75%. Unfortunately, ammonium thiosulphate dissolves the silver grains slowly, in addition to its more rapid attack on the silver bromide crystals. In very thin emulsion layers its use is acceptable, as it is in contact with the emulsion for so short a time that the developed grains are probably not significantly eroded. But, in the thicker emulsion layers, where the fixation times may be very long indeed and a more rapid action would be very welcome, the silver grains themselves may be so reduced in size that they are no longer visible if a rapid fixer is used. If plain hypo is used for fixation, it is important that it should be freshly made up, and discarded after use.

Hardeners may be added to acid fixing-baths. These reduce the swelling of the gelatin in the subsequent washing, and hence its liability to mechanical injury. In most microscope work it is possible to avoid physical damage to the emulsion during washing, and to protect it subsequently under a cover-glass, so that hardeners are only an added complication. With thick emulsions, hardeners in the fixing-bath can considerably increase the time needed for fixation. They may have a place in the processing of autoradiographs of large objects, such as chromatograms.

In a nuclear emulsion, which is usually 50% by volume silver bromide crystals, very few of which become converted into silver grains, it is obvious that removal of the undeveloped crystals by fixation will reduce the volume of the emulsion considerably. After fixation, dehydration in alcohols, and mounting in a conventional histological medium under a coverslip, an emulsion layer may be as little as one-third its original thickness during exposure. This may make the recognition and following of particle tracks in thick emulsion layers very difficult, and it is often useful to impregnate the emulsion with glycerin or some similar agent in order to re-swell it to a thickness nearer that during exposure, if track recognition is necessary.

By the process of fixation, then, the silver bromide remaining in the emulsion after development is brought into solution. Afterwards, the products of fixation must be washed out of the emulsion layer. This is usually done in running tapwater. A flat, shallow dish and rapidly running water will give the shortest washing time. At this stage the gelatin usually swells considerably, and great care must be taken not to damage or displace the emulsion. If the emulsion layer is very thin, as in electron microscope work, or not very firmly attached, as when an impermeable layer of polyvinyl chloride covers the specimen, washing must be extremely gentle. It may help to keep the specimen horizontal while washing, or to use several changes of stationary water. It is a good idea to keep the temperature of all the solutions used to process autoradiographs within a few degrees of each other, preferably at or below 20°C. Hardening the emulsion during fixation may reduce the swelling of the emulsion considerably. Alternatively, it may help to fix the gelatin with formalin either before or after development.

It is worth noting that some of the silver–thiosulphate complexes are light-sensitive. It is sensible to work under safelighting until the washing process is complete, since an earlier return to full lighting may result in the desposition of a fine, dark precipitate in the gelatin. The local composition of the tapwater may have something to do with this light-induced precipitate, since many laboratories have never experienced it.

Even with prolonged washing, it is difficult to remove the last traces of thiosulphate from emulsion layers. Thiosulphate ions appear to become adsorbed to the surface of silver grains[8]. In most instances, this does not matter to the autoradiographer. If the emulsion is transferred after fixation to an acid medium, however, the developed grains may be eroded with the formation of silver sulphide. I have known the grains to vanish completely from an autoradiograph, causing considerable consternation, following the use of certain staining procedures, or even histological mounting media. It is possible to reduce thiosulphate adsorption by the silver grains by treating the emulsion with a solution of potassium iodide before fixation[8].

PRACTICAL CONSIDERATIONS

The preceding sections on the photographic process provide the theoretical basis for many of the autoradiographic techniques that will be described in more detail later. Several practical hints can be drawn from them, which are of such wide application that it is best to list them here.

First, the handling of emulsions requires that any possible source of latent images other than those due to radioactivity in the source must be very carefully excluded. They should never be heated above 50°C. The chemicals that come in contact with them should be of high quality: only distilled water should be used, at least until the stage of washing after fixation; vessels to hold the emulsion during melting should be of glass, plastic or high grade stainless steel, and should be scrupulously clean, and the same applies to spoons, forceps or stirring rods. Pressure should be carefully avoided on dry emulsion layers, particularly scratching or rubbing pressures. Drying of emulsion layers should be slow and gentle, to prevent the shrinkage of the gelatin from stressing the crystals. Lighting should be kept to the absolute minimum, and the manufacturer's recommendations on filters should be strictly followed. Safelighting should be switched off altogether when not needed, and exposure of the emulsions to full light be deferred until at least several minutes of washing have passed after fixation.

Next, the emulsion must be exposed to the radioactive source under the best possible conditions. We have seen that moisture and oxidising agents can promote latent image fading. Emulsions must be thoroughly dried before exposure, and protected during exposure from strange chemicals that might either produce latent images or result in their loss. Sealed containers of glass or plastic are the best. The temperature chosen for exposure should be controlled.

Since emulsions respond to a range of insults by the production or loss of latent images, the pattern of developed grains seen over a specimen cannot be interpreted in terms of the radioactivity it contains without two simple controls. The first is a similar but non-radioactive specimen exposed under identical conditions. The presence of silver grains above background levels in the emulsion over this control will alert you to the fact that some component of the experimental specimen may be producing silver grains by mechanisms other than radioactivity. The second control is a similar specimen exposed to an emulsion that has been fogged with light or radiation. Again, the conditions of exposure and development must be identical to the experimental ones. If the developed grain density is not uniformly high over the control specimen, the experimental ones also may have similar regions of rapid latent image loss. These two controls should be built into each and every autoradiographic experiment: without them, interpretation is impossible.

The choice of developers is wide and often not critical. For light microscope work, there is no advantage in using a physical developer. Most people work with recipes based on Metol or Amidol. Metol is an excellent general purpose developer; the restrictions on its use are that it works best in highly alkaline media, which may soften the gelatin unduly in some applications, and that it does not penetrate thick emulsion layers readily. Amidol is again an excellent general developer, which can be used in neutral or even slightly acid media. It is the developer of choice for all thick emulsion layers, such as used in track autoradiography. It is less stable when made up than Metol, and it only develops surface latent images, unless special formulae are used, so that it is more sensitive to latent image fading and chemography than Metol. A great range of other developers is available. Sugars, haemoglobin, and even old Burgundy wines have been used[9].

At the electron microscope level, the choice of developer becomes much more critical. It will be discussed in more detail later (p. 405).

Probably more important than the choice of a developer are the decisions about the conditions of development. It cannot be too strongly emphasised that there is no "correct" developing time for a given emulsion. Do not just adopt the conditions given in someone else's paper. They are seldom fully specified, and there is no guarantee that they are applicable to your experiment. Determine the optimum conditions for yourself: it is very simple. Prepare a series of identical autoradiographs from specimens similar to the experimental ones. Choose a developer, a dilution for it, and a temperature at which to work which can be accurately controlled: this temperature should be below 22°C. Either avoid agitating the developer altogether, or fix up some reproducible means of agitation. Develop autoradiographs for increasing lengths of time, e.g. 1, 2, 4, 8, 12 and 16 minutes. Prepare them for viewing in the manner you will use for your experimental series. Then, either by quantitative methods or even just by observation, determine in your conditions the time that gives the best signal-to-noise ratio (Fig. 3).

If grains are small and few even at the longer times, the rate of developing is too slow. Try increasing the concentration of developer or its temperature. If the plateau of development seems too short for comfort, and background levels are building up very early in the series, try reducing the concentration or temperature of development. I personally prefer development times in the range of 6–15 minutes. It seems slicker to reduce this to 1 or 2 minutes, but the control of development becomes less reproducible at such short times, and the extra few minutes are a very small fraction of the total experimental time.

We have seen that the conditions of viewing determine the definition of a developed grain. If your conditions of viewing change, the same series of autoradiographs at varying development times can be re-examined under the new optical conditions to pinpoint the best time.

SPECIAL TECHNIQUES

A number of specialised techniques have been used in autoradiography over the years. The more important of them are reviewed briefly below, together with one or two problems in interpretation that are posed by aspects of the photographic process.

(a) *Hypersensitisation*

Herz[10] has suggested the use in autoradiography of emulsions in which the sensitivity has been increased by treatment with triethanolamine (TEA) before exposure. Barkas[11] discusses a number of experiments into the mechanism by which TEA increases emulsion sensitivity, without reaching any firm conclusions. This treatment is capable of giving an increase of approximately 30% in the number of developed grains per 100 μm of particle track for high energy electrons in Ilford nuclear emulsions. Unfortunately hypersensitised emulsions develop background fog very fast at room temperature[11]. This technique has been used at times by the particle physicists, but not to any significant extent in autoradiography.

(b) *Intensification of the latent image*

This is often referred to as latensification. The technique involves depositing another metal, usually gold, on the silver deposits present at sensitivity specks immediately before development[12]. This is done by soaking the emulsion in a specially prepared solution of a gold salt. The effect is to increase the size and catalytic potential of each silver deposit, in other words, to increase its probability of development. As we have seen, normal chemical development falls short of developing every crystal with a silver deposit in it, because the probability of development of a two-atom nucleus, for instance, is close to that of a crystal without a silver deposit at all. Latensification increases the probabilities of development of all crystals with silver deposits without affecting crystals without deposits. It thus permits many crystals with what would have been latent sub-images to be developed in conditions that do not produce high backgrounds[13].

Gold latensification was first introduced into autoradiography in an attempt to increase the efficiency in electron microscope work, where the conflicting demands of resolution and efficiency make it imperative to get the utmost out of the exposed emulsion layer. It is now a routine technique in many laboratories for electron microscope work. It also has valuable applications to light microscope autoradiography, permitting an appreciable reduction in exposure times.

Rechenmann and Wittendorp[13,14] have studied gold latensification and its applications to autoradiography in considerable detail. They have found that gold treatment increases the kinetics of development in all the combinations of emulsion, isotope and developer they have examined. But the effect of gold is not merely to achieve the same grain densities in a shorter development time: even at long times, when the numbers of developed grains should have reached a plateau (Fig. 4), there is an increase in the total number of grains developed by comparison with emulsion layers without gold latensification. In their hands, this increase is consistently achieved with remarkably low fog levels.

The gold treatment is clearly transforming latent sub-images into latent images. The next question is how stable are the sub-images during exposure? Rechenmann and Wittendorp[13] have shown that, like latent images, they can be very stable in suitable exposure conditions.

In their work, gold latensification is more effective with less highly sensitised emulsions: with K2 the improvements in grain yield are much higher than with K5 (Fig. 6). Gold treatment is more effective with emulsions with smaller crystal sizes. It is more effective with β particles of higher energies. All these conditions are likely to produce a higher proportion of rather small silver nuclei in hit crystals, i.e. a high proportion of latent sub-images to latent images. Finally, the increases in efficiency seen with any combination of isotope and emulsion vary with the redox potential of the developer that is used after the gold treatment.

Gold latensification is not recommended by Eastman Kodak for their new product, 129–01, since the sensitisation process given to this emulsion in manufacture includes treatment with gold salts. It appears that the more highly sensitised Ilford emulsions have also received gold treatment in manufacture, which may account in part for the reduced effectiveness of gold latensification with these emulsions.

If the background is to be held at very low levels, great care must be taken to reduce every source of potential latent sub-images to the lowest possible level. Safelighting must be kept low, and drying of emulsion layers should be very slow and gentle.

Gold latensification is discussed in relation to a number of techniques later (pp. 341, 365, 372, 405, 406, 409).

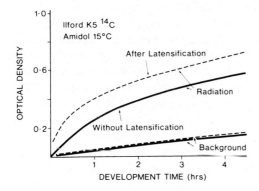

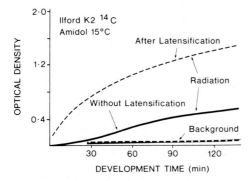

Fig. 6. The increases in efficiency resulting from gold latensification of layers of Ilford K2 and K5 emulsions. The emulsion layers received standard exposures to a source of carbon-14 and were then developed in Amidol at 15°C. The optical densities after gold treatment (-------) are uniformly higher than those without gold (———), while the background levels remain unchanged. (a), K2 emulsion. (b), K5 emulsion. (Redrawn from Rechenmann and Wittendorp, 1976)

(c) *Exposure in a magnetic field*

This technique was originally introduced to try to increase the efficiency of autoradiographs with tritium. The idea was to bend the particle tracks in a magnetic field so that low energy β particles emitted in the direction away from the emulsion would be made to curve round into the emulsion. Initial reports claimed very substantial increases in efficiency[15,16]. Caro[17], however, showed that the magnetic field used was quite insufficient to affect the trajectories even of the low energy β's of tritium, and experiments by Zessack and Fromme[18] seemed to confirm his view.

It seems possible to me that the observed improvements in efficiency could have been due to an effect on the position and size of the silver deposits in hit crystals. It has been demonstrated frequently in the photographic literature that electric fields are capable of concentrating photolytic silver at one end of an exposed crystal[1]. Such concentration in a crystal hit by a β particle might result in the collection of all the metallic silver at the sensitivity speck nearest one pole, rather than its distribution between several such specks. Its probability of surviving exposure without fading, and its probability of development would both be increased. A recent Russian paper[19] suggests that an electric field can have opposite effects on different emulsions, those with a relatively large number of sensitivity specks responding with an increase in the number of developable grains, while those with very few specks may even show a decreased response.

(d) *Dose dependence*

Salpeter and Szabo[20] have drawn attention to a puzzling phenomenon in electron microscope autoradiographs. The efficiency in grains per β particle falls as the density of β particles increases. This effect was seen at very low densities of β particles, so that it was not due to an increasing probability of crystals being hit by more than one particle. Nor were the higher densities produced by longer exposures to the same source, which might have raised the possibility of latent image fading affecting the experiment.

Dose dependence of the emulsion response has only been described in the monolayers of silver halide crystals used in electron microscope autoradiography. It is more severe with the developer Microdol-X than with D-19 or Elon-ascorbic acid. It seems at present likely that a large silver deposit in a crystal may in some way inhibit the formation or development of smaller deposits in nearby crystals. It is interesting that Microdol-X is a developer with a strong solvent action on the crystals. Dose dependence is of importance in quantitative work at the electron microscope level.

(e) *Silver bromide precipitation on the specimen*

Silk, Hawtrey, Spence and Gear[21] have described a method which involves depositing silver on the specimen, and converting this to the bromide in situ with bromine vapour. From what has been said earlier about the functions of gelatin in nuclear emulsions, it seems unlikely that a layer of silver bromide of this nature would act as a reliable and reproducible recording medium. The autoradiographs they published are not very striking, their experiments, which seem poorly controlled, do not agree with the results of other electron microscope techniques, and their method has not been repeated by anyone else. From first principles, this would seem an unlikely approach to take.

Recently, a rather more sophisticated method has been described for making a silver bromide emulsion in the laboratory[22]. This involves the formation of silver particles in gelatin on the specimen, and their subsequent bromination after treatment with collodion to restrict redistribution of the silver into fewer larger particles. A mean crystal diameter of 100 Å is claimed, giving very good resolution in electron microscope autoradiography. The important factors of sensitivity and reproducibility are not mentioned. Knowing the care and skill required in the manufacture of nuclear emulsions, I would be very surprised if a "do-it-yourself" recipe such as this were to prove sufficiently reliable.

(f) *Direct deposition autoradiography*

Normandin[23] has suggested a system of identifying radioactive sites in tissue sections by depositing metals, silver or gold, directly on them without the intervention of a photographic emulsion, and his system has been used by Hemmings and Williams[24], in a study of protein transport in the placenta. Epon sections of tissue are "exposed" in the light for 20 seconds to solutions of gold or silver chloride, and the sites at which the metals are deposited are taken to be radioactive.

It will be seen in the next chapter that the disintegration of a radioactive atom not only produces a β particle, but also an activated nucleus, which has undergone transmutation to another element. No attempt has been made to date to make identification of the nucleus, rather than the β particle, the basis for localising radioactivity. Although Normandin does not claim to do this in his confusing article, it seemed a possible explanation of his results. Rapid calculation based on the grain density he demonstrates show that he achieves an impossibly high figure in 20 seconds; however, and, in view of other curious features in the article, it seems best to avoid his attempt to "develop" non-existent autoradiographs.

(g) *Modifications to the developed grains*

Developed and fixed autoradiographs may be treated in many ways to modify the silver grains. It is possible, for instance, to produce grains that are coloured[1]. This may have some application to the autoradiography of two isotopes in the one specimen[25]: this is often carried out with two emulsion layers, the second applied and exposed after the first has been processed. Grain counting then follows in each emulsion separately (pp. 306–308). It may help to have the grains in one layer coloured.

It is possible to bleach out developed grains in a system of image reversal. Sisefsky[26] used this technique in a study of the radioactivity of airborne particles. The reversal process left each radioactive particle in the centre of a clear area in the emulsion, the diameter of which could be related to its activity. This system permitted the particles to be studied microscopically, without interference from overlying silver grains.

Finally, a number of methods have been described for making silver grains radioactive. Stevens[27] has suggested this as a method of manufacturing very small standard sources from photographic step wedges. This approach has been used, in conjunction with a Geiger tube fitted with a small counting window, as a rapid method of grain counting[28]. Thackray, Roman, Hetherington and Brian[29] have described its use for the intensification of grossly underexposed photographs, the radioactive negative or print being autoradiographed to give a workable picture. In the latter two applications, a very great amplification factor is introduced by making the silver grains themselves radioactive.

(h) *Fluorography, or scintillation autoradiography*

If a radioactive source is placed in contact with a scintillator, either solid or liquid, the particles emitted will have some of their energy converted into photons. These may then be recorded on a photographic emulsion.

This procedure was first suggested by Wilson[30], to improve the very low efficiency with which tritium is recorded by X-ray films. These generally have an anti-abrasion coating over the emulsion layer, thick enough to prevent all but a tiny fraction of the β particles from tritium reaching the emulsion. Conversion of some of the particles' energy to photons, which have a much larger range in matter, resulted in higher efficiencies of detection.

The photographic consequences of this conversion from β particles to photons are complex and interesting. Photons, having a longer range, are distributed over a wider volume of emulsion, and there is a high probability of single photons producing single photolytic electrons in individual crystals. As we have seen, the single atom silver deposits that result are very unstable, and many such photons will go unrecorded. This is the basis of low-intensity reciprocity failure in photography. β particles, on the other hand, tend to produce several free electrons simultaneously in their passage through a crystal, and thus a stable silver deposit. Recording of β particles in nuclear emulsions show no phenomenon comparable to low-intensity reciprocity failure. So the number of silver grains can be linearly related to the number of β particles at low densities; the same is not true of photons.

Two approaches from the photographic literature have been adopted to improve the stability of the silver deposits from single photon events. First, the rate at which the single silver atom reverts to a silver ion and a free electron is temperature dependent[31,32]. Exposure at $-70°C$ or below slows this rate tremendously, increasing the chances of a second photon hitting the crystal before loss of the first silver atom, and we have seen that a two-atom nucleus is stable. An added bonus is that light output from many scintillators is improved at low temperatures[33]. Considerable improvements in efficiency have resulted from low temperature exposures of scintillation autoradiographs (Fig. 7)[34].

The second procedure to render these single photon events stable has involved exposing the whole emulsion layer to a flash of light of carefully controlled intensity before autoradiographic exposure[35]. The theoretical background to this is that two- and three-atom nuclei of silver at sensitivity specks are stable, but have a very low probability of development. They are, however, quite efficient collectors of subsequent photolytic electrons. So the light flash is controlled to give a minimal increase in background on routine development, but a greatly improved efficiency for the collection of single photon events during the autoradiographic exposure (Fig. 7). If correctly chosen, a pre-exposure to light can eliminate low-intensity reciprocity failure, giving a linear response between photons and silver grains at low densities[35].

Presumably these two approaches are to some extent competing, and the increases in efficiency given by each are not additive if both are used simultaneously.

Scintillation autoradiography has been widely applied to the autoradiography of macroscopic specimens, and the detailed techniques are dealt with on p. 289. The efficiency of recording of photons on blue-sensitive X-ray

Exposure conditions	Tritium content (nCi)
No scintillator at 22°C	800
Scintillator at 22°C	87
Scintillator at $-70°C$	1.4
Controlled light exposure followed by scintillator at $-70°C$	0.14

Fig. 7. The lowest detectable amounts of tritium in bands 1×10 mm in size on polyacrylamide gels in a 24 hour exposure to X-ray film, using scintillation techniques. (Results taken from Bonner and Laskey, 1974, and Laskey and Mills, 1975)

film is reasonably good, and this method has a great deal to offer in situations where the longer range of the photons permits much higher fluxes to reach the emulsion.

More recently, a number of reports have claimed greatly increased efficiencies in light and electron microscope autoradiography from the use of scintillators impregnated into specimen, emulsion or both (e.g. ref. 36). These should be regarded with reserve. Nuclear emulsions are not sensitised to light during manufacture, and are very inefficient at recording photons, even though their spectral sensitivity does roughly correspond with the emission wavelength of the commonly used scintillators, such as PPO. If the β particles from the source can reach the emulsion, no increase in efficiency can possibly be expected from converting part of their energy into photons. If, of course, β particles directed away from the emulsion can be converted into photons, some of which reach the emulsion, a slight increase in efficiency of the order of $\times 1.5$ can perhaps be achieved[37,38]. Claims of increases in efficiency by 40–80 times have been made[39]. There is no theoretical basis for expecting such improvements. I have been unable to confirm these results in my own experiments at the light microscope level, and have had many reports from other laboratories to the same effect. The likely explanation for the greatly increased efficiencies claimed is that emulsions that have been very thoroughly dried by immersion in a dioxane-based scintillator before a short exposure have been compared to inadequately dried emulsions exposed for several weeks. Latent image fading in the latter group could introduce a very considerable difference in efficiencies. The attempts to apply scintillation autoradiography at the light and electron microscope levels are discussed in more detail on p. 366 and p. 341, respectively.

(i) *Xerography, and the use of Polaroid film*
Dobbs[40] has discussed the use of xerography in autoradiography. While interesting, I can see no immediate application for these techniques. A number of others have reported the use of Polaroid film for the autoradiography of chromatograms[41,42]. It may have a place where darkroom facilities are not available, on expeditions, for instance, or in the autoradiography of macroscopic specimens. The techniques using X-ray films are so simple, however, that I cannot see widespread use of Polaroid film in the near future.

(j) *Eradication of unwanted latent images*
It is possible to remove existing latent images in a nuclear emulsion by exposing it to conditions which favour latent image fading[11]. This may be valuable if emulsion has been accidentally exposed to radiation or to any

other factor which could cause unwanted background. The techniques available for doing this are discussed in more detail on p. 127.

(k) *The removal of unwanted emulsion layers*
This may be necessary if the emulsion has been ruined, perhaps by exposure to light, and the specimen is difficult to replace. It has also been used, after photography of the autoradiographic trace, to permit more accurate examination of the underlying specimen. The gelatin of the emulsion can simply be removed by digestion for 2–5 minutes in 1% potassium hydroxide at room temperature. This will often leave developed grains still in contact with the specimen. These can be removed by treatment with 7.5% potassium ferricyanide for 15 minutes, 3 changes of 30% sodium thiosulphate for 5 minutes each, and thorough washing in running water.

REFERENCES

1 C.E.K. Mees and T.H. James (Eds.), *The Theory of the Photographic Process*, 3rd ed. Macmillan, New York, 1966.
2 J.W. Mitchell (Ed.), *Fundamental Mechanisms of Photographic Sensitivity*, Butterworth, London, 1951.
3 C. Waller, *1st European Symposium on Autoradiography*, Rome, 1961.
4 B.M. Kopriwa, *Histochemistry*, 44 (1975) 201.
5 A.W. Rogers and P.N. John, in L.J. Roth and W.E. Stumpf (Eds.), *The Autoradiography of Diffusible Substances*, Academic Press, New York, 1969.
6 A.W. Rogers, *J. Microscopy*, 96 (1972) 141.
7 M.G.E. Welton, *J. Phot. Sci.*, 17 (1969) 157.
8 G.I.P. Levenson and C.J. Sharpe, *J. Phot. Sci.*, 4 (1956) 89.
9 M. Abribat, *Sci. Ind. Phot.*, 15 (1944) 204.
10 R.H. Herz, *Lab. Invest.*, 8 (1959) 71.
11 W.H. Barkas, *Nuclear Research Emulsions, Part 1*, Academic Press, New York, 1963.
12 T.H. James, *J. Colloid Sci.*, 3 (1948) 447.
13 R.V. Rechenmann and E. Wittendorp, *J. Microscopy*, 96 (1972) 227.
14 R.V. Rechenmann and E. Wittendorp, *J. Microscopic Biol. Cellulaire*, 27 (1976) 91.
15 C.G. Harford and A. Hamlin, *Nature*, 189 (1961) 505.
16 T. Nakai, *J. Cell Biol.*, 21 (1964) 63.
17 L.G. Caro, *Nature*, 191 (1961) 1188.
18 U. Zessack and H.G. Fromme, *Zeitschr. Wissenschaft. Mikrosc.*, 67 (1966) 225.
19 V.I. Kalashnikova, A.A. Kolyubin and B.D. Lemeshko, *Instrum. Exp. Tech. (USSR)*, 16 (1973) 420.
20 M.M. Salpeter and M. Szabo, *J. Histochem. Cytochem.*, 20 (1972) 425.
21 M.H. Silk, A.O. Hawtrey, I.M. Spence and J.H.S. Gear, *J. Biophys. Biochem. Cytol.*, 10 (1961) 577.
22 B.H.A. Van Kleeff, W.E. de Boer, R. Kokke and T.O. Wiken, *Exptl. Cell Res.*, 54 (1968) 249.

23 D.K. Normandin, *Trans. Amer. Microscop. Soc.*, 92 (1973) 381.
24 W.A. Hemmings and E.W. Williams, *J. Microscopy*, 106 (1976) 131.
25 E.O. Field, K.B. Dawson and J.E. Gibbs, *Stain Technol.*, 40 (1965) 295.
26 J. Sisefsky, *J. Phys. Sci. Instrum.*, 6 (1973) 74.
27 G.W.W. Stevens, *J. Microscopy*, 106 (1976) 285.
28 I. Haikal and O. Bobletter, *J. Chromatogr.*, 76 (1973) 191.
29 M. Thackray, D. Roman, E.L.R. Hetherington and H.H. Briand, *Intern. J. Appl. Radiation Isotopes*, 23 (1972) 79.
30 A.T. Wilson, *Biochim. Biophys. Acta*, 40 (1960) 522.
31 U. Lüthi and P.G. Waser, *Nature*, 205 (1965) 1190.
32 S. Prydz, T.B. Melö and J.F. Koren, *J. Chromatogr.*, 59 (1971) 99.
33 S. Prydz, T.B. Melö, J.F. Koren and E.L. Eriksen, *Analyt. Chem.*, 42 (1970) 156.
34 W.M. Bonner and R.A. Laskey, *Eur. J. Biochem.*, 46 (1974) 83.
35 R.A. Laskey and A.D. Mills, *Eur. J. Biochem.*, 56 (1975) 335.
36 B.G.M. Durie and S.E. Salmon, *Science*, 190 (1975) 1093.
37 H.A. Fischer, H. Korr, H. Thiele and G. Werner, *Naturwissenschaften*, 58 (1971) 101.
38 L.-A. Buchel, *Proc. Fifth Int. Congress Histochem. Cytochem.*, Bucharest, 1976, p. 397.
39 W. Sawicki, K. Ostrowski and E. Platkowska, *Histochemistry*, 52 (1977) 341.
40 H.E. Dobbs, *Intern. J. Appl. Radiation Isotopes*, 14 (1963) 285.
41 D.D. Jackson and M. Kahn, *Intern. J. Appl. Radiation Isotopes*, 20 (1969) 742.
42 C.O. Tio and S.F. Sisenwine, *J. Chromatogr.*, 48 (1970) 555.

CHAPTER 3

The Response of Nuclear Emulsions to Ionising Radiations

It might be thought unnecessary in a book of this sort to describe the particles and electro-magnetic radiations emitted by radioactive isotopes. It is certainly beyond the scope of this book to deal with their physical characteristics comprehensively. But biologists using radioisotopes frequently have gaps in their knowledge. I have been shown an electron microscope autoradiograph by a research worker who assured me that the coiled thread of metallic silver forming each developed grain was in fact the complete track of a β particle. The particles of major interest to the biologist will be reviewed briefly, concentrating on their characteristics as they can be seen in nuclear emulsions.

RADIOACTIVE ISOTOPES

Each atom consists of a dense, positively charged nucleus and a number of negatively charged electrons in orbit around it, like planets around the sun. The chemical properties of each atom are determined by the number of electrons around it, which depends in turn on the number of positive charges in the nucleus. The nucleus contains two types of particle: the proton, which carries a single positive charge, and the neutron, which is of similar mass, but is uncharged. The naturally occurring atoms of each element must clearly have the same number of protons in the nucleus. They also have a constant number of neutrons, and if for any reason the ratio of protons to neutrons is altered, the atomic nucleus becomes unstable. For each element, then, a number of different isotopes may exist, as the ratio of protons to neutrons is varied, but only one of these isotopes is normally stable — that which is naturally occurring. The unstable or radioactive nuclei disintegrate, ejecting charged particles until they reach a new balance of neutrons to protons which is stable.

Each radioactive isotope has well-defined characteristics. The manner in which disintegration takes place, in other words the types of radiation or particle emitted, their energies, and the daughter nucleus which is produced, are constant for each atom of that isotope. The probability or time-scale of disintegration is also characteristic for the isotope: this is usually expressed as the half-life, or time over which half the nuclei of that isotope will disintegrate. This time is independent of the total number of radioactive nuclei present, so that for a given mass of radioactive material, half the unstable nuclei will disintegrate in one half-life, half of the remainder in the next, and so on. Half-lives may range from fractions of a second to thousands of years.

The radioactivity of a given mass of material is measured in units called Curies: 1 Curie equals 3.7×10^{10} disintegrations per second. This is a statement of the rate at which unstable nuclei are disintegrating. Obviously, to achieve the same disintegration rate, an isotope with a long half-life will require far more unstable nuclei than one with a short half-life.

The events which take place when an unstable nucleus disintegrates affect the atom itself, and the surrounding atoms. If a nucleus attempts to change its ratio of protons to neutrons by ejecting a charged particle, it is clear that its own residual charge must alter. Since this charge determines the number of orbital electrons, and hence the chemical nature of the atom, disintegration produces an atom of another element. The charged particle is ejected with a certain amount of energy, and both the particle and the energy must in the last analysis be absorbed by surrounding atoms. The patterns of energy loss by α and β particles will be considered in the next section.

Any mass of a radioactive isotope is a diminishing asset, decaying at the rate indicated by its half-life. But a specific chemical compound labelled with a radioactive isotope is even more vulnerable. At each disintegration a molecule is disrupted by an abrupt switch of one atom to a different chemical state, while surrounding molecules are bombarded by ejected particles. In some cases, compounds may be produced which will catalyse the breakdown of molecules in which radioactive disintegration has not yet taken place. Radiation decomposition is a complex phenomenon, and is discussed at length by Bayly and Weigel[1], and by Evans[2].

In the following section, the interactions of charged particles with nuclear emulsions will be considered. This is basically the same as the behaviour of these particles in any other medium, except that the density of the medium affects the probability of interaction for a given length of particle track. Nuclear emulsions have a high density, about 3.8 during exposure; in tissues and in most histological embedding media, which have a density about 1.1, track lengths will be correspondingly greater and the density of events along the track more widely spaced.

PARTICLES AND RADIATIONS

The previous chapter has presented the mechanism by which a silver bromide crystal responds to photons of light falling on it. The absorbed energy of the photon raises one electron to a higher energy level, and this enables the electron to move through the crystal structure until it is trapped at a sensitivity speck, resulting in the conversion of a silver ion to elemental silver. This minute deposit of silver within the crystal forms the latent image, which catalyses the conversion of the crystal to metallic silver in a chemical developing solution. Ionising radiations can also impart energy to the silver bromide crystals through which they pass, producing a series of latent images.

In considering the effects of ionising radiations on nuclear emulsions, the radiations themselves must be separated into two categories. The first of these is the group of charged particles, including the α and β particles which form the main radiations of interest to autoradiographers. These particles, by virtue of the charge they carry, can exert an effect at a distance on the electrons in orbit about the atomic nuclei of silver and bromine which make up the crystal. If positively charged, they can attract electrons out of orbit; if negatively charged, their passage near to an electron may be enough to repel it out of orbit. In passing through matter, the charged particles lose energy in a closely spaced series of charge effects on the electrons that come within their field. In addition, they will also lose energy in less frequent interactions with positively charged atomic nuclei. This loss of energy in many little packets results in the production of many latent images distributed along the course of the particle through the emulsion. On development, these silver grains mark out the track of the particle.

Unlike photons, charged particles have a very low probability of producing single atoms of metallic silver at sensitivity specks. Instead, they tend to convert a number of silver ions to elemental silver almost simultaneously in each hit crystal. The effects of this on the stability of the silver deposited at sensitivity specks has already been described (p. 17).

The second category includes the uncharged particles, such as the neutron, and the electro-magnetic radiations, such as the X and γ rays. Uncharged particles only lose energy by direct collision with electrons or nuclei. Since such collisions are relatively unlikely, these particles may travel for considerable distances through nuclear emulsion without leaving a latent image to indicate their passage. The path taken by such particles through emulsion cannot be traced, though the infrequent collisions in which they are involved may produce recognisable patterns. In much the same way, the electro-magnetic radiations lose energy only in direct "collisions", imparting energy to orbital electrons in relatively few, widely spaced events, such as the processes of photoelectric absorption or of Compton scattering, or the production of a positron and an electron within the field of an atomic nucleus. The details of these interactions may be found in textbooks of atomic physics. Their result, from the viewpoint of the user of nuclear emulsions, is that electro-magnetic radiations leave no recognisable track, but produce sporadic electron tracks throughout the emulsion.

Clearly, it is the charged particles which produce recognisable tracks that are most usefully studied by means of nuclear emulsions, and a few of these will now be considered in more detail.

THE α PARTICLE

α Particles are relatively massive. Each one is identical to the nucleus of a helium atom, consisting of two protons and two neutrons. It therefore carries two positive charges. As one might expect, such large particles are not common in the radioactive disintegration of the lighter atomic nuclei. They are emitted by a few isotopes of elements with a high atomic number and, in particular, from the naturally occurring series of actinium, uranium, and thorium. Each isotope which emits α particles does so at one or more specific energy levels: in other words, all the α's emitted by one isotope have the same initial energy (or, sometimes, two distinct initial energies), and this energy of emission is characteristic of that isotope. All α particles have initial energies greater than 4 MeV, and very few have energies greater than 8 MeV, so that they form a relatively homogeneous group when one comes to consider their tracks in nuclear emulsion.

Each α particle has a rest mass nearly 8000 times that of an electron. It is not surprising, therefore, that a collision between these two particles will knock the electron off its course, without noticeably influencing the path of the α particle, like a collision between a cannonball and a tabletennis ball. Even in a collision with the nucleus of a hydrogen atom, an α particle will only be slightly deviated from its path.

By virtue of its two positive charges, an α particle will attract orbital electrons over a considerable distance, in atomic terms, while there will be mutual repulsion between it and other atomic nuclei which it may approach. An α particle dissipates energy very rapidly indeed in these effects on electrons, so that its range in matter is short, despite its high initial energy. Another way of describing the same thing would be to say that α particles cause tremendous havoc to the electron shells of the atoms through which they pass. In terms of the photographic process, this implies that relatively large nuclei of metallic silver will be formed at sensitivity specks in every silver halide crystal through which the particles pass.

So the track of an α particle in nuclear emulsion will be very dense, with every crystal along its path activated, and will be quite straight, apart from a slight chance of deviation toward its termination, if it collides with an atomic nucleus after it has already been slowed down. The track will also be relatively short: in Ilford G5 emulsion, all α's of initial energies between 4 and 8 MeV will have ranges between 15 and 40 μm. The track will be fairly wide, as the energised electrons resulting from the many collisions along the path of the α particle will themselves often travel short distances through the emulsion, causing latent images in adjacent crystals. These secondary electrons are often called δ rays, and they give the α track an irregular, fringed edge at very high magnifications.

This high rate of energy loss results in silver deposits forming at sensitivity specks in every crystal hit, and each deposit having a high probability of containing a large number of atoms of elemental silver. An emulsion of low sensitivity will be sufficient to record the tracks of α particles and, similarly, mild development will make these grains visible.

Fig. 8 shows two α tracks, radiating outwards from a single source. These tracks are quite characteristic. They are easy to record. They may be extrapolated back towards the specimen to indicate the exact source with

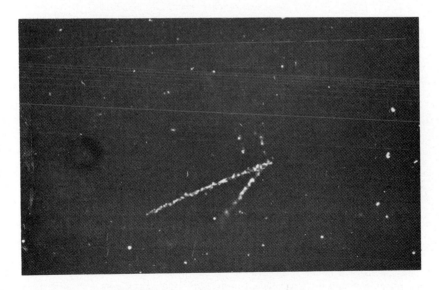

Fig. 8. Two α-tracks and one β-track coming from the same point, recorded in Ilford G5 emulsion. Note the width and high grain density of the straight α-tracks, and the irregular grain spacing of the tortuous β-track. ($\times$ 720)

great precision. They are easy to count, and quantitative procedures are simple and reliable. It is a great pity that so few of the isotopes of interest to the biologist emit α particles. Fig. 9 shows an autoradiograph of the spleen of a patient who had been given an intravenous dose of a thorium-containing compound several years previously. The tracks are so unmistakeable that their distribution can be plotted with great precision. In the course of autoradiography, α tracks will often be seen forming part of the background. α-emitting isotopes occur as impurities in glass, and also in the emulsion itself.

β PARTICLES

β Particles are really electrons of nuclear origin. They have the same mass as electrons, and the same single negative charge. The ejection of a β particle from the nucleus is the commonest way in which artificially induced radioactive nuclei achieve stability, and most of the radioisotopes of interest to biologists fall into this group.

Unlike α particles, the β particles emitted by one particular isotope do

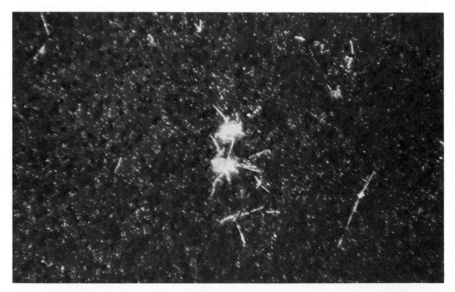

Fig. 9. Photomicrograph of the spleen from a patient injected 14 years previously with a thorium-containing compound. Crystalline deposits of thorium salts can be seen, with many α-tracks radiating out from them. Autoradiograph prepared by dipping in Ilford G5 emulsion: photography by combined transmitted and incident light. (× 170). (Tissue kindly supplied by Dr. H. Levi)

not all have the same initial energy. They show a spectrum of energies, ranging from a maximum value down to zero. The maximum energy (E_{max}) is characteristic of the particular isotope. The shape of the energy spectrum varies, too, from one isotope to another. Some, like phosphorus-32, are bell-shaped, with relatively few β's at the high and low ends of the energy spectrum and a peak in between; others, like calcium-45 or carbon-14, have the peak probability of emission nearer the low energy part of the spectrum. Several spectra are illustrated in Fig. 10. Marshall[3] presents useful data on the calculation of energy spectra for isotopes emitting β particles.

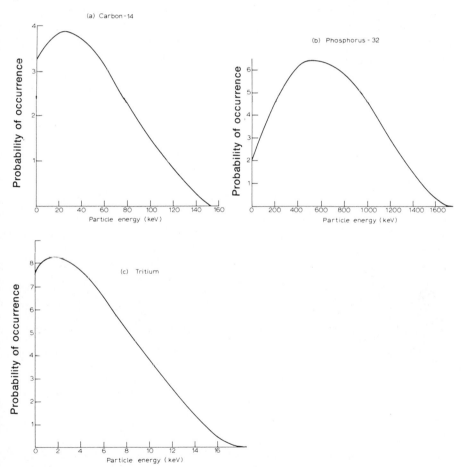

Fig. 10. Energy spectrum of 3 β-emitting isotopes. (a), Carbon-14, $E_{max} = 155$ keV. (b), Phosphorus-32, $E_{max} = 1.7$ MeV. (c), Tritium, $E_{max} = 18.5$ keV.

The range of maximum energies is far greater for β particles than for α's. Tritium has an E_{max} of 18 keV, while for some isotopes the figure is over 3 MeV. It is clear at once that β tracks will be far more variable in length than the tracks from α particles.

This variability becomes even more pronounced when one considers the passage of a β particle through matter. Having the same mass and charge as an orbital electron, a β particle may impart sufficient energy to it to eject it from its orbit but, since we are now dealing with a collision between two table tennis balls, the β particle itself will be deflected from its course. This mutual repulsion between the β particle and the electrons near which it passes is the principal means by which the β particle loses energy. It produces a random, buffeting effect on the path of the β particle, with large numbers of small deviations of course, at times summating to give a curved path, at times cancelling out to approximate the path to a straight line (Fig. 11). Very occasionally, the collision between the β particle and an electron imparts sufficient energy to the latter to send it careering off as a δ ray with

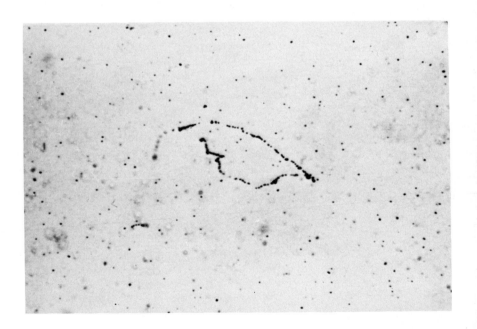

Fig. 11. Photomicrograph of a β-track recorded in Ilford G5 emulsion. The large, densely packed grains at the termination of the track can be readily recognised; by contrast, the grains are smaller and more widely spaced at its origin. The acute angle at the extreme left represents a nuclear collision: several other nuclear collisions can be seen in the terminal part of the track. The initial energy of this particle was about 190 keV. ($\times$ 700)

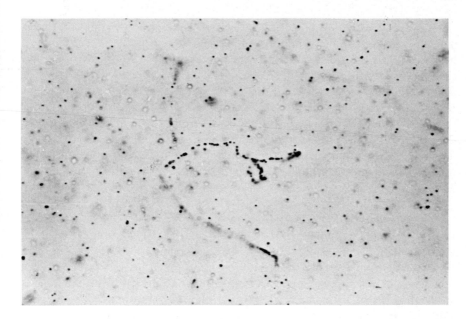

Fig. 12. Part of a β-track recorded in Ilford G5 emulsion. The track appears to branch, giving two endings, both with large, closely packed grains. This is a collision with an orbital electron, which has had sufficient energy imparted to it to produce a recognisable track. By convention, the longer track is considered to be the β particle, the shorter one the electron, or δ track. ($\times 700$)

a track that can be recognised (Fig. 12). Since the β particle and the δ ray are both electrons, they cannot be distinguished from each other by their behaviour. By convention, the one with the shorter track is called the δ ray, and the other is considered to be the original β particle. Where a recognisable δ ray occurs, the β particle undergoes an abrupt change of course and loses a considerable amount of energy to the δ electron; the two tracks separate at practically 90°.

A β particle may also lose energy in large packets by passing close to an atomic nucleus. The positive charge of the nucleus accelerates the β particle as it approaches, may cause it to change direction, and exerts a braking effect on it as it departs. The energy lost by the β particle in this process appears in X-rays (Bremsstrahlung) which are not recorded by the emulsion as a recognisable track. The net effect of a nuclear collision, as far as it can be seen in the emulsion, is that there is a sudden and often considerable change of course, a sharp angle in the track, which is usually quite distinct from the gradual curves that result from random buffeting by orbital electrons (see Fig. 11).

The rate at which β particles lose energy is much lower than that of α particles, so their ranges in emulsion are much greater. The term "range" has been used in two rather different contexts in the literature. In some cases the point-to-point distance travelled by the β particle is meant, in others, the net distance which it travels from the source. Since the path of a β particle is so tortuous, and may even twist back towards its origin, these two measurements may be very different. In order to prevent confusion, the point-to-point distance travelled by the β particle is best referred to as its track length, and its penetration into matter as its radius, since this is the radius of the sphere around the origin of the particle that contains the entire track. It is clear that the radius cannot be greater than the track length.

The track length of a β particle of initial energy 20 keV is about 3 μm in Ilford G5 emulsion; for an initial energy of 6 MeV, the length becomes about 10 mm. For an α particle of 6 MeV, the comparable figure is 26 μm. It is obvious that the rate of energy loss is so much lower for β particles that one cannot expect anything like the same density of latent images along their path. In fact, the β particle may pass through many crystals without imparting sufficient energy to them to create a latent image. Other crystals will have the minimum energy given them to render them developable: still others will receive considerable amounts of energy, resulting in the deposition of much more metallic silver within them than the bare minimum needed for development. So the track of a β particle through nuclear emulsion consists of silver grains, some small, some larger, with many gaps. It is never quite straight, but always shows small deviations, and there will be occasional abrupt changes of direction (nuclear collisions) and, more rarely, places where the track branches into two (δ ray formation). The distribution of small and large grains and of gaps along the track is more or less random, and there will sometimes be several grains closely spaced, sometimes regions where the grains are widely separated (Figs. 11 and 12).

But the distribution of grains along the track is not completely random. As can be seen from Figs. 11 and 12, the grains tend to be larger, and more closely spaced, towards the end of a track. This part of the track also tends to be more tortuous than the rest of it. Both these phenomena have the same cause. Over a wide range of energies above 500 keV, the rate at which a β particle loses energy is practically constant. As the β particle is slowed down to below 500 keV, the rate at which it loses energy increases, gradually at first, then more rapidly until, right at the very end of the track, the rate of energy loss is about 8 times as high as it was at 500 keV. Inevitably, the number and size of latent images per micron of track also increase towards the end of the track.

The higher the sensitivity of the emulsion, the more grains will be recorded per micron of track. A relatively insensitive emulsion can record the last few microns of a β track, where the rate of energy loss is high; but it would fail to yield more than a few widely spaced grains earlier in the track where the β particle is still highly energetic, and losing its energy more slowly. These grains represent crystals of silver halide in which, by chance, sufficient energy has been dissipated for the formation of a latent image, and they might well be so widely spaced that they could not be identified as a continuous track. Similarly, the development of an emulsion becomes critically important with energetic β particles. Powerful development will lift more latent images to the level of visible grains in the track. This will at the same time increase the frequency of background grains, making a high grain density in the track necessary for its recognition.

By comparison with α particles, then, β particles are much more difficult to record. Technically, the emulsions required must be much more sensitive, and development must be more strictly controlled. The long and irregular β tracks are more difficult to recognise, and cannot be extrapolated back with certainty to indicate their point of origin. It is impossible to know, when a β particle enters the emulsion, whether one has the complete track of a β of X KeV, or the final portion of the track of a very much more energetic particle. Most of these difficulties are inherent in the properties of the β particles themselves, and so cannot be side-stepped.

Many of the factors governing the recording of β particles in nuclear emulsions have now been investigated. Zajac and Ross[4] and Levi, Rogers, Bentzon and Nielsen[5] present most of the experimental data available for β particles of initial energies less than 400 keV, and the agreement between these two sets of results is good.

From a combination of these results, Levi et al. have calculated equations describing many of the basic properties of β particles in this energy range. The track length in Ilford G5 emulsion is related to the initial energy of the particle by the equation,

$$\log \bar{L} = 1.59 \log E - 1.51$$

where $\bar{L}$ is the mean track length in microns, and E is the initial particle energy in keV. The numbers of grains produced by β particles, under plateau conditions of development, are related to their initial energies by the equation

$$\log \bar{G} = 1.19 \log E - 0.74$$

where $\bar{G}$ is the mean number of developed grains per track (Fig. 13). The relationship between the number of grains in the track and the track length is given by (Fig. 14),

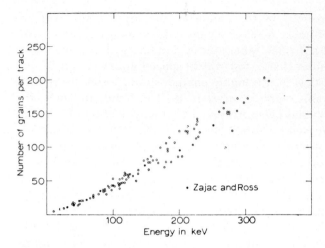

Fig. 13. The relationship between the number of grains per track and the initial energy of the β particle. Open circles represent individual tracks recorded in Ilford G5 emulsion: solid circles represent mean values for groups of particles in Kodak NT-4 emulsion, determined by Zajac and Ross (1949). (NT-4 was formerly produced by the Eastman Kodak Company). (From Levi et al., 1963)

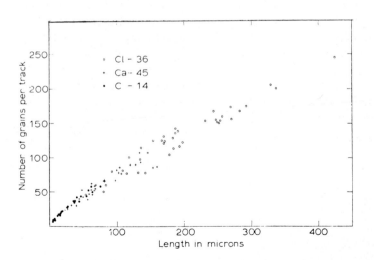

Fig. 14. The relationship between the number of grains in the track and the grain-to-grain track length for individual β-tracks, recorded in Ilford G5 emulsion. (From Levi et al., 1963)

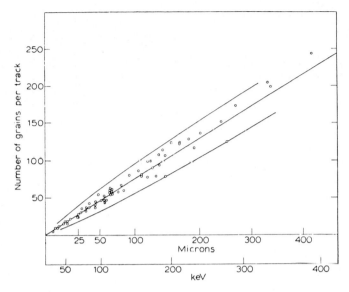

Fig. 15. The numbers of grains in β-tracks related to the track length and to the presumed initial energy of the β particles. Open circles represent individual tracks recorded in Ilford G5 emulsion: the lines represent the mean and 95% confidence limits of measurements made by Zajac and Ross (1949) on groups of monoenergetic electrons recorded in Kodak NT-4 emulsion. (NT-4 was formerly available from Eastman Kodak Company). (From Levi et al., 1963)

$$\log \bar{G} = 0.747 \log \bar{L} + 0.385.$$

There is considerable scatter about the mean value for both the track length and the number of grains per track, the coefficient of variation in both cases being about 20%.

From these data, it is possible to calculate the probable initial energy of a β particle from either its track length or the number of grains in its track in Ilford G5 emulsion. These relationships are summarised graphically in Fig. 15.

The track radius, defined as the radius of the smallest sphere centred at the origin of the β particle which contains every silver grain in the track, is a much more variable parameter, as might be expected from the very irregular track patterns which β particles characteristically produce. In the rather narrow range of initial energies between 20 and 150 keV, the relationship between track length and radius can be expressed as (Fig. 16)

$$\log \bar{R} = 0.816 \log \bar{L} - 0.042.$$

It is thus possible to calculate the radius of the sphere about a point

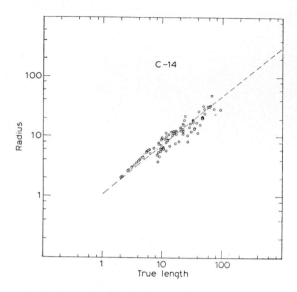

Fig. 16. The relationship between track radius and the grain-to-grain track length for β particles from carbon-14. The radius is that of the smallest sphere of G5 emulsion, with the first grain of the track as its centre, that contains every grain in the track. This relationship is energy dependent and should not be used for other isotopes. (From Levi et al., 1963)

source which will contain a given percentage of the tracks originating from the source. For instance, a sphere of radius 20 μm will contain every silver grain in 90% of the tracks originating from a point source of carbon-14 at its centre. The remaining 10% of the tracks will leave this sphere of G5 emulsion at some point. The corresponding radius around a point source of calcium-45 is 45 μm.

If one considers the individual silver grains produced by β particles originating from a point source of carbon-14 surrounded by emulsion on all sides, 29% of them will lie within a radius of 5 μm from the source, 50% within 9 μm, 75% within 17 μm, and 90% within 25 μm.

The measurement of radioactivity in terms of the number of disintegrations taking place within a source in a given exposure time is technically very difficult, and will be discussed in detail in Chapter 12. But it is clearly possible, on the basis of this mass of data on the behaviour of β particles in Ilford G5 emulsion, to count the number of tracks or even the number of individual silver grains within a given volume of emulsion, and to estimate from this the disintegration rate within a labelled source at a known distance from the volume examined.

It cannot be emphasised too strongly that nuclear emulsions provide a recording medium for charged particles which is versatile, sensitive, and reliable. Biologists have hardly begun to exploit the full potentialities of this detector, as can easily be realised by looking at the highly sophisticated measurements made by physicists with similar basic techniques[6-8]

Apart from the β-emitting isotopes used experimentally, the autoradiographer will meet background β particles from a number of sources. Potassium-40 is a naturally occurring β-emitter encountered in glassware, and carbon-14 will be found in the gelatin of the emulsion. In addition, all the stray background ionising radiations that man is heir to can produce δ rays, or secondary electron tracks.

These, then, are the particles which will concern the autoradiographer in practically every experiment. β Particles are often likened to drunks, starting out with a long stride and a spurious confidence, weaving through matter with increasingly frequent changes of direction, and hesitant, shorter steps until the original impetus is lost and the track ends. Whatever recording system is used, from a monolayer of minute crystals to a layer of emulsion 100 μm or more in thickness, the characteristics of β particles must be understood if the patterns of silver grains produced by them are to be interpreted sensibly.

RAYS

There is little more to be said about these. As stated on p. 45 they betray their presence only by infrequent secondary electrons, which may produce background grains or even tracks, indistinguishable from those of β particles. γ rays are only a source of additional background to the autoradiographer, unless he is working with macroscopic specimens at very low resolution.

For this latter type of work, it is clear that the emulsion should be highly sensitised, and its crystals should be large, to give the maximum blackening for the incident radiation. It is often possible to increase the efficiency of recording for γ rays significantly by placing a layer of material of high atomic number, such as lead, directly against the emulsion, sandwiching the latter against the specimen. This dense material acts as an intensifier, increasing the probability of secondary electrons, which are the actual particles registered by the emulsion.

OTHER IONISING RADIATIONS

(a) *Extra-nuclear electrons*
Some isotopes undergo rearrangements of the neutrons and protons in their nuclei without the ejection of a charged particle, losing energy as a

low energy γ photon. This may have a significant probability of losing its energy to one of the orbital electrons in the inner K or L shells. This electron will leave the atom, behaving like a β particle in every respect. Since it does not originate from the nucleus directly, but comes from the electron shell, it is not called a β particle but an internal conversion electron. The gap left in the inner electron shell is filled by one of the outer electrons of the same atom, and the transition is accompanied by the emission of low-energy photons which, since they do not come from the nucleus but from the electron shell, are called X-rays, not γ rays. These X-rays may, in turn, eject an outer electron from orbit, a so-called Auger electron.

The decay schemes of such isotopes may be very complex, with a number of electrons of specified energies and probabilities. Iodine-125 is an example of such an isotope. It produces 1.64 electrons per disintegration; their energies and probabilities of emission are given in Appendix 4.

The interest of such extra-nuclear electrons for the autoradiographer is that they tend to have very low initial energies, in the range of those emitted by tritium. They thus share with tritium the advantage of giving very good resolution in most autoradiographic systems. Iron-55[9], chromium-51[10] and technetium 99m[11] are further examples. A useful discussion of such isotopes is presented by Forberg, Odeblad, Söremark and Ullberg[12].

(b) *Positrons*

A few isotopes produce positrons, which have the same mass as electrons, but carry a single positive charge: sodium-22 is an example. These lose energy in repeated buffetings by orbital electrons, in rather the same way that β particles do. The typical tail of the β track is not produced, however, since the positron ends its brief existence by annihilating with an orbital electron, giving off two quanta of γ rays of 511 keV each. Positron emission and electron capture are related phenomena, and may occur in the same disintegration. In electron capture, one of the innermost electrons is captured by the nucleus. This event in itself gives rise to no detectable trace in nuclear emulsion.

(c) *Cosmic rays*

These consist of highly energetic charged particles which bombard the upper atmosphere continually. At ground level cosmic radiation is a mixture of secondary and subsequent products of this bombardment. Some components of this radiation are easily recognised in thick layers of nuclear emulsion, by their straight tracks of great length, marching through the emulsion with occasional δ tracks originating from them (Fig. 17).

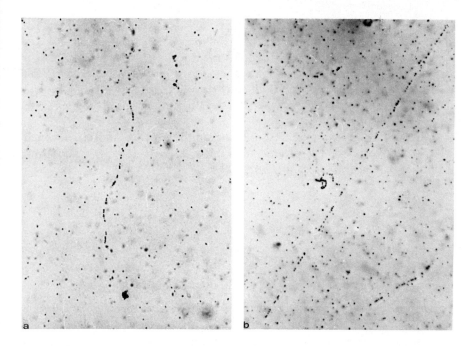

Fig. 17. Portions of the tracks of (a) a β particle and (b) a high energy cosmic ray. The latter is quite straight: such tracks can often be followed for very long distances through thick emulsion layers. A δ ray can be seen coming from it. By contrast, the β-track has many changes of direction as a result of buffeting by orbital electrons, producing a characteristic, irregular track, even in the absence of nuclear collisions (Fig. 11). Tracks recorded in Ilford G5 emulsion. ($\times 540$)

(d) *Neutron activation*

A number of elements have no satisfactory radioactive isotope, the only known ones having half-lives so short that their commercial production and use are impossible. There has been some interest recently in the possibility of irradiating tissue sections with high neutron fluxes in the attempt to produce these isotopes in situ, recording their presence either during or immediately after the irradiation. Boron-10 decays under high neutron fluxes to lithium-7 with the emission of an α particle, with an extremely short half-life[13]. There are obvious difficulties in the use of nuclear emulsions in this type of autoradiography, however, as the high levels of neutrons, γ rays and secondary electrons tend to produce very high background levels. These experiments have turned more and more to plastic films as detectors: the latter will record α or heavier particles, but not electrons or photons[14]. An excellent discussion on the uses of these solid state detectors is presented by Thellier, Stelz and Wissocq[15].

Neutron activation is also used in the localisation of plutonium in tissues[16]. The radioactive isotope has such a long half-life that its detection in minute quantities is very difficult. Under neutron irradiation the plutonium undergoes fission, and the fragments can produce tracks in emulsions or solid state detectors.

These are examples of the technique of radioactivation analysis, in which the radioactive isotope is produced in the specimen itself by irradiation, usually with neutrons.

REFERENCES

1 R.J. Bayly and H. Weigel, *Nature*, 188 (1960) 384.
2 E.A. Evans, *J. Microscopy*, 96 (1972) 165.
3 J.H. Marshall, *Nucleonics*, 13 (1955) No. 8, 34.
4 B. Zajac and M.A.S. Ross, *Nature*, 164 (1949) 311.
5 H. Levi, A.W. Rogers, M.W. Bentzon and A. Nielsen, *Kgl. Danske Videnskab. Selskab. Mat.-Fys. Medd.*, 33 (1963) No. 11.
6 P. Demers, *Ionographie*, University Press of Montreal, 1958.
7 C.F. Powell, P.H. Fowler and D.H. Perkins, *The Study of Elementary Particles by the Photographic Method*, Pergamon, London, 1959.
8 W.H. Barkas, *Nuclear Research Emulsions, Part I*, Academic Press, New York, 1963.
9 D.M. Parry and N.M. Blackett, *J. Cell Biol.*, 57 (1973) 16.
10 P. Ronai, *Intern. J. Appl. Radiation Isotopes*, 20 (1969) 471.
11 R.F. Barth, J. Clancy and J.M. Pugh, *J. Microscopy*, 109 (1977) 211.
12 S. Forberg, E. Odeblad, R. Söremark and S. Ullberg, *Acta Radiol. (Ther.)*, 2 (1964) 241.
13 R.H. Stinson, *Can. J. Bot.*, 50 (1972) 245.
14 K. Becker and D.R. Johnson, *Science*, 167 (1970) 1370.
15 M. Thellier, J. Stelz and J.-C. Wissocq, *J. Microscop. Biol. Cell.*, 27 (1976) 157.
16 W.S.S. Jee, R.B. Dell and L.G. Miller, *Health Physics*, 22 (1972) 761.

CHAPTER 4

The Resolution of Autoradiographs

In the previous chapter, we have seen that any isotope that emits β particles does so over a wide spectrum of energies, and that the path taken by each β particle is unpredictable, and may have many abrupt changes of direction. The initial course taken by a β particle ejected from the source appears to be randomly determined, so there is an equal probability of the track starting off in any direction.

If one considers a point source emitting β particles, which is surrounded on all sides by a layer of nuclear emulsion thicker than the maximum track length of the particles, clearly the source will lie at the centre of a sphere of developed grains, and the density of grains will decrease as one proceeds outwards from the source towards the perimeter of the sphere (Fig. 18a). The diameter of the sphere will be determined by the greatest track radius, as defined in the last chapter, which is related to the maximum energy of β particles emitted. The distribution of silver grains within the sphere will be a complex function, influenced by the shape of the energy spectrum of the isotope, amongst other factors.

For the purposes of calculation, it is usually assumed that the density of silver grains falls off as one proceeds away from the source as the inverse of the square of the distance. In fact, the density decreases more rapidly than the inverse square law would suggest, as shown by Levi, Rogers, Bentzon and Nielsen[1].

If one slices such a sphere of emulsion in half (Fig. 18b), the source now has silver grains on one side only. This is the situation when the source is supported on a glass slide, and covered with an emulsion layer thicker than the maximum range of the emitted particles. This model, where the emulsion is on one side of the source only, and is thick relative to the ranges of emitted particles, is used in track autoradiographs, and also very frequently in the autoradiography of tritium at the light microscope level: it will be discussed in more detail later. One half of the space previously occupied by emulsion has now been replaced by glass, which is considera-

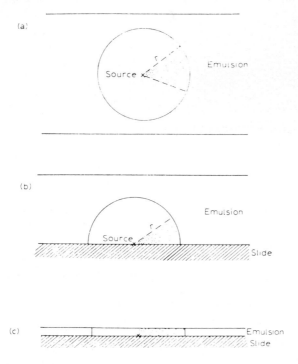

(a)

(b)

(c)

Fig. 18. Diagrams illustrating the three source–emulsion relationships most commonly used in autoradiography. (a), The source is completely surrounded by emulsion to a thickness greater than the maximum track radius (r). (b), The source is mounted on a support, such as a glass slide, and covered on one side only by a thick layer of emulsion (c), The typical grain density autoradiograph, with a layer of emulsion that is thin relative to the track radius on one side of the source.

bly less dense than emulsion. The shape of the hemisphere of grains in the remaining emulsion is distorted slightly in consequence. Particles that enter the glass will have a longer path than they would have done in emulsion. If they are scattered back across the glass–emulsion interface, they will produce grains in the emulsion which have a wider distribution about the source than is seen when the source is completely surrounded by emulsion.

If one now considers a thin section through the centre of the sphere of emulsion (Fig. 18c), this approximates more nearly to the usual histological autoradiograph. The source, supported on a glass slide, is covered on the other side by a thin layer of emulsion, and many of the particles emitted will travel through the emulsion into the air beyond. If the emulsion is viewed perpendicularly to the plane of the slide, the source appears as the centre of a circular area of silver grains. Once again, altering the density of the medium that originally occupied the rest of the sphere from that of emulsion to that of air has the effect of increasing the diameter of this circle, and lowering the grain densities within it. The ranges of β particles in air are so much longer than in emulsion, and those β's which are scattered back into the emulsion re-enter over such a wide area, that they effectively raise the background by a minute amount, which is usually not detectable.

The three models illustrated in Fig. 18 cover practically every type of autoradiograph. In track autoradiography, sources may be suspended in emulsion or covered on one side only with a relatively thick emulsion layer, to give the sphere or hemisphere type of geometry. A tritium source covered by an emulsion layer 3 or more microns thick is another example of the hemisphere model. Most autoradiographs produced for the light or electron microscope are examples of the grain density model, where the emulsion layer is thinner than the maximum track length. Since the grain density model is the most widely used, it is not surprising that most of the studies of resolution have been made on this type of autoradiograph.

DEFINITIONS OF RESOLUTION

If one considers the type of preparation shown in Fig. 18c, a plot of the grain density along a line passing through the source would look rather like the curve shown in Fig. 19.

The grain density is highest over the source, falling off symmetrically as the distance from the source increases on either side. It is very difficult to define the precise diameter of the circular area of silver grains around the source. The grain density falls away gradually to background levels and the end point is impossible to determine with certainty. Resolution is therefore usually defined in terms of some other function of the grain density, which is more convenient to measure in practice.

Unfortunately, different authors have chosen different functions of the grain density curve as the basis for definitions of resolution. The choice of definition has a considerable influence on the numerical value given to the resolution in any particular set of experimental conditions. Before comparing the resolution claimed by one author with that stated elsewhere, it is a good idea to make sure that they are both talking about the same thing. In particular, it is important to distinguish between definitions based on point sources, and those based on linear or extended plane sources; and between definitions based on grain density (the number of grains per unit area of emulsion) and on the percentage of the total grains produced by the source.

Let us view Fig. 18c from above. The point source has a high density of

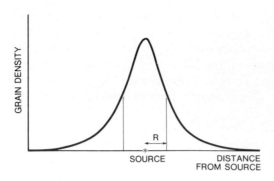

Fig. 19. A graph showing the distribution of silver grains, in arbitrary units of distance, around a point source in a grain density autoradiograph. R indicates the resolution, defined as the distance from the source at which the grain density falls to one half of that found over the source itself.

grains over it: concentric rings of emulsion taken further and further from the source have fewer and fewer grains in them. If one takes concentric rings of uniform width, one can count the number of grains in each ring, and plot this figure against distance from the source to produce a curve of grain distribution around the source, from which some definition of resolution can be derived. Alternatively, one can calculate the grain density in each ring. But since the area of each ring is much larger than the area of the ring immediately inside it, it is obvious that the curve of *grain density* will fall off much more steeply than the curve of *grain distribution*. A definition of resolution based on some function of the grain density about a point source will necessarily give smaller values for the resolution than a definition based on the numbers of grains within given distances of the source.

A linear source of radioactivity can be considered as an infinite number of point sources arranged along a line, and each point source will produce grains in just the same pattern as a single point source (Fig. 20). Of the grains produced by any one such point source, only those in a direction at right angles to the line will lie as far from the line as from their point of origin: those β particles emitted in the direction of the line will give grains near the line, however far the grains may be from their actual point of origin. If we divide the emulsion into bands of equal width, similar to the concentric rings around the point source, and count the number of grains in each band, this measure of resolution will fall off more rapidly than the grain distribution around a point source in identical conditions of autoradiography. But since the area of each band is the same, there is now

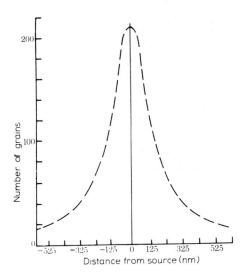

Fig. 20. The distribution of silver grains around a linear source of tritium, autoradiographed with Ilford L4 emulsion and developed with Microdol X. The HD value is the distance from the line which delimits the area containing half the grains produced by the source. (Data from Salpeter et al., 1969)

no difference between curves of grain distribution and curves of grain density.

Clearly, the definition of resolution determines the actual, numerical value to be attached to it in any given set of conditions.

In photography, resolution is usually defined in terms of the distance that must separate two objects before they can be distinguished from each other. If two sources were to lie very close to each other, the silver grains produced by the one would overlie the other, and vice versa, giving a density curve with a single peak like that shown in Fig. 20. As the sources are moved further apart, this peak would broaden and become lower, and finally the grain density midway between the two sources would be lower than the two peaks over the sources. Gomberg[2] examined this situation, defining resolution as the distance which must separate two sources of equal strength if the grain density between them falls to half that seen over each source. The same model has formed the basis for an extensive theoretical treatment of resolution by Bleecken[3–5].

A widely accepted definition of resolution was proposed by Doniach and Pelc[6]. With a grain density curve similar to that in Fig. 19, they defined the resolution as the distance from the source at which the grain density is one half that seen directly over the source. They presented a series of calculations of resolution based on a point source, and on a linear source at right angles to the plane of the emulsion.

Nadler[7] used a rather similar definition. He worked with extended planar sources, and calculated the distance from the edge of the source at which the grain density fell to one-half of that above the edge itself. Hill[8] used the same definition in his experimental study of resolving power with tritium, as did Creese and MacLagan[24].

Bachmann and Salpeter[9] defined resolution as the radius of the circle about a point source which contains half the silver grains produced by that source, the half-radius (HR). This definition is a useful one when one comes to the analysis of autoradiographs at the electron microscope level. Here, one is faced with a number of scattered grains, and the problem is to identify their source. If a circle of the same size as the resolution circle is placed around each grain, there is by definition a 50% probability that the point of origin of the β particle lies within it.

In addition to the references given above, theoretical treatments of resolution in the grain density type of autoradiograph have been presented by Gross, Bogoroch, Nadler and Leblond[10], Lamerton and Harriss[11], Caro[12], and Pelc[13].

Recent work by Salpeter, Bachmann and Salpeter[14] on resolution at the electron microscope level has been based on a linear source, with measurements of grain distribution in bands of equal width parallel to the source. Such "hot lines" have proved convenient to make, and the collection of grain numbers within the parallel bands at increasing distances from the line is relatively simple. It is essential to continue counting from bands up to a considerable distance from the line, however. If counting stops before grain densities have reached background levels, not all the grains produced by the line will be recognised, and the estimated resolution will be too small. Salpeter and her co-workers[14-16] have carried out these measurements under many different conditions, and determined the pattern of distribution of developed grains around the line, and the distance from the line that contains half the grains produced by it — the half-distance (HD) (Fig. 20). Although changing the thickness of the specimen, or the crystal diameter of the emulsion, produced considerable changes in the HD, they found that the *shape* of the curve of distribution of grains around the line was similar in all cases. This observation allowed them to produce a generalised curve of grain distribution in units of HD which has since been shown to fit autoradiographs at the light microscope level also[17]. A mathematical treatment of resolution[14] has also been presented.

This generalised curve of grain distribution (Fig. 21) is a valuable step forward. It is a relatively simple matter to calculate the predicted grain

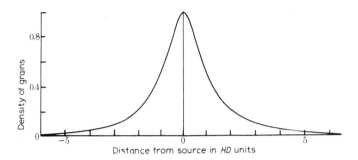

Fig. 21. The distribution of silver grains around a linear source of tritium, measured in units of HD. This generalised curve describes the distribution found in all the experimental conditions examined by Salpeter et al., 1969.

distributions around sources of various sizes and shapes, once the distribution around a linear source is known. In Fig. 22, predicted grain densities are shown for hollow circular sources and for solid discs of various radii, and many other shapes are calculable. One only needs to determine the relevant value of HD for one's own autoradiographs to be able to predict grain distributions around any labelled biological structure that can be approximated to a fairly simple geometrical shape. The concepts of HD and HR will be discussed several times in the rest of this book, particularly in connection with the interpretation of autoradiographs.

FACTORS THAT GOVERN THE RESOLUTION OF GRAIN DENSITY AUTORADIOGRAPHS

The mathematical treatment of resolution appeals to some people. It is a complex affair, and every treatment has to start from simplifying assumptions. Since these are likely to be different in each paper, there is an endless field for argument and for the exercise of mathematical expertise. Fortunately for the poor biologist, all the theoretical models proposed to date agree in indicating that certain geometrical factors in the preparation of the autoradiograph have an important bearing on how good or bad the resolution will be. What is more, experimental test systems, such as the test charts developed and used by Stevens[18] and Bleecken[19], have confirmed the effects of altering these factors on the resolution.

Whatever definition of resolution you adopt, and whatever system of autoradiography, the following factors will help to determine the resolu-

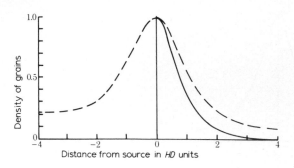

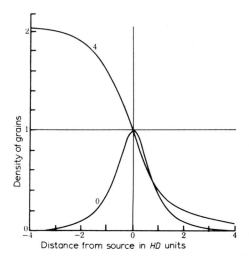

Fig. 22. Generalised curves showing the computed distributions of silver grains around sources of different shapes. Positive values on the abscissa lie outside the edge of the source, negative values inside the edge. (a), The solid line represents a point source, the dotted line a hollow circular source of radius 4 HD units. (b), Line O relates to a point source, line 4 to a solid disc of radius 4 HD units. (From Salpeter et al., 1969)

tion you obtain. They can be divided into factors operating in the source itself, and factors concerned with the nuclear emulsion.

(a) *Factors in the source*

(i) *The choice of isotope.* Clearly, if the maximum energy of the β particles coming from the source is increased, there will be particles producing silver grains at greater distances from the source. The whole curve of distance versus grain density shown in Fig. 19 will be broader, and the resolution will be poorer. In grain density autoradiographs, there is a very definite improvement in resolution to be gained from using an isotope with lower energies. If, as with iodine, several isotopes are commercially available, it is common sense to use the one with the lowest energy of emission if high resolution is needed. The great value of tritium in autoradiography lies in its very low maximum energy (18 keV).

(ii) *The distance between source and emulsion.* If the source is separated from the emulsion, instead of being in direct contact with it, this is similar in principle to examining a thin section which does not pass through the centre of the sphere of emulsion described in Fig. 18. The grains lying over the source when this section is viewed from above are now at a distance from it, instead of being immediately adjacent, and they have a correspondingly lower probability of being hit. This section through our notional sphere of silver grains has a more uniform grain density over it, without a sharp peak corresponding to the position of the source itself. If we use the definition of resolution based on the distance at which the grain density is half that over the source, the resolution will be poorer; the HR will also be greater.

Curiously enough, the further away from the source that the emulsion layer lies, the further up the notional sphere of silver grains one proceeds, until the circle containing all the grains from the source begins to get appreciably smaller. Theoretically, examining a tangential section through the upper pole of the sphere of grains will give very good resolution. In practical terms, the efficiency of looking at an emulsion layer so far from the source is so low that this sytem is not deliberately used in practice. There are a very few situations with tritium, however, and maybe with other isotopes with extra-nuclear electrons of low energies, in which increasing the distance between source and emulsion may improve the resolution, provided the density of the intervening layer is reasonably high.

It is sometimes necessary to interpose between source and emulsion a layer of inert material, such as polyvinyl chloride, to prevent chemography. Any such layer should be kept as thin as possible, to minimise its adverse effects on resolution.

(iii) *The thickness of the source.* If the source, instead of being the dimensionless point source that we have been considering so far, has an appreciable thickness, the situation that results is really a special case of separation of source from emulsion. The rod-shaped source, lying perpendicular to the plane of the emulsion, can be regarded as a series of point sources. The uppermost of these will have the characteristic curve of resolution seen in Fig. 19. The next below will, in effect, be separated from the emulsion by the first layer of source, and its resolution will be correspondingly worse, and so on down the rod. The resultant resolution of the complete rod will be a composite of the resolutions at each level, and will be worse than that of a point source in contact with the emulsion.

Once again, while that statement is true for grain density autoradiographs of the type illustrated in Fig. 18c, hemisphere autoradiographs of the type in Fig. 18b may, in certain circumstances, show an improved resolution with increased section thickness. With tritium and similar isotopes of very low energies, an effect may operate, like the one described above in considering source–emulsion separation, to give a slight improvement in resolution from a thicker source. The effect is discussed in Appendix B of a paper by Salpeter, Fertuck and Salpeter[16].

Though I have described an increasing thickness of source as making the resolution worse, this effect cannot be entirely separated from a consideration of the density of the source. If the material which makes up the source is of high density, the β trajectories through it will be shorter and more tortuous than with a source of lower density. β Particles will enter the emulsion over a circle of smaller diameter, and the resolution will be better. Gupta, Moreton and Cooper[20] claimed that heavy metal staining of electron microscope sections labelled with tritium reduced the HD by up to 30% compared to unstained sections. Salpeter[21] did not find a significant reduction in resolution from staining of tritium-labelled sections but, in later work[16], found the HD reduced by about 20% with iodine-125. There is no logical reason to predict a different effect of heavy metal staining with isotopes as similar in energies of emission. The disagreement between these workers is really on the significance of the difference. Gupta's work suggests an improvement large enough to take into account in experiments with tritium; Salpeter's work, on the other hand, suggests that the difference is not large relative to the other experimental variables of the technique.

In general, however, increasing source thickness leads to poorer resolution.

(b) *Factors in the emulsion*

(i) *The thickness of the emulsion.* The effect of increasing the thickness of the emulsion layer is rather similar to increasing the thickness of the source. A thin section of emulsion through the centre of the sphere of emulsion of Fig. 18 will give grain densities which decrease rapidly with increasing distance from the source. A section of emulsion parallel to this, at some distance from the source, will have grain densities that are still maximal over the source itself, but which decrease relatively slowly as one moves away from the point over the source. Superimposing several of these sections of emulsion at different levels, the resultant grain density does not fall off nearly so abruptly with distance from the source as is the case with a thin emulsion layer in contact with the source.

In the majority of autoradiographs, possible increments in emulsion thickness are small by comparison with the maximum range of the particles in emulsion. In the few cases where this is not so, an increase in emulsion thickness may just superimpose the upper pole of the hemisphere of grains on the rest of the hemisphere, with little significant change on the total resolution, or even a very slight improvement. Obviously, increases in emulsion thickness above the maximum range of the particles in emulsion will have no effect at all on resolution.

(ii) *The length of exposure.* In a situation such as that shown in Fig. 19, it is clear that the probability of a crystal being hit by a β particle is far higher directly over the source than it is several microns away from it. It is equally true that the probability of a crystal over the source receiving two hits from different β particles is much higher. Such a crystal, unfortunately, cannot produce two developed grains in consequence.

If exposure is held to a relatively short time, so that no crystal has a significant chance of a double hit, the grain distribution will show the sharp peak characteristic of Fig. 19. If exposure continues until the probability of double hits over the source becomes considerable, the grain density over the source will not increase proportionally to the increase occurring away from the source, resulting in a broader, flatter curve, and poorer resolution. Ultimately, all the crystals over the source will be hit, and further increases in exposure will only give more grains in the areas of lower grain density away from the source.

Complete saturation of the emulsion in this way is fairly obvious. But deviations from a linear response in the volume of emulsion immediately over the source occur at much lower grain densities than is commonly realised. The resolution is often better in autoradiographs with low grain densities, quite apart from the advantages to be gained, from the point of view of statistics, in working with many lightly labelled sources rather than a few heavily labelled ones (see p. 299).

(iii) *The size of the silver halide crystals.* We have seen in Chapter 2 that the latent image produced in a silver halide crystal as a consequence of the passage of a β particle will lie at a preformed sensitivity speck, and not necessarily on the path of the crystal. The developed grain that results will have a point of contact with the parent crystal at this latent image site, but may not coincide with the position of the crystal.

If the size of the crystal is smaller, the position of the developed grain must correspond more closely to the trajectory of the particle. This will result in improved resolution.

(iv) *The size of the developed silver grains.* There is some evidence that a developed silver grain may not correspond in position with the halide crystal from which it grew[9,22]. Obviously the two must make contact at the latent image, if nowhere else, but the fully developed grain is considerably larger than the parent crystal and may lie to one side or other of the latter. The smaller the size of the developed grain, the nearer will its centre be to the latent image. Where the highest possible resolution is needed, in electron microscope autoradiography, developing methods that produce small grains have a small but significant influence on the overall resolution.

(v) *The sensitivity of the emulsion.* On p. 52, it was explained that β particles have a lower rate of energy loss per micron of track at higher energies, up to about 500 keV. In other words, at higher energies they have a smaller probability of losing enough energy in a silver halide crystal to make it develop. If the emulsion has a low sensitivity, or if development is incomplete, an even higher threshold of energy loss per crystal must be reached before a developed grain will result.

Immediately over the source, β particles will enter the emulsion at the complete spectrum of energies of the isotope being used. At a distance from the source, only relatively low energy portions of the ends of the tracks will be encountered. If the emulsion is so insensitive that high energy particles do not have a reasonable probability of causing developed grains along the initial part of the track, relatively few grains will appear over the source. This is equivalent to requiring more than one hit per crystal to give a grain over the source while, in the areas away from the source, one hit per crystal will be enough. This would clearly distort the shape of the curve of grain density against distance, giving poorer resolution. Although this is likely to be only a minor effect by comparison with other factors that have been listed, it is worth taking into account in experiments where the very highest resolving power is required.

In summary, these are the eight factors which determine the resolution of an autoradiograph. Their relative importance varies from one experiment to another. The following sections will illustrate how they apply to various experimental situations.

RESOLUTION AND THE AUTORADIOGRAPHY OF MACROSCOPIC SPECIMENS

The techniques described in Chapter 14 are largely based on the use of X-ray film. In these techniques, many of the above variables are already determined. The thickness of the emulsion layer is fixed by the manufacturer, with the one exception that it is sometimes possible to sharpen up an autoradiograph by removing the layer of emulsion furthest from the specimen after photographic processing (p. 332). Crystal size and sensitivity and the size of the developed grains have little or no significant effect on resolution at the macroscopic level. The important variables in this group of techniques are the choice of isotope, source–emulsion separation and the thickness of the source. The length of exposure may also be significant.

Isotopes of very low energy, such as tritium, pose technical problems with X-ray film (p. 317). Carbon-14 and sulphur-35 give excellent results, however, and better resolution than is obtainable with phosphorus-32, for instance.

The emulsion layer should be in direct contact with the specimen, separated from it only by the anti-abrasion coating of the X-ray film itself. Separation of specimen from emulsion considerably reduces the clarity and sharpness with which radioactive structures in the specimen can be identified.

The thickness of the source may be determined by other factors, but there is a bonus to be had from preparing a thin specimen. Resolution improves considerably.

Overexposure produces a very dramatic loss of resolution, and care should be taken to work with degrees of blackening over the specimen that lie within the range of variation of optical densities produced on the film by the radioactive ladder (Fig. 23), if resolution is at all important.

In many preparations for macroscopic viewing, such as chromatograms, high resolution is of very little significance compared to the ability to detect low levels of activity.

RESOLUTION IN GRAIN DENSITY AUTORADIOGRAPHY AT THE LIGHT MICROSCOPE LEVEL

Two different situations must be considered here: the very low energy isotopes of which tritium is the prime example, and the much larger group of radioactive isotopes of higher maximum energies, including carbon-14,

74

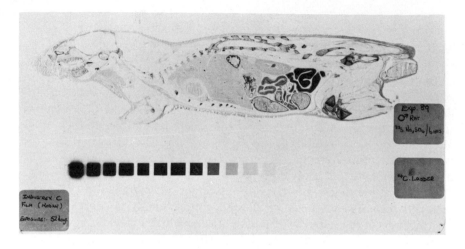

Fig. 23. A whole-body autoradiograph of a rat 6 hours after the injection of [³⁵S]sodium sulphate, prepared with Kodak Industrex C X-ray film. The ladder represents known concentrations of radioactivity in doubling dilutions. (Material prepared by Dr. S.A.M. Cross)

sulphur-35 and phosphorus-32. In addition, we shall look briefly at the resolution possible with autoradiographs of freely diffusible substances.

(a) *Tritium*

With a maximum energy of 18.5 keV, the β particles from tritium seldom travel more than 2 μm in biological specimens or 1 μm in nuclear emulsions. This is the most significant single fact in considering resolution with tritium at the light microscope level.

As can be seen from the data in Fig. 24, increasing section thickness from 0.4 to 1.0 μm has a negligible effect on the HD. Increasing emulsion thickness from a monolayer of L4 to the 3.5 μm of AR-10 has a similar, very small effect on resolution. Increasing the crystal size from 0.14 μm (L4) to 0.26 μm (NTB2) again increases the HD only slightly.

Source–emulsion separation reduces the efficiency of the autoradiograph without affecting resolution much. Since the particles from tritium have a very high rate of energy loss due to their low initial energies, emulsion sensitivity cannot effectively discriminate between the start and end of their trajectories, thus influencing resolution. At the light microscope level, the effect of developed grain size on resolution can be ignored.

Since all the hit crystals will lie within a very small radius of the source, the probability of double or even multiple hits on crystals immediately over the source will be correspondingly high. This can increase the HD up to a

Isotope	Section thickness (μm)	Emulsion	Developer	H.D. (μm)
^{3}H	0.5	L4	D-19	0.32
	0.9	L4	D-19	0.35
	0.45	NTB-2	Dektol	0.37
	0.45	AR-10	D-19	0.35
^{14}C	0.5	L4	D-19	0.7
	0.9	L4	D-19	1.1
	0.45	NTB-2	Dektol	0.7
	0.45	AR-10	D-19	2.0

Fig. 24. HD values for various combinations of section thickness, emulsion and isotope in grain-density autoradiography at the light microscope level. Liquid emulsions were applied as dense or overlapping monolayers. (Data provided by Dr. M.M. Salpeter)

value of perhaps 0.5 μm where the autoradiograph is heavily overexposed.

It seems that, whatever one does, the HD for tritium at the light microscope level will be between 0.3 and 0.4 μm.

(b) *Isotopes of higher maximum energy*

For grain density autoradiographs at the light microscope level, β-emitting isotopes with a maximum energy of 100 keV or more present totally different problems to those of the very low energy group, as seen in Fig. 24.

First and foremost, the lower the maximum energy the better the resolution[19]. Carbon-14 will give a lower HD than phosphorus-32 under identical conditions.

Source–emulsion separation degrades the image severely, as does an increasing thickness of source. The effect of increasing the thickness of the emulsion layer is similar.

Since a larger number of crystals are within the range of the β particles emitted, problems from double hits do not become significant until far higher doses of β's have been received by the emulsion. The size of crystals and of developed grains are relatively unimportant, but the sensitivity of the crystals becomes quite important.

For the best results, a low maximum energy, a very thin specimen in direct contact with a very thin layer of highly sensitised emulsion, and an exposure timed to keep small the probability of double hits on crystals immediately over heavily labelled areas will give the highest resolution.

Some idea of the HD values likely with various combinations of experimental variables will be gained from Fig. 24.

(c) *The autoradiography of diffusible materials*

When the labelled source is freely diffusible in the biological specimen, movement of radioactivity from its position during life can contribute to the final distribution of silver grains. In a poorly controlled specimen, diffusion during freezing and subsequent freeze-drying can give a horrible spread of radioactivity. If freezing is rapid and subsequent handling avoids thawing at any stage, there is no reason why the resolution obtained with diffusible materials should not be comparable with that of fixed and embedded histological specimens[23,24]. An interesting observation has been made by Clarkson and Sanderson[25], who studied cryostat sections of plant material. After freeze-drying, a section of 10 μm nominal thickness would remain at its original thickness over cellulose cell wall, but would dry down to a very thin layer of almost molecular dimensions over the lumen of conducting vessels. This effect can obviously introduce considerable variations in resolution into a single section. While animal tissues are usually rather more homogeneous, the same effect must occur to some extent over cytoplasm and extracellular fluid. Possible movements of labelled molecules during freezing and freeze-drying will become much more critical as techniques evolve for studying freely diffusible molecules by electron microscopic autoradiography.

RESOLUTION IN ELECTRON MICROSCOPE AUTORADIOGRAPHY

(a) *Theoretical studies*

Autoradiographic resolution has received more intensive and anxious attention from electron microscopists than from anyone else. The discrepancy between the resolving power of the electron microscope itself — in the range of 5–10 Å — and that of the photographic emulsion as a detector for β particles is so great that the measurement of autoradiographic resolution, the study of the factors that influence it and the devising of methods to manipulate those factors have occupied a lot of time and effort. Since all studies at this level start with the thinnest possible emulsion layer, a packed monolayer of crystals of small diameter, they all are concerned with the grain density model (Fig. 18c). In fact, our understanding of the factors affecting resolution in this system at the light microscope level owes a great deal to studies in electron microscope autoradiography.

The first major attempt to tackle this problem was made by Caro[12].

Starting with a thin biological specimen in direct contact with a monolayer of silver halide crystals he identified three factors which contribute to the scatter of silver grains around a radioactive source. The first is the geometric relationship between source and emulsion, which determines the pattern of distribution of the paths of the β particles through the emulsion layer. Next is the fact that a latent image forms in a hit crystal at a preformed sensitivity speck, and not necessarily on the path of the particle itself. The third factor is the uncertain relationship between the centre of the developed grain and the latent image. By employing a physical developer, p-phenylenediamine, Caro reduced his last source of uncertainty considerably, producing tiny, comma-shaped grains which indicated the site of the latent image with considerable precision.

Caro[12] did not evaluate separately the second and third sources of error, but concentrated on the geometric error, which appeared to contribute the major part of the uncertainty in identifying the source of radioactivity in the specimen. He broke this geometric error down into three components: (1) the possibility that a β particle will have sufficient energy to reach a crystal at a given distance from it; (2) the effective size of the crystal, taking into account its shielding from the source by adjacent crystals; and (3) the solid angle subtended to the source by the unshielded portion of the crystal. Assuming a point source of tritium 500 Å away from a packed monolayer of crystals 1000 Å in diameter, Caro found that the grain density at a distance X from the source (D) could be approximated to the grain density over the source itself, (D_0), by the equation

$$D = D_0 e^{-1.6X}.$$

Using the same approach, Caro[12] was able to calculate predicted grain densities around labelled bacteriophages, and found experimentally a reasonably good agreement with prediction. He also calculated predicted grain densities for two thicknesses of source and two diameters of crystal (Fig. 25). These curves show that reducing the crystal diameter from 1000 Å to 100 Å (far smaller than any available emulsion with useful sensitivity) would only improve the resolution by a relatively small amount. By contrast, reducing the specimen thickness from 500 Å to 100 Å would have a much greater impact on resolution.

This paper focussed attention on many points of significance. Caro[12] drew attention to the undoubted benefit that would result from smaller crystals in terms of increased information per unit area of emulsion, for instance, even if the predicted resolution, strictly defined, did not seem to improve very dramatically. He defined resolution as twice the distance at which the grain density falls to half its value directly over a point source,

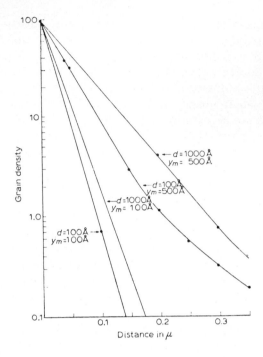

Fig. 25. The expected distribution of silver grains in an electron microscope autoradiograph of a point source labelled with tritium. d indicates the diameter of the silver halide crystals forming a monolayer over the source: y_m indicates the vertical distance from the source to the crystals. The improvement in resolution from reducing the thickness of the source is greater than that predicted from a smaller crystal size. (From Caro, 1962)

and on this basis predicted a value of 860 Å for a tritium source 500 Å away from a packed monolayer of 1000 Å crystals.

Bachmann and Salpeter[9] produced the next significant treatment of resolution in electron microscope autoradiography. Their method of autoradiography differed slightly from Caro's: in particular, they introduced a carbon layer of 50–60 Å between specimen and emulsion. The three factors identified as contributing to the scatter of grains about a point source by Caro[12] were again recognised, but this time an attempt was made to evaluate the influence of crystal and grain diameters on resolution. The photographic error (E_p) due to the combination of these two factors was taken to be

$$E_p = \sqrt{\frac{a^2}{5} + \frac{b^2}{12}}$$

where a is the diameter of the silver halide crystal and b the diameter of the developed grain. Fig. 26 shows their calculated values of E_p for two emulsions and several developing routines. It appears from their calculations that the diameter of the crystal is more important than that of the developed grain in its influence on resolution.

Emulsion	Developer	Diameter of halide crystals	Size of developed grains	Mean photographic error	Efficiency
Ilford	Microdol X	1000–1600	2000–3000	900	1 : 10
L4	p-Phenylenediamine	1000–1600	400– 700	600	1 : 13
Kodak	Dektol	300– 550	800–1500	400	1 : 37
NTE	Gold latensification and Dektol	300– 550	800–1500	400	1 : 12
	Gold latensification and Elon–ascorbic acid	300– 550	400– 600	280	1 : 9

Fig. 26. The effects of different development routines on the photographic component of resolution and on the efficiency, for monolayers of two emulsions exposed to tritium. Distances are in Ångstrom units. (From Bachmann and Salpeter, 1965)

Bachmann and Salpeter[9] next considered the distribution of β particle trajectories in the plane of the emulsion, or the geometric error (E_g). Fig. 27 shows their model. D is the centre of the developed grain, X the lateral distance between the source (S) and D, and θ the angle between the vertical and the path of the β particle. The distance between S and the emulsion is d, which is a sum of $t_e/2 + t_s/2 + t_i$ where t_e is the thickness of the emulsion layer (the diameter of the silver halide crystals), t_s the source thickness and t_i the thickness of the intervening layer of carbon. The density of β particles crossing the emulsion plane at any point lies between the limiting values of $\cos^3 \theta$ and $\cos^2 \theta$, and the fraction of the total emitted β particles that cross the emulsion plane in a circle of radius $\tan \theta$ is $1 - \cos \theta$. The geometric error was defined as the value of X at which $1 - \cos \theta = 0.5$ or the radius of the circle around the source through which half the β particles reaching the emulsion would travel. The resolution which was defined as the HR, or radius of the circle around a point source which contains half the silver grains produced by it, was calculated to be

$$HR = \sqrt{E_p^2 + E_g^2}.$$

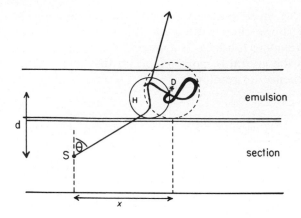

Fig. 27. Diagrammatic model of an electron microscope autoradiograph. The source, S, lies in the mid-plane of the section. The β particle leaves the source at an angle, θ, taking a straight path to the emulsion layer where it hits a crystal, H. The centre of the developed grain is at D. The distance from source to emulsion is taken as d, from the mid-plane of the section to the mid-plane of the emulsion. The projected distance from the source to the middle of the developed grain is x. (Kindly provided by Dr. M.M. Salpeter)

In Fig. 28, the HR is calculated for two types of specimen. It will be seen at once that the resolution appears to be much worse than that calculated by Caro[12] and given above. But it must be remembered (see p. 64) that the

	Specimen I	Specimen II
Section thickness	350	1000
Carbon layer	50	50
Emulsion	NTE	L4
Emulsion thickness	600	1300
Development	Gold–Elon	Microdol X
Photographic error	280	900
Geometrical error	720	1650
Total error	770	1850

Fig. 28. Factors affecting resolution calculated for two different types of electron microscope autoradiograph. Specimen I has been prepared to give the highest resolution: NTE emulsion has now been superseded by Kodak 129-01. Specimen II represents the most common preparation with Ilford L4 emulsion. The total error approximates to the value for HR. It is clear that geometrical factors contribute more than photographic factors to the total error. Distances are in Ångstrom units. (From Bachmann and Salpeter, 1965)

grain density about a point source falls off much more rapidly than the *grain distribution*. If Bachmann and Salpeter's calculations are used to estimate the geometrical error in terms of grain density, similar to Caro's, a value much nearer to his will be found.

This theoretical treatment illustrates well the contributions to resolution that are made by such factors as specimen thickness, and diameters of halide crystal and developed grain. The predicted values of HR for tritium range from 770 Å with Kodak NTE and the best conditions that can probably be obtained at present with sectioned biological material, to 1850 Å for a thicker section covered with a monolayer of L4.

In a later paper, Salpeter, Bachmann and Salpeter[14] looked for experimental verification of these predicted values of HR. This paper is basically empirical. It proved more convenient to study the grain distribution around a radioactive line source than to construct point sources, so a "hot line" was autoradiographed under a number of conditions, and the shortest distance from the centre of each developed grain to the line measured, up to 2 μm on either side. The equivalent concept to the HR for a point source, the HD (or distance from the line within which 50% of all the grains from it lie) has been found, and is listed for the various conditions studied in Fig. 29. Values of HR are not identical with the HD: HD = HR/1.7. When this correction is made, it will be seen that the theoretical predictions suggest a slightly better resolution (by 5–30%) than was found in practice. Once again, Fig. 29 shows clearly the improvements in resolution that follow from reducing the section thickness, and the diameters of halide crystal and developed grain. It was on the basis of these distribution curves that Salpeter, Bachmann and Salpeter[14] derived their universal curves for grain density studies (see pp. 66, 67).

So an HD of about 800 Å can already be obtained with sectioned material, and if only the section thickness and the halide crystal diameter could be considerably reduced, further improvements in resolution would follow. Unfortunately resolution cannot be considered in isolation from all the other factors that make an autoradiograph interpretable; the conflicting requirements imposed by efficiency, adequate contrast and total radioactivity in the specimen frequently prevent the highest resolution from being achieved. These problems are discussed more fully on p. 109.

(b) *Practical considerations*

As experience has accumulated in autoradiography at the electron microscope level, it has become clear that some factors affect resolution more severely than others, and that the list varies with the energies of particles emitted by the isotope[16].

With carbon-14 and sulphur-35, the relatively energetic β particles have

Isotope	Emulsion	Section Thickness	Development	HD
^{3}H[a]	L4 monolayer	120	Microdol X	165
			p-Phenylenediamine	145
		50	Microdol X	145
			p-Phenylenediamine	130
	NTE double layer	120	Dektol	125
		50		100
	NTE monolayer	120		100
		50		80
^{14}C[b]	L4 double layer	100	Microdol X	285
		50		235
	L4 monolayer	100		230
		50		180
	NTE double layer	100	Dektol	250
	NTE monolayer	100		200
^{55}Fe[c]	L4 monolayer	100	Microdol X	130
^{125}I[d]	L4 monolayer	100	Gold-EAS	92
		50		84
		50	p-Phenylenediamine	82
	NTE monolayer	100	Dektol	55
		50		52
^{203}Hg[e]	L4 monolayer	50	D-19	180

[a] Salpeter et al., 1969
[b] Salpeter and Salpeter, 1971
[c] Parry and Blackett, 1973
[d] Salpeter, Fertuck and Salpeter, 1977
[e] Williams, 1977

Fig. 29. Experimentally determined HD values taken from the literature for various combinations of isotope, specimen, emulsion and developer at the electron microscopic level. Distances are all in nanometres.

ranges which are very large by comparison with the thicknesses of sections and emulsion layers that are useful in electron microscopy. The HD increases linearly with increasing section thickness. Increasing the emulsion from a monolayer of L4 crystals to a double layer produces a 40% increase in HD. An emulsion of high sensitivity is needed to give acceptable yields of silver grains, and this factor is usually much more important than crystal size, which has a relatively small effect on the HD. Separation of source from emulsion must be kept to a minimum.

These factors of source or emulsion thickness and source–emulsion separation, are so much more important than the other factors affecting resolution that they virtually determine the HD. With isotopes of even higher maximum energies, these factors will be even more significant.

With tritium, the increase in HD with increasing section thickness is less dramatic than with carbon-14. The HD's obtained with single and double layers of L4 are virtually identical, for reasons discussed on p. 71. Source–emulsion separation also is less serious than with carbon-14, though still affecting the resolution significantly. As these factors become less important, the influence of crystal diameter on the HD increases. Changing from a monolayer of NTE (crystal diameter 0.06 μm) to one of L4 (0.14 μm) produces a 20% increase in the HD. At the same time, it is reasonable to accept a lower sensitivity of crystal. A further slight improvement in HD results from fine-grain developing techniques.

With iodine-125, most of the autoradiographic image at the electron microscope level is due to extranuclear electrons of 3–4 keV. There is no significant effect on the HD due to variations of the thickness of section or emulsion. Going from the small crystals of NTE to the larger ones of L4 increases the HD by 80%. Again, a slight improvement in HD results from fine-grain development procedures.

As far as the staining of the section with heavy metals is concerned, one would predict no significant effect on the HD with carbon-14 or sulphur-35. With tritium, the resolution may be improved slightly[20,21]. With iodine-125, a reduction of around 20% will be seen in the HD[16].

THE RESOLVING POWER OF TRACK AUTORADIOGRAPHS

The grain density model of autoradiography that has been discussed above is by far the most commonly employed. In some experimental situations, however, there are advantages in using much thicker emulsion layers, and studying the tracks of the β particles, rather than variations in the density of developed grains. This corresponds to the sphere and hemisphere concepts of Fig. 18 and it has interesting consequences for the resolution obtainable.

Let us consider a source mounted on a glass slide, and covered with nuclear emulsion to a thickness greater than the maximum range of the particles emitted — such a situation with carbon-14, for instance, would require a thickness of about 50 μm[1]. If the track density is kept fairly low, by exposing for a short time, it is possible to recognise the individual tracks and to distinguish their beginning from their termination. Each track can, therefore, be traced back to the point at which it entered the emulsion. It is possible to plot the distribution of points of entry around a source, just as the distribution of individual grains was plotted in Fig. 19. The resolving power of this system then becomes, by analogy, the radius of the circle around the source which contains one-half of the points of entry of β particles originating from the source.

If the source is assumed to be very small, is in direct contact with the emulsion, and is mounted on a plane surface, it is clear that about half the particles emitted will enter the emulsion directly from the source, the other half being directed into the glass slide (Fig. 30). A small proportion of those fired into the emulsion will subsequently be scattered back into the slide, leaving a curved section of track visible. Similarly, some of the tracks entering the glass slide will be scattered into the emulsion at a distance from the source. But if one considers all the tracks that appear in the emulsion, the majority enter directly from the source, and present no particular problem in recognition. The problem of high resolution in track autoradiographs becomes one of defining these points of entry with the greatest possible precision.

In this situation, some of the recommendations for obtaining high resolution in grain density autoradiographs require modification. To begin with, there is little improvement in resolving power on using a lower energy isotope. In fact, with high energy emitters, such as phosphorus-32, the tracks tend to be much straighter near their origin than with lower energies, such as carbon-14. It is easier to extrapolate a straight track back towards its origin than a crooked one. With particles of higher energy there

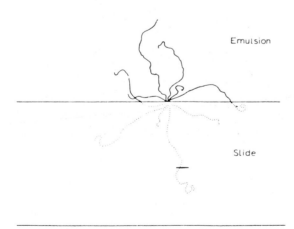

Fig. 30. The distribution of β particle trajectories around a source mounted on a glass slide and covered with a thick layer of emulsion. Half the particles enter the emulsion directly from the source, the other half entering the glass. Scattering across the glass–emulsion interface occurs in both directions, but particles cross this plane randomly over a considerable area.

will be a smaller percentage of very tortuous short tracks, with which the uncertainty as to the precise point of entry may be considerable.

The separation of source from emulsion will still reduce the resolving power considerably, and increasing thickness of the source will have a similar effect. The thickness of the emulsion layer now is an advantage, however: it does not need critical control, provided only that it is thick enough to enable one to determine which is the beginning of the track and which the end. Background particles entering the emulsion from the air surface should be recognised and rejected with confidence.

The dimensions of the silver halide crystals cannot, unfortunately, be reduced very far, as it is usually not practicable to view a thick emulsion layer except with conventional transmitted light, but their sensitivity becomes a matter of critical importance. If, at the start of a high energy track, the grains are on average 4 μm apart, there may be tracks which appear to start quite a way from the source. If the sensitivity is so much higher that the average grain spacing becomes 1 μm, the first grain in each track will correspond much more closely to the point of entry of the β particle.

The exposure time will have to be kept short as, if there are many tracks criss-crossing in one microscope field, the problems of identification become impossibly complex.

Given a highly sensitised emulsion and correct development, the resolving power of track autoradiography is probably comparable to that of conventional thin emulsion layers for carbon-14 and sulphur-35, and considerably better for higher energy isotopes such as phosphorus-32.

Alpha tracks. These are straight and short, and very easy to extrapolate back into the source (Figs. 8 and 9). The only problem they raise in considering resolution is that, with a nuclear emulsion of high sensitivity, the tracks become so broad, due to the δ rays, that it is a cylinder that is extrapolated back rather than a line. For the best resolving power, an emulsion of low sensitivity and fairly small grain size offers the best precision — for instance, Ilford K0.

RESOLUTION AND THE INTERPRETATION OF AUTORADIOGRAPHS

It should be quite clear by now that the resolution of an autoradiographic system is a measure of the scatter of developed grains over and around radioactive sources in the specimen. It is *not* a statement of the minimum size of source that can be identified. It may be quite possible to

86

determine the source in a complex specimen when the structures within the specimen are smaller than the HD. The use of the HD value in interpreting the grain distribution in autoradiographs is considered in greater detail in Chapter 11.

REFERENCES

1 H. Levi, A.W. Rogers, M.W. Bentzon and A. Nielsen, *Kgl. Danske Videnskab. Selskab, Mat.-Fys. Medd.*, 33 (1963) No. 11.
2 H.J. Gomberg, *Univ. of Mich. Project* AT (11-1)-70, No. 3, 1952.
3 S. Bleecken, *Naturforsch.*, 23B (1968) 1339.
4 S. Bleecken, *Naturforsch.*, 23B (1968) 1350.
5 S. Bleecken, *Naturforsch.*, 23B (1968) 1478.
6 I. Doniach and S.R. Pelc, *Brit. J. Radiol.*, 23 (1950) 184.
7 N.J. Nadler, *Can. J. Med. Sci.*, 29 (1951) 182.
8 D.K. Hill, *Nature*, 194 (1962) 831.
9 L. Bachmann and M.M. Salpeter, *Lab. Invest.*, 14 (1965) 1041.
10 J. Gross, R. Bogoroch, N.J. Nadler and C.P. Leblond, *Am. J. Roentgenol. Radium Therapy, Nucl. Med.*, 65 (1951) 420.
11 L.F. Lamerton and E.B. Harriss, *J. Phot. Sci.*, 2 (1954) 135.
12 L.G. Caro, *J. Cell Biol.*, 15 (1962) 189.
13 S.R. Pelc, *J. Roy. Microscop. Soc.*, 81 (1963) 131.
14 M.M. Salpeter, L. Bachmann and E.E. Salpeter, *J. Cell Biol.*, 41 (1969) 1.
15 M.M. Salpeter and E.E. Salpeter, *J. Cell Biol.*, 50 (1971) 324.
16 M.M. Salpeter, H.C. Fertuck and E.E. Salpeter, *J. Cell Biol.*, 72 (1977) 161.
17 M.M. Salpeter, G.C. Budd and S. Mattimoe, *J. Histochem. Cytochem.*, 22 (1974) 217.
18 G.W.W. Stevens, *Brit. J. Radiol.*, 23 (1950) 723.
19 S. Bleecken, *Atompraxis*, 14 (1968) No. 8, 1.
20 B.L. Gupta, R.B. Moreton and N.C. Cooper, *J. Microscopy*, 99 (1973) 1.
21 M.M. Salpeter, *J. Histochem. Cytochem.*, 21 (1973) 623.
22 B.M. Kopriwa, *Histochemistry*, 44 (1975) 201.
23 T.C. Appleton, *J. Histochem. Cytochem.*, 14 (1966) 414.
24 R. Creese and J. MacLagan, *J. Physiol.*, 210 (1970) 363.
25 D.T. Clarkson and J. Sanderson (personal communication).

CHAPTER 5

The Efficiency of Autoradiographs

We have seen in the previous chapters that β particles can cause developable latent images in nuclear emulsions. Since the β particles cause the developed grains, it is reasonable to assume that a simple relationship exists under any given set of conditions between the number of β particles leaving the specimen during exposure and the number of silver grains produced by them. In the ideal case, there should be a direct proportionality between the radioactivity of the specimen and the number of silver grains produced by it.

Unfortunately, many factors can influence the efficiency with which an emulsion records β particles. The great advantage which autoradiography has for the biologist is the retention of structural patterns in the specimen; but this implies immediately that variations in shape and density between one component and another within the specimen will also be retained, with all the effects on efficiency which result. In Geiger or scintillation counting, care is taken to reduce every sample to a uniform specimen so that the efficiency of counting is the same for all samples. With the pulse counting techniques, each sample is presented to the same detector; in autoradiography, each sample is covered with very many halide crystals, each an individual detector. So the concept of efficiency assumes great importance in autoradiography. It is very easy to count silver grains and to compare the numbers counted over various specimens in the cheerful assumption that the efficiency of autoradiography has been the same in every case. It is often depressingly difficult to autoradiograph several specimens with the same efficiency.

In this chapter, the various factors which can influence the efficiency of autoradiography will be considered. The interpretation of autoradiographs is a more complex matter, involving the resolution, or scatter of silver grains about the source, as well as the efficiency; this will be considered in a later chapter.

87

DEFINITIONS OF EFFICIENCY

Just as resolution has been variously defined by different autoradiographers, so there are several definitions of autoradiographic efficiency. A common one relates the number of silver grains produced in a layer of nuclear emulsion to the number of particles entering it. This is quite a convenient definition for experiments into efficiency. It is relatively simple to take an extended, uniformly labelled source and to determine the number of particles per unit area per unit time that leaves its surface, using some type of electronic pulse counter. The same source can then be exposed to an emulsion layer, and the number of developed grains counted. The efficiency can then be calculated directly.

In the majority of radioisotope techniques, however, efficiency means something rather different. It is the number of events recorded, as pulses in the case of a Geiger counter, related to the number of radioactive disintegrations taking place in the source. This is a more complex concept, as it introduces many variables in the source itself. If the source is relatively thick by comparison to the range of the β particles emitted, many of them will never reach the recording device at all.

In this book, I shall use "efficiency" to mean the emulsion response relative to the number of radioactive disintegrations taking place in the source during exposure. The term "grain yield" will be used for the mean number of grains produced per β particle entering the emulsion. The grain yield should be a constant for any given grain density autoradiograph; the efficiency, however, can vary enormously from point to point within a single specimen.

This definition of efficiency must be interpreted in different ways with different autoradiographic systems. In a grain density autoradiograph prepared for examination under the light or electron microscope, the number of grains is the emulsion response to be related to the disintegration rate. When the emulsion is examined by the unaided eye, however, individual silver grains are not seen. The intensity of blackening of the film becomes the parameter that is related to the number of β particles emitted by the source during exposure. The effects of this substitution of blackening for grain number will be discussed below in relation to macroscopic specimens. When one comes to track autoradiographs, the unit to be observed and counted is not the individual silver grain, but the arrangement of grains to form a track. In this case, efficiency becomes the ratio between the number of tracks produced in the emulsion and the number of radioactive disintegrations taking place in the source.

FACTORS AFFECTING THE EFFICIENCY OF GRAIN DENSITY AUTORADIOGRAPHS

(a) *Factors in the source*

(i) *The choice of isotope.* The initial energies of the β particles emitted by the specimen influence the efficiency of grain density autoradiographs mainly from two effects. The first is that of distance travelled. The higher the energy of the particle, the further it is likely to penetrate through specimen and emulsion. The second is the rate of energy loss. As we have seen (p. 52), the lower the energy of a β particle, the greater the rate of energy loss. In photographic terms, this is the same as saying the low energy particles produce larger deposits of silver at the latent image specks of hit crystals. These two effects determine the interaction of particle energy with each of the other variables listed below, and will be discussed in relation to each of them.

(ii) *The distance between source and emulsion.* Separating the source from the emulsion will, in general, reduce the efficiency. The degree to which this effect operates will depend largely on the initial energy of the β particles. With tritium, for instance, the particle track is so short anyhow that even the thinnest inert layer will screen off a significant fraction of β particles from the emulsion. Perry, Errera, Hell and Durwald[1] have shown that 675 nm of formvar are sufficient to reduce the grain count over a tritium-labelled cell by 85%. With higher particle energies, this effect becomes less important until, with phosphorus-32, separation by several microns of material of density about 1.1 will not produce a significant reduction in efficiency. There is no appreciable difference, for instance, between a film that has been exposed in direct contact with a paper chromatogram containing phosphorus-32, and one that has been separated from the source by a single thickness of paper.

Curves for the external absorption of β particles are available in the literature, relating the density of the intervening layer (measured in mg/cm^2) to the proportion of β particles which penetrate through to the other side. Maurer and Primbsch[2] present such a curve for tritium, based on measurement with biological material; Pelc and Welton[3] have published a theoretical curve which is practically identical.

(iii) *The thickness of the source.* A flat, planar source which is only a few Ångstrom units thick in contact with a nuclear emulsion, will produce a given number of β particles entering the emulsion per hour. A source twice as thick, but otherwise identical, will project β particles into the emulsion at twice the rate. But if one continues this process of increasing the thickness of the source, plotting source thickness against counting rate (Fig. 31), the counts will soon fall below the line which indicates a direct proportionality between the two. Above a given thickness, adding a further layer of radioactive material to the bottom of a source does not produce the expected increase in counting rate, since this layer is separated from the emulsion by the overlying layers, which absorb some of the β particles coming from the lowest layer before they can be counted. With still greater source thicknesses, the departure from linearity becomes more and more marked until the addition of further radioactive material has no effect whatever on the count rate, which remains constant.

This phenomenon is known as self-absorption, and the general curve shown in Fig. 31 is found with any isotope emitting β particles, though the precise thicknesses vary with the energy spectrum of the isotope. In the initial part of the curve, where the increase in count rate is nearly enough linear for all practical purposes, the source is said to be infinitely thin. On the plateau, the source is said to be infinitely thick. At infinite thinness, the observed count rate is proportional to the amount of radioactivity in the source; at infinite thickness, the count rate is proportional to the concentration of radioactivity within the source, but is independent of its absolute amount.

It is obvious that the thickness of the source must affect the efficiency, except at infinite thinness. How important is this factor with various isotopes?

With phosphorus-32 and tissue sections 5 μm thick, one is working at

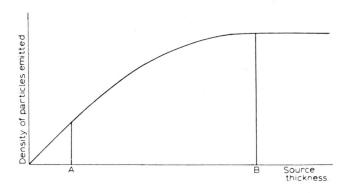

Fig. 31. The effect of increasing source thickness on the density of β particles leaving the surface of the source. Up to point A, source thickness is nearly proportional to particle density, and the source is "infinitely thin". Above point B, increases in source thickness have no effect on particle density, and the source is "infinitely thick".

infinite thinness. With carbon-14, a simple and direct method of calculating self-absorption has been presented by Hendler[4]. Using his figures, and assuming the density of fixed biological material to be 1.3, a section 1 μm thick would absorb 5.5% of the emitted β particles or, stated in the more usual terms, there would be 94.5% transmission of the particles emitted in the direction of the emulsion. At 5 μm, there is 82.5% transmission and at 10 μm, 70.3%. These figures compare very well with data presented by Dörmer and Brinkmann[5], who cut sections of uniformly labelled [^{14}C]methacrylate at various thicknesses and autoradiographed them under carefully controlled conditions. Their graph for self-absorption with carbon-14 is shown in Fig. 32. They found slightly less self-absorption than is predicted above, but their material had a density of around 1.1. From their data it is clear that sections up to 1 μm can be considered as infinitely thin, and that self-absorption corrections are not required in this range. This agrees with data from Housley and Fisher[6], who found no difference in efficiency between sections at 0.3 and 1.0 μm. Above 1 μm, corrections for section thickness become increasingly important.

Tritium represents the extreme case where self-absorption is the most important single factor in considering autoradiographic efficiency.

Falk and King[7] used uniformly tritiated methacrylate to investigate self-absorption, and their results are shown in Fig. 33. There is no evidence of an initially linear relationship between thickness of methacrylate and observed grain densities, suggesting that infinite thinness had already been

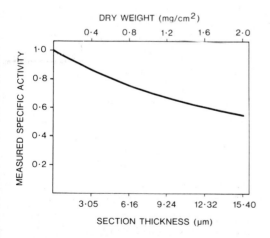

Fig. 32. Self-absorption from [^{14}C]methacrylate sections of constant specific activity but increasing thickness, determined by autoradiography under constant conditions with Kodak AR-10 stripping film. (Data from Dörmer, 1973)

AUTORADIOGRAPHIC EFFICIENCY AS A FUNCTION OF SECTION THICKNESS
WITH A SOURCE UNIFORMLY LABELLED WITH TRITIUM

Section thickness (μm)	Grain count per 100 μm^2	Efficiency (%)
0.5	1.22 ± 0.11	15.72
1.0	1.72 ± 0.09	11.08
2.0	2.18 ± 0.10	7.04
3.0	2.61 ± 0.10	5.62
4.0	2.99 ± 0.15	4.82
5.0	3.25 ± 0.15	4.19
6.0	3.28 ± 0.15	3.52
7.0	3.20 ± 0.18	2.94
8.0	3.25 ± 0.12	2.62
9.0	3.26 ± 0.14	2.34
10.0	3.16 ± 0.12	2.03

Fig. 33. Grain densities over sections of [^{3}H]methyl methacrylate, autoradiographed with Eastman Kodak NTB-2 emulsion under constant conditions. The efficiencies have been calculated on the basis of the known specific activity of the methacrylate. (Data from Falk and King, 1963)

left behind at 0.5 μm. After 3 μm, there is little increase in grain density, and infinite thickness is reached at 5 μm. These sections should have approximately the same density as fixed biological material. The dramatic effects of self-absorption on the efficiency of these autoradiographs can be clearly seen, with values ranging from over 15% at 0.5 μm to 2% at 10 μm. Very similar results were obtained by Oja, Oja and Hasan[8], using sectioned rat liver labelled with tritium.

Pelc and Welton[3] have computed self-absorption curves for tritium which match well with Falk and King's[7] data. Careful analyses of self-absorption with tritium are presented by Maurer and Primbsch[2] and by Perry et al.[1] with biological material. Maurer and Primbsch[2] showed that the density of nucleolus, nucleus and cytoplasm differed, even in fixed and sectioned material, from 0.018 mg/cm^2 for nucleus to 0.095 mg/cm^2 for nucleolus. So not only is self-absorption critically important with tritium, but it can and does vary from point to point within a biological specimen. As an example, self-absorption reduces the number of β particles reaching the emulsion to 40% relative to an infinitely thin specimen if the radioactivity is distributed in fixed and sectioned nuclear material 3 μm thick; comparable figures for cytoplasm would be 2 μm, and for nucleolus 0.6 μm.

An interesting discussion of these problems of density and thickness of the source was published by Perry[9].

Dörmer[10] has also drawn attention to the varying efficiencies that can occur within the same specimen with tritium, in an excellent review article on quantitative aspects of autoradiography. He presented data showing the cytoplasm of cells in the erythroid series to be significantly denser, in autoradiographic terms, than otherwise comparable precursors of white blood cells, due to the accumulation of iron in haemoglobin.

At what thickness does self-absorption become critical with tritium? Vrensen[11] has published figures which suggest significant self-absorption between polymer-embedded sections 400 and 1000 Å thick. In a subsequent study by Salpeter and Szabo[12], an alternative explanation for Vrensen's data was suggested: this was dose dependence, which is discussed in detail on p. 36.

We have already seen (p. 70) that heavy metal staining improves the resolution of electron microscopic sections labelled with iodine 125 and, less significantly, with tritium. It is likely that the increased scattering and absorption of β particles by heavy metal atoms responsible for this effect would also influence efficiency. In short, with stained biological material, even a thickness of 1000 Å is probably too thick for self-absorption to be ignored.

(iv) *Backscattering.* When a source is mounted on a support, such as a glass slide, some β particles will inevitably enter the support, only to be scattered back into the emulsion again. This scattering of particles back into the emulsion or specimen is largely a function of the atomic numbers of the elements making up the support: the higher these are, the greater the backscattering. So a section or smear on a glass slide will have more particles scattered back towards specimen and emulsion than an otherwise similar specimen on a plastic slide. If the support has a rough surface rather than a smooth one, backscattering is very slightly increased.

Of course, these backscattered particles may not affect the efficiency at all if the specimen is thick enough to absorb them before they reach the emulsion. So the degree to which backscattering affects efficiency is linked to source thickness and to β particle energy.

There have been few attempts to measure this effect directly. Dörmer[10] reviews the topic briefly, and it is discussed in electron microscope work by Salpeter and Salpeter[13].

(b) *Factors in the emulsion itself*

(i) *The thickness of the emulsion.* In a situation where the emulsion thickness is less than the maximum length of the β track, it is obvious that increasing the thickness of the emulsion will increase the total number of developed silver grains. For high energy β-emitters like phosphorus-32, and emulsion layers in the range of 1–20 μm, it is reasonable to assume that there is a linear relationship between the thickness of the emulsion layer and the efficiency of the system. For carbon-14, the first 2 μm of emulsion will effectively reduce the number of β particles entering the next layer, and so on, and the effect of increasing the emulsion thickness will not be linear.

This factor, then, also varies in importance with the maximum energy of β particle from the isotope under examination. With tritium, this is so low that the maximum track length in nuclear emulsion is about 3 μm, and the maximum radius, or penetration into the emulsion, probably about 1 μm. In this case, it is irrelevant how thick the emulsion layer is, provided it does not drop below 2 μm. As will be seen in Chapter 13 (p. 292), this means that liquid emulsion techniques, in which the emulsion thickness is not strictly controlled, can be used for quantitative measurements with tritium. With isotopes of higher energies, however, more accurate control of emulsion thickness is essential if grain counts are to be compared between one source and another.

At the electron microscope level, emulsion layers are so thin that the efficiency of recording for even tritium can be significantly affected by changes in emulsion thickness. Vrensen[11] drew attention to the increase in efficiency resulting from an emulsion layer two crystals thick, by comparison with the more usual monolayer. Salpeter and Szabo[12] have estimated that this increase is of the order of 20% with L4 emulsion. For iodine-125, this use of a double layer does not produce a significant increase in efficiency[14].

(ii) *The dimensions and packing of the silver halide crystals.* If one compares a layer 4 μm thick of an emulsion with silver halide crystals 0.5 μm in diameter with a similar layer of crystals of equal sensitivity, but of diameter 0.1 μm, it is obvious that a β particle travelling through the latter layer will hit many more crystals. The layer of emulsion made up of small crystals closely packed can carry more information per unit volume, and produce more developed grains per β particle. Provided the decrease in crystal size is not achieved at the expense of sensitivity, such an emulsion is more efficient than one with larger crystals.

(iii) *The sensitivity of the emulsion.* In Chapter 3 (p. 50), the track characteristic of a β particle travelling through nuclear emulsion was described. Not every crystal along the trajectory of the particle will have sufficient energy imparted to it to produce a developed grain, though every crystal hit will have some energy dissipated in it. If an emulsion is highly sensitised, less energy is needed to make an activated crystal develop into a grain. In such an emulsion, the grains will be closely spaced along the track of the particle. With lower sensitivities, only those grains that have had a large amount of energy dissipated within them will be developed.

Tritium is once again a special case. The energies of these particles are so low that they have a small probability of affecting more than one grain anyhow. At these low energies, the rate of energy loss is very high, and a relatively insensitive emulsion will record tritium with almost the same efficiency as a highly sensitised one — and often with a lower background. This is the rationale behind the use of Ilford K2 instead of K5 for tritium, or NTB-2 instead of NTB-3, a procedure which will be mentioned again on p. 297.

(c) *Factors operating during exposure*

(i) *Instability of the latent image.* It frequently happens that similar autoradiographs are exposed for different times, and that the efficiency in the longer exposure is found to be lower than in the shorter. As we saw in Chapter 2, latent images can regress during exposure, an effect that is favoured by the presence of excessive moisture and of oxidising agents. Generalised latent image fading can be due to incomplete drying of the emulsion, and the presence of atmospheric oxygen: in this case, it tends to affect large areas of emulsion, often being more severe over the section, where presumably drying was less thorough. It may vary from slide to slide, and will tend to be worse with longer exposures. This effect should be excluded from any quantitative experiment by the use of suitable control slides (see p. 302).

Of all the factors affecting efficiency, this one is probably the cause of the most disagreements between laboratories. Many autoradiographers are used to operating with considerable latent image fading: this gives a very favourable signal-to-noise ratio in the final autoradiograph, at the expense of a longer exposure (p. 124). It can also produce widely differing efficiencies of recording, unless the effect is recognised. If unduly low efficiencies are found in the literature, look first at the conditions of drying and exposure for an explanation.

Latent image fading over the specimen, due to a chemical interaction between parts of the specimen and the emulsion, can also occur. This negative chemography will produce wildly varying efficiencies over a single section in extreme cases (Fig. 34), and should also be excluded by control slides in which a high and uniform density of latent images has been created *before* exposure begins.

96

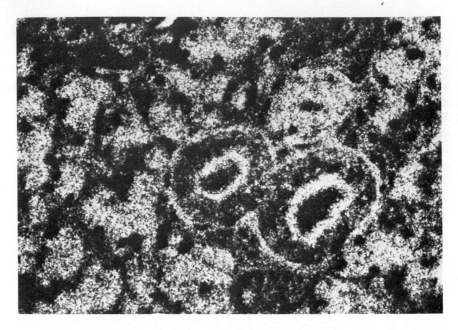

Fig. 34. An emulsion layer fogged by light and "exposed" under conditions used for autoradiography in contact with a cryostat section of rat submandibular gland for several days before development. Light areas indicate latent image fading, which is anatomical in distribution, occurring particularly over the cytoplasm of acinar cells and over two adjacent profiles of collecting duct. (× 750). (From Rogers and John, 1969)

(ii) *The temperature of exposure.* Several papers report variation in efficiency with the temperature of exposure of the autoradiograph[15,16]. Under constant conditions of development, the higher the temperature of exposure the higher the efficiency. It seems that the latent images formed at lower temperatures are smaller or more disperse. Particle physicists have frequently exposed nuclear emulsions at low temperatures, however, and successfully achieved reasonable efficiencies. The secret lies in slightly increasing the development. I have been able to expose routinely at − 70°C and, by adjusting the development, have found no significant loss of efficiency by comparison with exposure at 4°C, nor has the background been significantly higher.

(iii) *The duration of exposure.* Each halide crystal can only produce one developed grain. If many crystals are hit by more than one β particle, the efficiency will clearly be lower than if an identical autoradiograph is exposed for a shorter time.

Perry[9] has an excellent discussion on the probability of double hits. Biologists like myself are often surprised to find that this probability is finite, though very small, for each β particle after the first that travels through an emulsion layer. It does not become large enough to require correction factors until about 10% of the available crystals have already been hit. With grain density autoradiographs and isotopes of the maximum energy of carbon-14 or higher, it is reasonable to consider crystals at all depths in the emulsion as available. With tritium this clearly is not so, and double hits are a far more serious problem, as a point source in contact with the emulsion can only affect the few crystals that lie within 1 μm radius. If n_{max} is the total number of crystals that can be affected by a source, the observed grain count n_{obs} is related to the number that should have been seen in the absence of double hits, n_{true}, by

$$n_{obs} = n_{max} \left[1 - \left(\frac{n_{max} - 1}{n_{max}} \right)^{n_{true}} \right].$$

(iv) *Dose dependence.* Salpeter and Szabo[12] have drawn attention to a strange phenomenon in electron microscope autoradiography. At very low densities of hit crystals in the emulsion, the efficiency is far higher than at higher densities. This dependence of the efficiency on the dose received by the emulsion layer has not, to my knowledge, been seen in the thicker emulsions used by light microscopists. It is clearly related to the process of development, being more severe with Microdol than with either D-19 or Elon-ascorbic acid after gold latensification. It has been found with tritium, iodine-125 and carbon-14 (Fig. 35).

Interestingly enough, it shows no very direct correlation with the number of β particles entering the emulsion layer, when one compares carbon-14 with the other two isotopes. There is, however, a good correlation between the density of developable crystals in the emulsion layer and this effect[6]. It looks as if the presence of a rapidly developing latent image may suppress in some way the development of other hit crystals in the immediate neighbourhood. The developer that shows this effect most strongly, Microdol, has a significant solvent action on the silver halide crystals.

(v) *Other factors during exposure.* Reference has already been made to the use of scintillation compounds to try to improve efficiencies during autoradiographic exposure (p. 38). In the case of macroscopic specimens labelled with tritium and recorded on X-ray film, this is an established and very valuable technique, improving the efficiency of autoradiography very considerably. Although a number of papers[17-19] have recently appeared claiming improvements in efficiency in light microscope autoradiography

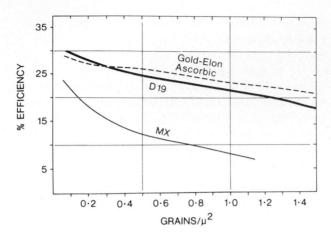

Fig. 35. A graph illustrating dose dependence. Monolayers of Ilford L4 were irradiated with known doses of tritium from a layer of radioactive gelatin of known thickness. The efficiency falls as the density of β particles reaching the emulsion increases. The exposure conditions excluded a significant contribution to this reduction in efficiency from either latent image fading or multiple hits. (From Salpeter and Bachmann, 1972)

by factors up to × 80, I remain at present unconvinced that this is a real effect. The striking thing about these papers is not the high efficiencies claimed for scintillation autoradiography, but the low efficiencies achieved in the absence of scintillation fluid. This, combined with the very variable increases claimed, makes me wonder if some or all of the improvement claimed is not due to better control of latent image fading in the presence of scintillation fluid. A number of laboratories, my own included, have been unable to reproduce these results (p. 365).

Claims for increase in efficiency from exposure in a magnetic field have been described elsewhere (p. 36). Here again the situation is unclear, though the explanation advanced for the way in which the improved efficiency is caused is almost certainly wrong.

(d) *Factors operating during development*

(i) *The choice of developer.* Development is a process of amplification, operating to increase the deposits of metallic silver at sensitivity specks until they reach the level of recognition. From the material in Chapter 2, it is obvious that this process determines the numbers of recognisable grains that are produced from labelled sources.

It is possible to achieve fairly similar efficiencies in many experimental situations with different developers, provided each is used optimally. There are a few situations, however, in which the choice of developer has a considerable influence on efficiency. Developers which have no solvent action are restricted to surface latent images: the usual recipes for Amidol produce such a developer. If negative chemography is reversing surface latent images preferentially, the use of such developers will exaggerate the loss, producing significantly lower efficiencies than other developers with the ability to reach internal sensitivity specks[20].

At the electron microscope level, however, strongly solvent developers such as Microdol are more liable to the dose dependent variations in efficiency described above[12] (p. 36). They are best avoided.

Physical developers in general seem to produce lower efficiencies in electron microscope autoradiography than chemical developers[21]; they also have a reputation for being rather less reliable.

The work of Rechenmann and Wittendorp[22] on development emphasises the importance of selecting the correct reducing potential for optimal development.

(ii) *The conditions of development.* The curve of development against time (Fig. 4) illustrates how important it is to select the optimal conditions for development and to control the variables that could affect it. While it is obviously possible to increase the efficiency by more active development, this is only possible at the expense of higher backgrounds. Attempts to squeeze the highest efficiencies out of autoradiographs by overdevelopment are self-defeating, since the recognition of lightly labelled sources may become more difficult as the background rises. Although the conditions of development affect efficiency, then, the ideal conditions are those which give the best signal-to-noise ratio, and the variables affecting development should be controlled at this level rather than manipulated to produce higher efficiencies.

(iii) *Intensification of the latent image.* The basis for this is discussed on p. 35. Briefly, small deposits of metallic silver at sensitivity specks may have other metals, usually gold, deposited on them before development. Such intensified deposits have a higher probability of development: unhit crystals, forming the main constituent of background in a good autoradiograph, have an unaltered probability of development. The result is to improve the signal-to-noise ratio, sometimes dramatically (Fig. 6).

Gold latensification is most widely used in electron microscopic autoradiography, where high efficiencies are particularly important, but Rechenmann and Wittendorp[22] have recently drawn attention to the value of this step at the light microscope level also. Latensification is most effective where the proportion of hit crystals failing to reach the status of developed grains is highest; in other words, where the ratio of latent sub-images to latent images is high.

These, then are the major factors that determine the efficiency of an autoradiograph. While the factors in the emulsion itself and those operating during exposure and development can and should be controlled, the preservation of structural patterns in the specimen will often produce significant variations in efficiency in the same autoradiograph, making quantitative interpretations difficult.

THE EFFICIENCY OF AUTORADIOGRAPHS OF MACROSCOPIC SPECIMENS

Specimens such as chromatograms or sections through large objects like whole mice or rat brains are usually viewed with the naked eye, or at very low magnifications. At this level, individual silver grains cannot be seen, and blackening in the film is the response that indicates radioactivity. This response is usually measured as optical density. Nuclear emulsions tend to have very small crystal diameters relative to other photographic materials, so that a high density of developed grains is needed to give a visible blackening. In X-ray emulsions the crystal diameter may be ten times greater than in a nuclear emulsion, so that the mass of silver formed by one developed grain is obviously very much more.

With isotopes of the energy of carbon-14 or higher, X-ray film is the obvious choice. Fig. 36 gives a rough idea of the radioactivity needed in a given spot of carbon-14 to give reasonable blackening in a stated exposure time. The large crystal size and the emulsion thickness of about 20 μm provide a recording medium which is very efficient, and which has sufficient resolution for most macroscopic work.

But X-ray films have a coating over the emulsion layer to reduce damage by handling. This anti-abrasion coating is sufficiently thick to screen off most of the β particles from tritium. The combination of this separation of source from emulsion with the self-absorption that is likely in a relatively thick specimen, makes the efficiency of this method far lower for tritium, probably by a factor of about 100 compared to carbon-14. In consequence, several techniques have been described as alternatives to simple apposition to X-ray film: these are discussed in more detail in Chapter 14. They include exposure to a special X-ray film without an anti-abrasion coat[23], or conversion of the β particle energy to photons by exposure in the presence of a scintillator. These steps are capable of improving the efficiency for tritium by factors of about $\times 12$ in the case of whole body sections exposed

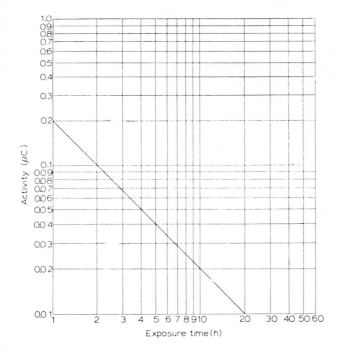

Fig. 36. A graph relating the activity of carbon-14 in μCi applied to a paper chromatogram to the approximate exposure time in hours needed to give a clearly discernible image on an opposed X-ray film. (From Tsuk et al., 1964)

to CEA ^{3}H-Film[23], and up to $\times 5000$ in the case of acrylamide gels exposed at $-70°C$ in the presence of a scintillator, after pre-exposure to a flash of light[24]. Note that the improvement offered by the ^{3}H-Film is limited to the β particles leaving the upper surface of the specimen; that potentially available from scintillation methods extends to disintegrations taking place in the full thickness of the specimen.

If the choice is available, it is better to avoid the use of tritium in favour of isotopes of higher energy, for the autoradiography of macroscopic specimens.

THE EFFICIENCY OF GRAIN DENSITY AUTORADIOGRAPHS FOR THE LIGHT MICROSCOPE

In light microscope autoradiographs with tritium, the most significant variables affecting efficiency are self-absorption and the separation of source from emulsion. The latter should be avoided wherever possible. A few figures from the literature will illustrate the dramatic influence of self-absorption on efficiency. Small bacterial cells such as *Escherichia coli* may be autoradiographed at an efficiency as high as 27% (ref. 25). Mammalian spermatozoa labelled in the DNA of the head, which is about 0.5 μm diameter, can give an efficiency of 17% (ref. 26). Sections of rat liver labelled more or less uniformly with a tritiated aminoacid give efficiencies around 5% at 5 μm thickness, and 2–3% at 10 μm (ref. 8).

Few if any of the β particles from tritium will penetrate further than 1 μm into the emulsion, and some data from Salpeter and Szabo[12] suggest that a packed monolayer will stop as many as 60% of the β particles from tritium. Provided only that the emulsion is thicker than 2 μm, variations in emulsion thickness have no effect on efficiency. Since the particles are all of low energy, there is little gain from the use of highly sensitised emulsions. The probability of double hits may be high with small sources, since only those crystals within a radius of 1 μm from the source will contribute to n_{max} in the equation on p. 97. From this point of view, there is an obvious advantage in using an emulsion with small crystal size.

Though it is usual to equate self-absorption with the thickness of the specimen, the results of Maurer and Primbsch[2] demonstrate how important variations in density can be in the specimen. Any extractive procedure that reduces specimen density without removing the radioactivity of interest will improve efficiency.

With isotopes of higher maximum energy, source–emulsion separation and self-absorption have progressively less effect on efficiency. As an example, if a bacterium labelled with carbon-14 has an efficiency of 60% in a given autoradiographic system, a section of density 1.3 and thickness 5 μm would register at about 50% efficiency, and increasing the thickness to 10 μm would reduce the efficiency to about 42%. With phosphorus-32, self-absorption can be effectively ignored in this range of source thickness.

On the other hand, variations in emulsion thickness now assume much greater importance. With phosphorus-32 one can assume a direct proportionality between efficiency and the thickness of the emulsion layer in grain density autoradiographs. With carbon-14, absorption in each micron of emulsion will reduce the number of β particles entering the next micron by about 10%. Sensitivity of the emulsion affects efficiency: with particle energies above about 500 keV the highest possible sensitivity will be needed, and even so many of the latent images will be relatively small, requiring considerable care in exposure and development. Since each source now can affect a far larger volume of emulsion than in the case of tritium, the probability of double hits significantly reducing the efficiency becomes less.

There have been few accurate measurements of efficiency with isotopes other than tritium at the light microscope level. Andresen, Chapman-Andresen, Holter and Robinson[27] and Dörmer and Brinkmann[5] present careful data on carbon-14, using Kodak AR-10. As a rough guide, grain density autoradiographs with carbon-14 should achieve an efficiency of 40–50% with sections 5 μm thick and emulsion layers in the range of 3–5 μm. The efficiency for phosphorus-32 in similar conditions might be between 30 and 40%. Recent work by Thiel and colleagues[37,38] presents efficiency values for iodine-125 and AR-10 stripping film, using smears of blood cells.

THE EFFICIENCY OF AUTORADIOGRAPHS FOR THE ELECTRON MICROSCOPE

A bewildering series of figures can be found in the literature for the efficiency of electron microscopic autoradiographs. Techniques vary slightly from one laboratory to another: this is particularly true of development and of the thoroughness with which conditions during exposure are controlled. Even taking the work of one laboratory, that of Dr. Salpeter, estimates of efficiency have changed with the years, partly as a result of correction factors for such effects as dose dependence. The following summary highlights the factors that are most significant in affecting efficiency, given the techniques which are emerging as the most widely accepted ones. Measurement is difficult at this level, since efficiency may be so crucially affected by small differences in variables which are themselves technically difficult to quantify. Any estimate of efficiency carries a standard deviation of the order of 10–20%.

Tritium. If one considers tritium-labelled specimens first, two methods of autoradiographing these at the electron microscope level exist, using the emulsions L4 and NTE (or its successor, Kodak special product type 129-01). Unfortunately, manufacture of the very useful NUC-307 emulsion has been discontinued.

Taking the emulsion L4 first, development with D-19 gives high efficiencies, but large, coiled silver grains. Elon-ascorbic acid after gold latensification gives smaller grains and roughly equivalent efficiencies, but involves more steps than D-19[21].

The Agfa-Gevaert physical developer based on Metol also gives similar efficiencies after gold latensification, and is reported to be simple and reproducible in use[28].

With tritium, source thickness is a major factor influencing efficiency. Up to about 1000 Å, with material of density 1.0–1.1, the specimen can be considered infinitely thin[12]. Above this, self-absorption begins to affect results significantly. Heavy metal staining of biological material in sections can be expected to reduce efficiencies by amounts which are on the borderline of significance, published estimates ranging between 7 and 30% of the expected efficiency in the absence of staining[29,30]. Separation of source from emulsion by a layer of evaporated carbon of 50–60 Å does not introduce a detectable loss of efficiency, but thicker layers or denser materials would begin to do so.

Increasing the thickness of the emulsion from a monolayer of L4 crystals to a double layer will increase efficiency by 10–20%[12].

Dose dependence will result in higher efficiencies of recording over lightly labelled sources[12]. Reductions in efficiency due to double hits begin to occur at relatively low grain densities around sources of small dimensions, since so few crystals lie within the range of the β particles. Both factors tend to compress the differences that can be detected between sources of differing radioactivity.

Exposures by the flat substrate method, on glass slides, will tend to have higher efficiencies than ones on coated grids, due to higher backscattering from the glass[13]. The magnitude of this effect has not been measured, but it is only likely to be significant with very thin sources, such as smears of labelled molecules.

Given all these variables, efficiencies in the range of 22–30% can be expected for tritium when a section of biological material is autoradiographed with a monolayer of L4, developed optimally with one of the above techniques.

Changing to NTE or Kodak 129-01*, all the factors operating in the source that are listed above for L4 will continue to operate. Now, with the smaller crystal diameters, going from a monolayer to a double layer will nearly double the efficiency.

It is customary to develop NTE with Elon-ascorbic acid after gold latensification. The new Kodak 129-01 cannot be latensified with gold, and Salpeter and Szabo[31] have found that Dektol is the best developer to use. Dose dependence is seen with this emulsion–developer combination, but it should be less sensitive to efficiency losses through double hits than L4, as more crystals lie within the range of the β particles from a point source.

The preliminary data suggests that Kodak 129-01 will be less liable to latent image fading than NTE. With care, L4 can give quantitative results

---

*This is the emulsion NTE-2 referred to by Salpeter and Szabo[31].

for up to one year's exposure or more, if the results are corrected for dose dependence[12]. NTE required exposure in an inert gas, such as helium, after careful drying, and even then fading and the build-up of background limited effective exposures to 3 months[32]. With the same conditions of exposure, it is hoped that Kodak 129-01 will be effective for much longer[31].

Efficiencies for NTE, developed optimally, were about 7% for a monolayer over a section of biological material. Efficiencies for Kodak 129-01 appear to be twice as high[31].

Iodine-125. This isotope is often assumed to be so similar to tritium in the energies of its particles that efficiency values for the two isotopes can be considered identical. This is far from the truth. In place of a continuous spectrum of particle energies, as with tritium, iodine-125 produces extra-nuclear electrons at specific energy levels. The majority of these have initial energies between 3 and 4 keV. This has interesting effects on the efficiency of autoradiography.

Source–emulsion separation and self-absorption will both be more significant for iodine-125: infinitely thin sections will need to be no thicker than about 400 Å. Heavy metal staining will reduce efficiencies by 10–15%[14].

Moving from a monolayer of L4 to a double layer produces no detectable increase in efficiency: with NTE and Kodak 129-01, the increase in efficiency from using a double layer is likely to be about 50%.

Dose dependence, being a function of the density of latent images in a given emulsion-developer situation, will be independent of the change from tritium to iodine-125; but the probability of double hits from very small sources will be even higher with the shorter ranges of the 3–4 keV electrons.

Finally, since efficiency is expressed as the number of developed grains per 100 disintegrations in the source, the complex decay scheme of iodine-125, yielding 128 electrons in the range 3–4 keV per 100 disintegrations, results in surprisingly high efficiencies. For a section of biological material and a monolayer of L4, optimally developed, values as high as 55–65% may be obtained[14]. With NTE, the comparable figure is around 15%, and this goes up to 30–40% for Kodak 129-01[31].

Isotopes of higher energies. Carbon-14, sulphur-35, calcium-45 and phosphorus-32 have all been studied at one time or another. With all of them, the standard autoradiographic preparation of a thin section covered by a monolayer of emulsion represents an infinitely thin source without measurable self-absorption. Source–emulsion separation by any layer

106

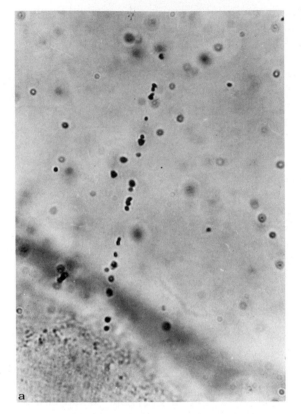

Fig. 37. Photomicrographs of β-tracks in Ilford G5 emulsion. The lower picture shows a track from phosphorus-32 in which the grain size and spacing are normal. The upper picture shows a similar track affected by latent image fading. The grains are smaller and more widely spaced. (× 850)

which is likely in such a preparation has no effect on the efficiency: the same is true of heavy metal staining.

Whatever the emulsion, going from a monolayer to a double layer will double the efficiency. Since many of the silver deposits at sensitivity specks will be very small by comparison with those produced by tritium, optimal development assumes great importance. Dose dependence will be encountered to much the same extent as with tritium, if the density of latent images is made the basis for calculation rather than the density of β particles hitting the emulsion. The probability of double hits diminishes dramatically, since a very much larger number of crystals lies within range of the β particles from a point source.

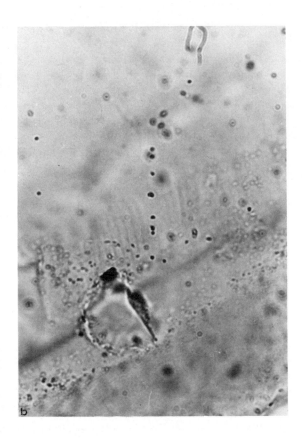

Emulsion sensitivity assumes considerable importance. While workable efficiencies can be obtained with L4, it is almost not worthwhile trying to use NTE: efficiencies with Kodak 129-01 are also likely to be many times lower than for L4.

Backscattering is likely to be more important than with tritium, and may well contribute significantly to efficiency values[13].

Efficiencies for carbon-14 and sulphur-35 in a section of biological material autoradiographed with a monolayer of L4 are likely to be 4–5%[13]. With small crystalline deposits of calcium-45, efficiencies of about 3% have been claimed[33]. With phosphorus-32, the efficiency for biological material in sections is likely to be much the same.

With carbon-14 and sulphur-35 and a monolayer of NTE, the efficiency is about 1%: this may well be 2% with Kodak 129-01. Efficiencies for calcium-45 and isotopes of higher energies may well fall below 1%.

THE EFFICIENCY OF TRACK AUTORADIOGRAPHS

If one now considers track autoradiographs, where the emulsion is thick relative to the maximum track length of the β particles being emitted, there are several factors that differ from those discussed above. The problems of self-absorption and emulsion sensitivity, however, are the same.

The key to the differences that do exist lies in the definition of a β track. This is usually accepted as four or more silver grains arranged in a linear fashion. This pattern is sufficiently improbable in random background grains to exclude the chance of counting fortuitous groups of individual grains as tracks. But all isotopes which emit β particles have a spectrum of energies extending right down to zero, and some of these particles will have insufficient energy to produce four silver grains in a row. With phosphorus-32, this fraction is so low that it is negligible. With carbon-14, 14% of the β particles will not produce a track[34]. With tritium, the fraction that does produce a track is very small[35].

So, if one considers the effect of particle energy on efficiency in track autoradiographs, the higher the maximum energy, the higher the efficiency. With particulate sources suspended in a nuclear emulsion, Levinthal and Thomas[36] have shown that 100% efficiency can be obtained with phosphorus-32. With phosphorus-32 in a 5 μm section, covered on one side by emulsion, the efficiency will be approximately 50%. With carbon-14 in a similar situation, the efficiency will be nearer 30% after allowing for self-absorption and the necessity to produce a track four grains long. The efficiency of track autoradiographs does not vary with emulsion thickness, provided always that the emulsion is thick enough to record the tracks adequately. Track autoradiographs are less critically affected by latent image fading. The exposure time is usually very short, as the interpretation of track patterns becomes very difficult if the track density is high. Further, latent image fading affects those crystals with minimal latent images first, and its initial effect on a β track is to reduce the grain density along the track, still leaving those grains which had more energy dissipated in them to indicate the passage of the particle. Thus the track may still be recognised and counted in the presence of mild latent image fading (Fig. 37).

Backscattering from the slide may contribute to the measured efficiency: the higher the isotope energy and the thinner the specimen, the greater the effect is likely to be.

The emulsion sensitivity becomes crucially important with energies as high as phosphorus-32, and the level of development also. Unless both are as high as possible, the grain spacing in the beginnings of the tracks of particles from the high energy end of the spectrum will be so wide that it will be very difficult to recognise the tracks at all.

THE CONFLICTING DEMANDS OF RESOLUTION AND EFFICIENCY

This and the preceding chapter have listed the factors determining resolution and efficiency in autoradiography. In some cases, control of one particular variable may improve both the resolution and the efficiency — the probability of double hits is an example.

In many cases, however, attempts to achieve the highest possible resolution produce emulsion responses so near background levels that the collection and interpretation of data become very difficult. This is particularly true in autoradiography at the electron microscope level. The demands of high resolution require the thinnest possible section and a monolayer of the emulsion with the smallest crystal diameter available, with no intervening layer. But such a section may contain very little radioactivity: such an emulsion layer may be relatively inefficient by comparison to other possibilities, and the absence of an evaporated carbon film between section and emulsion may favour the fading of latent images during exposure. Unswerving determination to produce high resolution may result in no autoradiographic image at all.

The technique that should be adopted for any particular experiment is nearly always a compromise between the conflicting demands of resolution and efficiency. It is probably best to begin with one or other of the compromises that have been well worked out and tried. Then, if either the resolution or the efficiency is not adequate for the experiment in question, the technique can be modified in specific ways by manipulating one or more of the variables discussed in these two chapters.

REFERENCES

1 R.P. Perry, M. Errera, A. Hell and H. Durwald, *J. Biophys. Biochem. Cytol.*, 11 (1961) 1.
2 W. Maurer and E. Primbsch, *Exptl. Cell Res.*, 33 (1964) 8.
3 S.R. Pelc and M.G.E. Welton, *Nature*, 216 (1967) 925.
4 R.W. Hendler, *Science*, 130 (1959) 772.
5 P. Dörmer and W. Brinkmann, *Histochemie*, 29 (1972) 248.
6 T.L. Housley and D.B. Fisher, *J. Histochem. Cytochem.*, 23 (1975) 678.
7 G.J. Falk and R.C. King, *Radiation Res.*, 20 (1963) 466.
8 H.K. Oja, S.S. Oja and J. Hasan, *Exptl. Cell Res.*, 45 (1967) 1.

110

9 R.P. Perry, in D.M. Prescott (Ed.), *Methods in Cell Physiology, Vol. 1*, Academic Press, New York and London, 1964.

10 P. Dörmer, in V. Neuhoff (Ed.), *Molecular Biology, Biochemistry and Biophysics, Vol. 14, Micromethods in Molecular Biology*, Springer-Verlag, Berlin, 1973.

11 G.F.J.M. Vrensen, *J. Histochem. Cytochem.*, 18 (1970) 278.

12 M.M. Salpeter and M. Szabo, *J. Histochem. Cytochem.*, 20 (1972) 425.

13 M.M. Salpeter and E.E. Salpeter, *J. Cell Biol.*, 50 (1971) 324.

14 H.C. Fertuck and M.M. Salpeter, *J. Histochem. Cytochem.*, 22 (1974) 80.

15 T.C. Appleton, *J. Histochem. Cytochem.*, 14 (1966) 414.

16 W. Sawicki, K. Ostrowski and J. Rowinski, *Stain Technol.* 43 (1968) 35.

17 W. Sawicki, K. Ostrowski and E. Platkowska, *Histochemistry*, 52 (1977) 341.

18 A.S. Mukherjee and R.N. Chatterjee, *Histochemistry*, 52 (1977) 73.

19 B. Durie and S. Salmon, *Science*, 190 (1975) 1093.

20 A.W. Rogers and P.N. John, in L.J. Roth and W.E. Stumpf, (Eds.), *The Autoradiography of Diffusible Substances*, Academic Press, New York, 1969.

21 B.M. Kopriwa, *J. Histochem. Cytochem.*, 15 (1967) 501.

22 R. Rechenmann and E. Wittendorp, *J. Microscop. Biol. Cell*, 27 (1976) 91.

23 B. Larsson and S. Ullberg, *Science Tools*, special issue, (1977) 30.

24 R.A. Laskey and A.D. Mills, *Europ. J. Biochem.*, 56 (1975) 335.

25 S. Bleecken, *Atompraxis*, 10 (1964) 1.

26 W.L. Hunt and R.H. Foote, *Radiation Res.*, 31 (1967) 63.

27 N. Andresen, C. Chapman-Andresen, H. Holter and C.V. Robinson, *C.R. Lab. Carlsberg Chim.*, 28 (1953) 499.

28 B.M. Kopriwa, *Histochemistry*, 44 (1975) 201.

29 M.M. Salpeter, *J. Histochem. Cytochem.*, 21 (1973) 623.

30 B.L. Gupta, R.B. Moreton and N.C. Cooper, *J. Microscopy*, 99 (1973) 1.

31 M.M. Salpeter and M. Szabo, *J. Histochem. Cytochem.*, 24 (1976) 1204.

32 M.M. Salpeter and L. Bachmann, in M.A. Hayat (Ed.), *Principals and Techniques of Electron Microscopy, Vol. 2*, Van Nostrand Reinhold, New York, 1972.

33 G.J. Huxham, A. Lipton and B.M. Howard, *Austr. J. Biol. Med. Sci.*, 47 (1969) 299.

34 H. Levi, A.W. Rogers, M.W. Bentzon and A. Nielsen, *Kgl. Danske Videnskab. Selskab. Mat.-Fys. Medd.*, 33 (1963) No. 11.

35 H. Levi, *Scand. J. Haematol.*, 1 (1964) 138.

36 C. Levinthal and C.A. Thomas, *Biochim. Biophys. Acta*, 23 (1957) 453.

37 E. Thiel, P. Dörmer, H. Rodt and S. Thierfelder, *J. Immunol. Meth.*, 6 (1975) 317.

38 E. Thiel, P. Dörmer, W. Ruppelt and S. Thierfelder, *J. Immunol. Meth.*, 12 (1976) 237.

CHAPTER 6

Autoradiographic Background

THE CAUSES OF BACKGROUND

In every autoradiograph, silver grains appear in the developed emulsion which are not due to radiation from the experimental source, but to other causes. These grains constitute the background. The recognition and measurement of radioactivity depend on the comparison of the grain or track density over an experimental source with the density found over a source that is known to be unlabelled. Clearly, the amount and the variability of background determine the minimum level of radioactivity that can be recognised.

To most people starting work with radioactive isotopes and autoradiography, background seems to be synonymous with cosmic radiation. When I first began this type of work, I went to great trouble to obtain permission from the National Coal Board to expose my slides at the bottom of a coal mine. This added a touch of the bizarre to the whole procedure, but I was rather crestfallen to find that the background was considerably higher in these slides which had been so carefully shielded from cosmic radiation than in control slides exposed in the refrigerator in the laboratory. In fact, cosmic rays form a comparatively insignificant component of background in most instances.

The major factors that contribute to autoradiographic background will vary from one laboratory to the next, from one technique to another, even between experiments carried out in the same laboratory with the same technique. The more important causes of background are discussed below, together with suggested ways of investigating their relative importance in any given experiment.

(a) *Development and background*

In Chapter 2 we have already seen that if the strength, the temperature, or the duration of development are progressively increased, more and more silver grains will be developed, regardless of the degree of exposure of the emulsion to radiation. Taking development time as an example (Fig. 4), beyond a threshold time, the number of silver grains lying along the tracks of particles which have passed through the emulsion does not increase significantly. At some development time longer than this, many tiny background grains appear, growing larger and more numerous as development is extended further. These effects have been studied, for instance, by Ahmad and Demers[1], and by Caro and Van Tubergen[2]. We have seen on p. 19 that some unhit crystals will always be developed, since the low probability of development of such crystals will be offset by their numbers. Their probability of development rises with increasing development. At many levels of overdevelopment, there will appear to be two populations of background grains: a few relatively large ones, which have begun to be developed early, and a far more numerous group that have not started the development process until later.

The optimal development time varies with the emulsion used, with the developer, and with many other factors, and there may be no alternative to finding it for the conditions that will be used in a particular series of experiments, using the methods outlined in Chapter 2.

A high background caused by overdevelopment can often be recognised by simple examination of the developed emulsion. The background grains will usually be randomly scattered throughout the emulsion, and noticeably smaller than the grains due to radiation. In cases where the background is very high, the processed emulsion may have a pinkish-grey colour when looked at with the unaided eye.

(b) *Background due to exposure to light*

The emulsions available for autoradiography vary considerably in their sensitivity to visible light, but all of them will show increasing background with increasing exposure to light. Appropriate safelighting conditions are recommended by the manufacturer for each emulsion, and these should be carefully followed. For the Ilford G, K and L emulsions, the light brown Ilford "S" safelight is satisfactory. This seems very bright to anyone used to working with Kodak AR-10, or the Eastman Kodak NTB emulsions, which require the dark red Wratten No 2 filter. The safelight filter only ensures that the wavelength of the light falling on the emulsion is that to which it is least sensitive. The light intensity is determined by the power of the bulb, which is usually taken to be 15 W, and by the distance between light source and emulsion. There is absolutely nothing to be gained by using very murky working conditions, and then carrying out each procedure 6 inches from the safelight in order to see what to do.

In effect, emulsions are at their most sensitive when adequately dried.

(Strictly speaking, the fading of the latent image is very rapid in a wet emulsion, giving a similar end result to a loss of sensitivity.) In the liquid emulsion techniques, for instance, a level of light intensity can be tolerated while the emulsion is molten and diluted with water that would be likely to cause an increase in background once the emulsion layers are fully dried. It is reasonable to use lower levels of lighting, or to work further from the safelight, when putting the dried autoradiographs away for exposure, or while developing them, than in the initial stages of preparing the emulsion.

Undue exposure to light can be simply recognised when it is at the stage of gross fogging. The silver grains occur throughout the emulsion, and there may be curious geometrical patterns visible, due to the partial shielding of one slide by the next one, or by some neighbouring object. The section itself may give a certain amount of protection to the overlying emulsion. The less obvious degrees of fogging can be difficult to recognise on examination of the emulsion alone. If light is suspected as a cause of high background, there may be no alternative to preparing and developing a series of emulsion layers on plain slides, without any biological specimen and varying the exposure to safelighting to see if there is any significant effect on the background.

Methods have been described for preparing and processing autoradiographs in total darkness[3]. If correct safelighting and handling procedures are employed, I have found no appreciable improvement in background levels from working in absolute darkness. If gold latensification is to be used, bringing to development numbers of crystals with silver nuclei too small for normal development procedures, then it does become important to reduce safelighting to an absolute minimum.

(c) *Background due to pressure*

Nuclear emulsions are sensitive to pressure, and beautiful fingerprints can be produced in them as patterns of developed grains. In some laboratories, this response to pressure is used routinely for "writing" numbers and letters on the dried emulsion layer with a pointed instrument. On development, they will appear clearly in black.

Obviously, scratches, fingerprints, and other gross insults to the emulsion must be avoided. But there are other examples of stress or pressure causing a high background which are not so immediately evident. Gelatin contracts on drying and, if this process is carried out too fast, or taken too far, developed grains will be produced. This type of artefact can show itself in two ways. Where the emulsion is in contact with the glass slide, it may be subjected to sliding, lateral stresses as it shrinks, producing background grains which are often arranged in curious patterns when viewed under low magnification. These grains tend to lie in the emulsion layer nearest to the

114

glass support. Very thin emulsion layers will tend to dry much more rapidly than thicker ones, and they will often have a higher background for this reason. In slides dipped in very dilute liquid emulsion, and dried in a vertical position, the background often increases sharply towards the top of the slide, where the emulsion layer is thinnest.

The other type of stress or pressure artefact is usually limited to the area over a specimen, such as a tissue section. The upper profile of a tissue section is usually very irregular: if the section was cut at a thickness of 5 μm, it will be 5 μm thick only in places. Over the blood vessels, for instance, and other tissue spaces, the thickness drops abruptly to zero, and the same is true of the edges of the section. With liquid emulsion

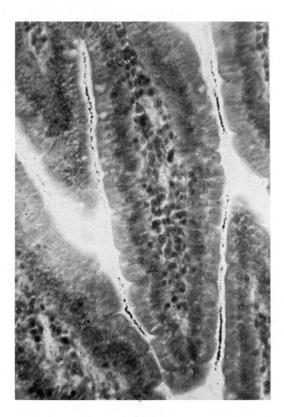

Fig. 38. Photomicrograph of a section of the small intestine of a mouse. The section was coated with a thin layer of Ilford K2 emulsion which was rapidly dried at 30°C in front of a fan. Exposure of one week was followed by routine development. Note the high background of random silver grains, and the dense lines of silver following the edges of the tissue, both resulting from the physical stressing of the silver halide crystals during rapid drying. (× 440)

techniques in particular, these irregularities fill with emulsion, producing many changes in thickness and occasional wedges, where the emulsion dips right down to the glass support. These places seem particularly vulnerable to stress artefacts on drying, and a line of silver grains, closely following such a change in contour in a section, should always be regarded with suspicion. If the line of silver grains is regular and confined to one plane in the emulsion, if it follows the change in contour very accurately, it is almost certainly an artefact. There are several papers in the literature which claim to demonstrate the localisation of radioactivity along the course of fibres of various sorts, in which the published microphotographs are beautiful examples of this type of artefact. Even the β particles from tritium have a measurable range in emulsion, and the silver grains from a uniformly labelled linear source would occur at many levels through the emulsion. They would be distributed at distances up to 2 μm on either side of the source, and would show some statistical variability in density from place to place along the source. Fig. 38 illustrates the very different appearance of the stress artefact.

Several precautions can be taken to avoid these pressure artefacts. The glass support should, wherever possible, be coated with a thin layer of gelatin (the subbing solution, described on p. 135) before the emulsion is applied. This not only improves the adhesion of emulsion to glass, but reduces the stress artefacts due to lateral displacements on drying. For routine work, emulsion layers should not be less than 3 μm thick.

The crucial step in controlling these pressure artefacts, however, comes in drying the emulsion after applying it to the specimen. Such artefacts may, for instance, be worse in the summer when the external temperature is high and the humidity low[4]. Another cause of seasonal variation may occur in the winter in very cold countries: in some buildings outside air is taken in, heated and distributed, resulting in very low humidities indeed in the darkroom. Many workers have their own drying routines, and these can vary greatly in detail. Most of them have certain features in common, which emphasise the factors that must be most carefully controlled. The first of these is the speed of drying. Sawicki and Pawinska[5] have made some interesting observations with Kodak AR-10. If the film is dried very slowly over 24 hours, by placing it in a closed but permeable box over calcium chloride in a desiccator, the observed background lies between 0.09 and 0.24 grains per 100 μm^2. If, under otherwise identical conditions, the film is dried rapidly in front of a fan, the background values are from 4 to 7 times higher.

With liquid emulsion techniques, the emulsion is heated in order to keep it molten for application to the specimen. Particularly if it is applied as a very thin layer, it will dry almost immediately. Caro and Van Tubergen[2] have drawn attention to the changes in distribution of silver halide crystals that can occur as a result of this rapid drying of a thin, warm emulsion, and have recommended cooling the emulsion to its gelling point prior to application. This procedure should be equally effective in reducing the stressing of the crystals during drying. Certainly, the emulsion should be cooled immediately it is on the specimen, and should be dried very gently in the gel state.

In quantitative work, it may be necessary to dry emulsions very thoroughly to prevent latent image fading. Messier and Leblond[6], using Eastman-Kodak emulsions and the dipping technique, found it necessary to expose their autoradiographs at extremely low relative humidities in order to prevent latent image fading. With Ilford emulsions, I have found that this sort of drying produces unacceptable high background levels. Ilford recommend a relative humidity of 40–45% during exposure, and the background is considerably less under these conditions, while fading of the latent image is not a problem for exposures up to about 3 weeks. There may well be genuine differences between the behaviour of the Ilford and Eastman-Kodak emulsions in this respect, and the high background from vigorous drying seen in the former may not occur with the latter. The whole problem of latent image fading and its control will be discussed in more detail later (p. 265).

One further step to reduce the probability of stress artefacts on drying has been suggested by Waller[7]. Emulsions, as supplied from the factory, normally contain various plasticising agents which tend to reduce the shrinkage of the gelatin on drying, and these agents may be leached out if the emulsion passes through aqueous media before exposure. Waller recommended that all aqueous solutions that come into contact with emulsions prior to exposure should contain added glycerol to a final concentration of 1%. The detailed descriptions of technique given in later chapters often include this addition of glycerol.

If it is necessary to expose an emulsion under extremely dry conditions or in vacuo, Ilford will provide their emulsions with added plasticiser.

(d) *Chemography*

In most of the autoradiographs that biologists prepare, emulsion comes into contact with biological material of some sort during exposure. Many reactive groups, in particular those that are reducing agents, are capable of producing a latent image in silver halide crystals by direct, chemical action.

Tissue that has been through the processes of fixation, dehydration, embedding in paraffin wax, sectioning, and subsequent dewaxing, is less likely to give rise to this type of artefact than fresh tissue sectioned on a cryostat, for instance. But in most situations the possibility of chemography exists, and one of the most striking things about this particular source of background grains is its unpredictability. One series of sections out of a score processed in apparently the same way will show it, where the others do not.

The opposite effect to this, negative chemography, may also occur. Certain reactive groups in the specimen may result in very rapid fading of the latent image in the adjacent crystals, resulting in an area of emulsion that is virtually incapable of registering the passage of a charged particle. Fig. 39 shows a dramatic example of this type of chemography. The material is muscle, fixed in formalin, embedded in paraffin, and autoradiographed by the dipping technique, using Ilford K2 emulsion. The photomicrograph shows a control slide which had been fogged by light before being exposed, together with the rest of the slides, for two weeks. The emulsion has been very effectively bleached, with complete loss of the latent image, in the regions over the section only. It is often almost impossible to trace down the variable that produces this effect in one block, but not another.

A further example of chemography is provided by Fig. 40. This shows an autoradiograph of a section of a human femur, obtained at post-mortem. In the course of a long illness, this patient received many injections of the short-lived isotope calcium-47. It had been suggested that this material might have been contaminated with traces of the longer-lived calcium-45, and that enough might have accumulated in her bones to give an autoradiograph. So a portion of femur was placed in contact with an X-ray film, and the autoradiograph obtained. The result looks quite impressive. There is obviously heavy blackening present, and it seems to follow an anatomical distribution. It is clear, however, that there are also some areas of this "autoradiograph" that are lighter than the background away from the bone. This indicates that these areas are being bleached by some compounds unknown, presumably originating in the tissue. If negative chemography is occurring, the blackening also may be an artefact. The bone was therefore dipped in a solution of nitrocellulose, coating it with an inert and relatively impermeable layer which was still thin enough for the majority of the β particles to penetrate. The autoradiograph that was then obtained, under otherwise identical conditions of exposure, was completely negative. It was obvious that all the blackening seen in the first picture was artefactual, due to chemography. Since it was blackening unrelated to radiation from the source, it was, by definition, background.

It is one of the most significant characteristics of chemography that it frequently has an anatomical distribution[8,9], giving rise to a very plausible "autoradiograph". It is quite common also to have this combination of blackening with bleaching in the same specimen. Neither observation should be very surprising, since presumably the chemical groups responsi-

118

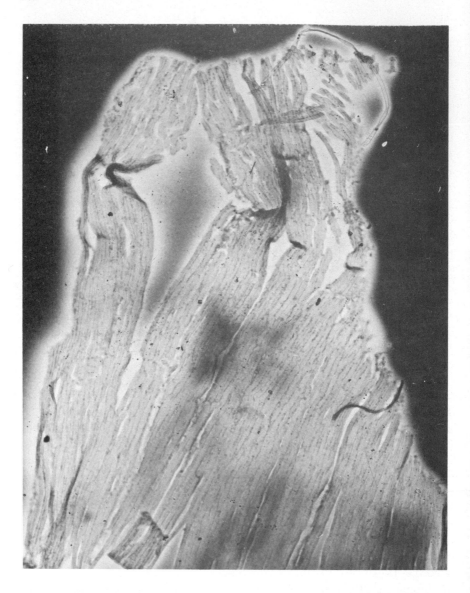

Fig. 39. Photomicrograph of a section of skeletal muscle: the tissue had been formalin fixed and embedded in paraffin wax. After coating with Ilford K2 emulsion, the specimen was fogged by light before being exposed in the dark for 18 days. In the absence of negative chemography, the emulsion would have been uniformly black on development. In fact, gross fading of latent images has occurred over and just around the section, indicating negative chemography. (×27)

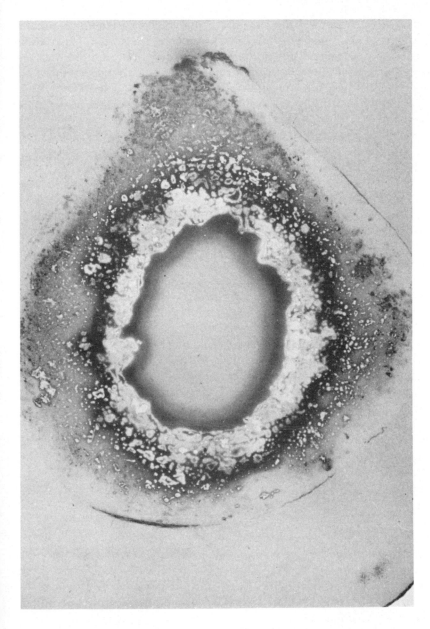

Fig. 40. An X-ray film that has been in contact with the cut surface of a human femur. Note the anatomical distribution of blackening, and also the areas that are lighter than background. This image is entirely due to chemography. A second exposure of the same specimen, separated from the film by a thin but impermeable membrane, failed to produce any blackening at all. (× 4)

ble vary in concentration from one part of a biological specimen to another in some sort of anatomical fashion. The spread and severity of this artefact may increase with increasing exposure time. It is obvious that this artefact is so serious a pitfall in the interpretation of an autoradiograph that adequate control measures must be taken in every single experiment. This point is discussed further in considering the design of autoradiographic experiments (p. 302).

Chemography depends on the diffusion of reactive molecules into the emulsion from the specimen, and their reaction there with the silver halide crystals. The rates of both these processes are temperature dependent, so it is possible to reduce the severity of this artefact by exposure at low temperature[10]. The resulting drop in efficiency reported by Welton[11] can be effectively restored by slight increase in either the temperature or duration of development.

The choice of developer may have a surprising influence on the severity of chemography[10]. Chemical effects on the silver halide crystals are presumably maximal at the crystal surface, and a developer that relies on surface latent images only will be much more sensitive to latent image loss from chemography than one that attacks internal latent images also.

It is sometimes possible, with experience, to suspect the presence of chemography by simple examination of the emulsion under high magnification. If the grain density is significantly lower over the specimen than away from it, negative chemography should be suspected. The silver grains due to positive chemography are often large, irregular, and sometimes clumped together. Their distribution over the specimen is often thicker and more even than one would expect from radioactivity, and there is no suggestion of their arrangement in segments of tracks. But these are not fully reliable criteria, and the only real protection against chemography lies in the adequate design of the experiment, with the correct use of control procedures.

(e) Contamination of the emulsion

Just as certain reactive groups in the specimen may affect the performance of the emulsion, so may traces of contaminating chemicals in the emulsion itself. Nuclear emulsions are carefully designed and their manufacture is finely controlled, and great care should be taken to keep them free from contamination.

Glass, certain plastics, and high grade stainless steel are the only materials that should be allowed to come into contact with nuclear emulsion, and they should all be scrupulously clean. Metallic ions are a frequent cause of trouble, particularly copper, and copper water baths and other metal objects, such as spoons or forceps, should be avoided at all costs. Distilled or deionised water should be used for all solutions.

It is often difficult to clean glassware that has held emulsion. The most satisfactory procedure I have found is to soak it first in normal sodium hydroxide, which digests away the gelatin. After washing, chromic acid will remove any remaining traces of silver salts. Thorough washing for several hours in running tap water, followed by distilled water, will be needed to remove the acid.

A different type of contamination may occur if the darkroom is not kept clean. This is particularly likely to happen if it is shared with amateur photographers. Developer and fixer get spilt on the floors and benches, and allowed to dry. Subsequent movement, or switching on a fan, stirs up dust containing all manner of strange compounds which may settle on emulsion layers which are in process of drying. This type of background can be more easily appreciated if two slides without specimens on them are coated with emulsion, and the one fogged by exposure to light. After leaving them in the darkroom with the fan on for an hour, they should be put away for several days, then developed and examined critically. It is often a salutory experience to note just how many strange specks are visible on the surface of the emulsion, and how likely they are to produce areas of blackening or bleaching.

Serious research with autoradiographic techniques requires a separate darkroom. Any spilt solutions should be mopped up at once and the area washed. Periodic cleaning of the darkroom and washing of the working surfaces is a good idea.

(f) *Environmental radiation*

It is obvious that extraneous sources of radiation will produce background in nuclear emulsions. Cosmic rays form only one component of this.

At ground level, cosmic rays form a wide spectrum of particles, from very high energy ones that penetrate long distances through matter, to low energy secondary electrons. Any shielding material interposed between the emulsion and the sky will screen off some of this radiation. The cosmic ray background will be lower in the basement of a 6-storey building than in an adjacent single-storey hut. A small box of lead bricks two inches thick will reduce the cosmic ray intensity still further.

Some rocks contain appreciable amounts of radioactive isotopes, as do the building materials made from them. But these naturally occurring sources of radiation are likely to be of academic interest only in most instances. X-ray machines, and laboratories using γ-emitting isotopes, are much more dramatic sources of background. On a smaller scale, the glass of the microscope slide contains potassium-40 and traces of α-emitting isotopes, while the gelatin of the emulsion itself has minute amounts of carbon-14 in it. If the isotope under study emits γ rays or β particles of very high energy, it is possible for adjacent slides to irradiate one another during exposure, producing very significant levels of background.

With track autoradiographs, many of these sources of radiation can be distinguished from radioactivity in the specimen by the pattern of track they produce. With thin emulsion layers, however, it is likely that only the densely ionising α particles can be satisfactorily recognised.

Background caused by environmental radiation may occasionally be recognisable from examination of the emulsion. The silver grains will be arranged along the tracks of individual particles, and these will be randomly distributed throughout the emulsion, without reference to the specimen.

Environmental radiation, then, provides the irreducible minimum of background to most experiments. But the other factors listed in this chapter usually are responsible for the majority of the background grains seen in an autoradiograph. Their effect is superimposed on the slow increase of background due to cosmic radiation. In all cases of unduly high background, these other factors, singly or in combination, are likely to be responsible. If the low levels of grains due to cosmic radiation are demonstrably the major component of the background in a series of autoradiographs, and this background is still unacceptably high, then it may become necessary to investigate methods of shielding the emulsion during exposure.

(g) Spontaneous background

The nuclear emulsions used in autoradiography are very highly sensitised products. In all of them, an occasional silver halide crystal will develop a latent image speck spontaneously. The more highly sensitised the emulsion, the more likely this event becomes. It is a good rule never to use an emulsion that is more highly sensitised than is strictly necessary. Low energy β particles from tritium, for instance, will be recorded in a less sensitive emulsion than that needed for β particles of much higher energies. α Particles will leave their characteristic tracks in even less sensitive media. The very highly sensitised emulsions, such as Ilford K5 or Eastman Kodak NTB-3, will have a higher rate of formation of spontaneous background, will be sensitive to a wider range of environmental radiations, and will be less tolerant to the raised temperatures necessary in melting and to the stresses of drying, than will K2 or NTB-2. The overall background of the former under similar conditions will always be higher, so that it seems only reasonable to reserve their use for experiments with β

particles of fairly high energy, which really require their higher sensitivity.

High temperatures increase the rate of formation of spontaneous background, so that emulsions are ideally stored before use and during exposure just above freezing point. The hydrated emulsions in gel form that are used in liquid emulsion techniques should not be frozen before application and drying.

The rate of formation of spontaneous background tends to increase with the age of the emulsion. Nuclear emulsions should always be used as soon after manufacture as possible though, with good conditions of storage, they usually remain in reasonable conditions for up to two months.

(h) *Causes of background specific for stripping film*

In the stripping film technique, the emulsion layer has to be stripped off its support before its application to the specimen. If this is done in conditions of low humidity, static electricity may be generated, with a crackling noise and even visible flashes of light accompanying the stripping process. This cause of background can often be recognised by the presence of dense lines of developed grains running parallel to each other across the developed emulsion. The control of this artefact lies in the control of temperature and humidity in the darkroom, which is discussed further in Chapter 15. It is an interesting idea to chain one's technician to the water pipes in the darkroom to try to earth the plate during stripping, but I am afraid it is not really effective.

Stevens has drawn attention to another source of high background in AR-10 stripping film[12]. The concentration of soluble bromide ions in the emulsion is carefully controlled during manufacture to achieve a good balance between sensitivity and the rapid growth of background. When the film is floated out on distilled water before picking it up on the specimen, diffusion into the water reduces the effective concentration of soluble bromide to the point that background may build up to unacceptable levels in exposures of 10 weeks or even less. By stripping the film on to distilled water containing 10 mg/litre potassium bromide and 50 g/litre glucose, background growth can be restrained without loss in sensitivity.

INFORMATION TO BE OBTAINED FROM THE EXAMINATION OF BACKGROUND

It is a good idea to take time for the systematic study of background levels in every batch of autoradiographs made. A great deal can be learnt about the techniques in use, even without making detailed grain counts. In particular, if slides from one batch of autoradiographs are exposed for

124

varying times, a lot of useful information can be gained from their examination.

Figs. 41 and 42 illustrate the results that may be found in generalised form, with grain count per unit area plotted against exposure time. In the ideal autoradiograph (line A), there are no background grains present at the start of exposure, and their number should increase slowly but linearly with time. This situation is almost never achieved, however, and nearly every series will show a certain level of background right at the beginning of exposure (line B).

If the background is unduly high at the start of exposure (line C), several possibilities exist. The emulsion could be old, or have a high level of background due to its previous history. Undue exposure to safelighting, or drying that is too rapid, may also be responsible.

If the increase in background with time is linear but rapid (line D) this suggests that the level of environmental radiation is too high. With higher energy isotopes, it may be that one slide is irradiating its neighbours during exposure.

In many instances, the initial rate of increase of background will level off after a short time (line E). In other words, an equilibrium will be reached

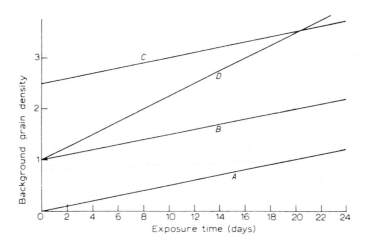

Fig. 41. Graphs illustrating the growth in background with increasing exposure times, in the absence of latent image fading. (A), The ideal situation, with no background at the start of exposure and a slow increase during exposure. (B), The more usual situation, with appreciable background present at the start of exposure. (C), High background levels at the start of exposure, suggesting old emulsion or faulty technique in preparing the autoradiograph. (D), Reasonable background levels at the start of exposure, but rapid increase during exposure, suggesting high levels of environmental radiation.

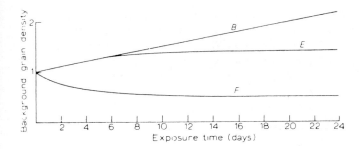

Fig. 42. Graphs illustrating the effects of latent image fading on background levels. (B), Normal situation, with low levels of background at the start of exposure and slow increase during exposure. In (E), slow fading of latent images produces deviation from linearity after a few days and a plateau of grain densities later in exposure. In (F), more severe fading produces an initial fall in background levels followed by a plateau at a very low level.

between the rate of formation of latent images on the one hand, and the process of latent image fading on the other. The extreme case of latent image fading (line F) produces an initial decrease in background, as the latent images formed in preparing the autoradiograph disappear. The curve then levels off as the equilibrium between formation and the fading process is reached.

THE MEASUREMENT OF BACKGROUND

In quantitative studies, the grain counts observed over the specimen will include background grains, and should be corrected by subtracting the mean number of background grains found in similar volumes of emulsion away from the radioactive areas in the specimen. This is often done by counting grains in areas of emulsion well away from the specimen. This is not a strictly correct procedure, though the error it introduces is often small.

From what has already been said about the various causes of background, it is clear that it may vary significantly from areas over the section to areas away from it. The uneven upper profile of the specimen may result in deviations from the normal emulsion thickness. Failure to dry the emulsion adequately before exposure may affect the area where there is a specimen beneath the emulsion more severely than the emulsion alone, resulting in more latent image fading over the section. But the biggest variable is chemography, which may result in considerable differences in density of background grains between the two sites.

The ideal way to estimate background is to expose an inactive, but otherwise identical, specimen to the same emulsion. Background counts can then be taken from emulsion over areas of non-radioactive specimen matched to the areas from which the experimental counts are obtained.

Up to the present, few autoradiographers have had any rational basis for selecting the number of areas to be counted in order to get a realistic estimate of the mean background. A paper by England and Miller[13] discusses how the effort of data collection can best be distributed between labelled sources and background, to achieve a desired level of statistical accuracy. The graphs from their paper can be referred to in the Appendix (p. 417).

In any comparison of the radioactivity present in two or more sets of sources, the mean background level is the statistic that should be looked for. In one very important group of experiments, however, this information is not sufficient. In many situations, the number of labelled cells is the parameter measured, rather than the level of radioactivity of the cells themselves. The percentage of labelled cells in various parts of the cell cycle after incorporation of tritiated thymidine is often found as a basis to estimate cell kinetics. The injection into an animal of cells labelled in vitro may also be followed by counts of labelled cells in various tissue compartments at later times. In both types of experiment, the definition of "a labelled cell" is obviously important. If one assumes that any cell with more than the mean background grain density is "labelled", random fluctuations in background will result in many unlabelled cells being wrongly accepted. The usual procedure seems to be to take a figure, such as 4 grains per cell, and to accept as "labelled" any cell which reaches or exceeds this level. This rough-and-ready rule of thumb is liable to generate misleading data in both directions — false positives and false negatives. The random, statistical fluctuations in background will produce occasional volumes of emulsion with high grain densities, up to 10–15 grains over areas the size of a single cell[14], in the absence of radioactivity. If large areas of tissue are being scanned to find the occasional "labelled" cell, this upper end of the distribution spectrum of background counts may make a very significant contribution to the data collected. On the other hand, if genuinely "labelled" cells have a mean grain count of 10–15 grains per cell, there may be a finite probability of finding radioactive cells with fewer than 4 grains over them. Ideally, one wishes to know the distribution of background counts over areas the size and shape of a cell, and to compare this with the distribution of observed counts over the cell population being examined. Stillström[15] first presented statistical methods for this type of analysis, and Bresciani and co-workers[16] have a computer program which does much the same sort of thing. This problem is also discussed in detail by England and Miller[13], and by Moffatt, Youngberg and Metcalf[14] who demonstrate the very considerable difference that a critical analysis can make to the percentage of labelled cells in experimental situations. While all the methods quoted for calculating the percentage of labelled cells agree well when background is low and the mean grain count per cell high, where there is considerable overlap between "labelled" and "background" distributions the methods give rather different results. Given adequate sample sizes, a very simple method of calculating the percentage of labelled cells has been presented by England, Rogers and Miller[17]. This is given in the Appendix. Where "labelled" and "background" distributions overlap, results calculated by this method agree well with those by Stillström's technique[15].

In photometric estimation of grain densities, the background reading contains an element independent of the presence or absence of silver grains, due to light scattered by the optical system of lenses, emulsion, specimen and slide. This will be discussed further in Chapter 10. With television scanning systems of grain counting, Mertz[18] has drawn attention to an interesting method of reducing the effective background in tritium autoradiographs. As was pointed out in Chapter 3 (p. 52), β particles of very low energy have a high rate of energy loss. The latent images created in crystals hit by tritium β's tend to be large and, in most conditions of development, the silver grains will also be large relative to the average background grain. In using the Quantimet image analyser, Mertz has found that he can count the great majority of tritium grains while rejecting most of the background ones by particle size analysis.

BACKGROUND ELIMINATION

Many of the causes of an unacceptably high background operate during exposure and development. These can and should be controlled by the autoradiographer. Occasionally, due to causes outside his control, the emulsion has a high background before the start of exposure. It may be, for instance, that the emulsion has been obtained from abroad. The combination of several days in transit at fairly high ambient temperatures with heavy exposure to cosmic radiation at high altitude in an aircraft may result in an unacceptable level of background. In this situation, several methods exist for wiping the slate clean again, so to speak, before exposure begins, based on the deliberate fading of the latent images. The formation of a latent image is a reversible process, and fading of the latent image is favoured, as we have seen, by the presence of oxidising agents, or of a high relative humidity.

The simplest method of background eradication is to fail to dry the emulsion thoroughly, so that exposure takes place at a fairly high relative humidity. Latent image fading proceeds together with the formation of new latent images by the radiation from the specimen. At the end of a reasonably long exposure time, such as two weeks, the probability of any latent images surviving from the start of exposure is remote. Only those produced in the last few days will stand much chance of development. The result will be an autoradiograph with relatively low background. For studies of the localisation of radioactivity, this method is quite acceptable. It is a poorly controlled way of achieving background eradication, however, and is out of the question in quantitative work, since it is quite impossible to estimate the effective duration of exposure.

Waller[7] recommends keeping the emulsion for a period of 1 hour at 100% relative humidity and a temperature of 37°C. This is sufficient to accelerate fading in any pre-existent latent images, and has the advantage that it does not alter the emulsion in any way, whereas the several methods that have been proposed involving chemical treatment of the emulsion may affect the sensitivity adversely.

Another technique has been described by Caro and Van Tubergen[2]. After preparation, the slides are placed in a closed glass container, at the bottom of which are several layers of filter paper, moistened with a few drops of 6% hydrogen peroxide. After 2.5–3 hours, the slides are removed and dried carefully, and exposed in the usual way. This treatment can give a very impressive reduction in background. The sensitivity of the emulsion does not seem to be affected by this procedure. Hydrogen peroxide is known to cause fogging of emulsions in certain circumstances, however, and I have had some complete failures with this technique, without being able to pinpoint any differences in procedure.

With track autoradiographs, the process of pouring the molten emulsion effectively eradicates the background. Any particle tracks in the emulsion become "scrambled" by the pouring, and are converted into scattered individual grains. The number of background tracks should always be practically nil at the start of exposure.

Several of the techniques available for background eradication are discussed in some detail by Barkas[19].

REFERENCES

1 I. Ahmad and J. Demers, *Can. J. Phys.*, 37 (1959) 1548.
2 L.G. Caro and R.P. Van Tubergen, *J. Cell Biol.*, 15 (1962) 173.
3 E.B. Barnawell, M.R. Bannerjee and F.M. Rogers, *Stain Technol.*, 45 (1970) 40.

129

4 J.M. Lachapelle, *Ann. Dermatol. Syphilis*, 99 (1972) 286.
5 W. Sawicki and M. Pawinska, *Stain Technol.*, 40 (1965) 67.
6 B. Messier and C.P. Leblond, *Proc. Soc. Exptl. Biol. Med.*, 96 (1957) 7.
7 C. Waller, The properties of nuclear emulsions in relation to autoradiography, *1st European Symposium on Autoradiography*, Rome, 1961.
8 C.R. Hackman and H. Vapaatalo, *Experientia*, 28 (1972) 492.
9 J.S. Daniels and S.A. Gilmore, *Brain Res.*, 98 (1975) 343.
10 A.W. Rogers and P.N. John, in L.J. Roth and W.E. Stumpf (Eds.), *Autoradiography of Diffusible Substances*, Academic Press, New York, 1969.
11 M.G.E. Welton, *J. Phot. Sci.*, 17 (1969) 157.
12 G.W.W. Stevens, *J. Microscopy*, 106 (1976) 113.
13 J.M. England and R.G. Miller, *J. Microscopy*, 92 (1970) 167.
14 D.J. Moffatt, S.P. Youngberg and W.K. Metcalf, *Cell Tiss. Kinet.*, 4 (1971) 293.
15 J. Stillström, *Intern. J. Appl. Radiation Isotopes*, 14 (1963) 113.
16 M. Benassi, R. Paoluzi and F. Bresciani, *Cell Tiss. Kinet.*, 6 (1973) 81.
17 J.M. England, A.W. Rogers and R.G. Miller, *Nature*, 242 (1973) 612.
18 M. Mertz, *Histochemie*, 17 (1969) 128.
19 W.H. Barkas, *Nuclear Research Emulsions, Part I*, Academic Press, New York, 1963.

PART 2: THE PLANNING AND INTERPRETATION OF AUTORADIOGRAPHIC EXPERIMENTS

CHAPTER 7

Histological Techniques and Autoradiography

The techniques of preparing a specimen for viewing with either light or electron microscope have evolved to produce a consistent and informative picture, without any direct regard for the quantitative retention or extraction of particular tissue components. It just happens that these techniques are very well suited to one of the most widely used groups of autoradiographic experiments, those based on the incorporation of [³H]thymidine into DNA. Here, fixation and embedding preserve labelled DNA in the section while removing any trace of labelled precursors or metabolites. But this match between histological processing and the needs of the autoradiographic experiment is quite accidental and is not true of most experiments.

In an autoradiograph one attempts to make a biochemical observation against the background of the microstructure of the specimen. Both components are essential — the biochemical specificity of the labelling and the preservation of the microstructure — and the autoradiographer will often have to work out new variants of established histological methods and demonstrate their biochemical validity in terms of his own experiment.

In addition, autoradiography, like many other quantitative techniques, makes more severe demands on the quality of the material than are necessary to simple viewing of a section. Variations in section thickness and in tissue shrinkage during processing can make dramatic differences to the numbers of silver grains per unit area.

Finally, most autoradiographs involve placing the specimen in close contact with a highly sensitised nuclear emulsion. Chemical artefacts due to the presence of particular reactive groups in the specimen often impose severe constraints on the histological process, forbidding the use of particular reagents and making it advisable to insist on a higher standard of purity and cleanliness of other reagents than is common or reasonable in routine histology.

Putting biochemistry in its place, back in the three-dimensional matrix of cells that makes up living tissues, is not easy.

133

CLEAN WORKING CONDITIONS

Before considering the stages of histological preparation of a specimen in detail, it should be noted that autoradiography requires a degree of cleanliness and care, at every stage in the handling of the specimen, that is seldom needed in the production of histological preparations alone. It is a useful, and often sobering, exercise to take some stained sections produced in the normal way by the methods in use in the laboratory, and to examine them critically, by transmitted and by incident light, for "silver grains". Frequently, many black specks of the approximate dimensions of a developed grain will be seen by transmitted light, and their number will almost certainly be more by incident dark-field lighting. This "background", in the absence of an emulsion layer, can be completely ignored in normal microscopy, but it becomes highly embarrassing in an autoradiograph.

It is a good idea to keep one complete set of glassware and solutions for use with autoradiographs only. These solutions should be covered, to prevent dust accumulating on their surface, and to prevent changes in concentration due to evaporation or imbibition of water, and should be renewed frequently. Wet slides should never be allowed to dry in a position where dust, or chemicals in use in the general laboratory, can settle on them, but should be dried in a protected, dust-free position.

Glassware used for emulsion work should be very carefully cleaned. Vessels that have contained liquid emulsion are often difficult to deal with. After washing again in hot water to remove as much emulsion as possible, they should be soaked in normal sodium hydroxide solution for several hours. After washing in hot water, they should be immersed in cleaning acid, washed in cold running water overnight, and finally rinsed in distilled water. Even so, glassware that is frequently used for emulsion work tends to acquire a characteristic hazy appearance, due to the surface deposition of silver.

Microscope slides require special cleaning. Even the commercially obtained "pre-cleaned" slides are usually too dirty. Slides should be soaked overnight in cleaning acid, made by dissolving 100 g potassium bichromate in about 850 ml water, and adding 100 ml concentrated sulphuric acid; the acid should be added very slowly, with constant stirring. After the acid bath, the slides are washed for several hours in cold running tapwater, followed by two changes of distilled water for 30 min each. They are then dipped once in the following solution at room temperature:

Gelatin	5.0 g
Chrome alum	0.5 g
Water to make	1000 ml

The slides should then be allowed to drain and dry while in a dust-free atmosphere.

This gelatin solution (or "subbing" solution) should be filtered immediately before use. It is unwise to try and keep it for more than about 48 h, even in a refrigerator. It is usually convenient to produce a large number of subbed slides at one time, and to keep them in clean, covered containers of plastic or glass until they are required. These subbed slides are usually clean enough for autoradiography, and the layer of gelatin provides good adhesion both for the section and the emulsion. It may be difficult to make good smears of cell suspensions on subbed slides, in which case gelatinisation should be omitted, and special care will be required in processing and mounting the autoradiographs to prevent movement or loss of the emulsion.

Solutions used in preparing material for autoradiography should be made up from reasonably fresh stocks of chemicals of known purity.

Emulsions are sensitive to many forms of chemical contamination, and I have known cases of chemography which have been traced back ultimately to old or low grade reagents. One of the most troublesome causes of positive chemography that I have ever been asked to investigate was finally traced to the xylene used in dewaxing the sections prior to autoradiography. It was low-grade xylene, supplied in large metal drums. Changing to laboratory-grade xylene in glass bottles removed the chemography completely. Tapwater should be avoided, and distilled or ion-free water used instead in preparing the solutions used in histology up to the stage of autoradiography.

Common sense must guide the degree to which cleanliness is pushed: it can become an obsession, slowing down work unreasonably. But one should be aware that the steps of preparing a tissue for autoradiography can introduce contamination, and include these steps in the investigation if the controls for chemography indicate that this artefact is present.

THE PRESERVATION AND EXTRACTION OF RADIOACTIVITY IN HISTOLOGICAL PREPARATION

In order to prepare a thin enough section of a solid tissue for observation in the light or electron microscope, it must first be embedded in some supporting matrix of sufficient hardness to allow sectioning. The materials in general use at the light microscope level, paraffin wax and the newer polymers such as araldite and Epon, are not miscible with water, and the embedding of tissues in them is preceded by the precipitation of much of the macromolecular moiety by histological fixation, followed by the withdrawal of the water present by soaking the tissue in increasing concentrations of alcohol. It is often assumed that this type of processing leaves in the embedded tissue the proteins and nucleic acids, while removing the ions and many other small molecules, together with much of the fat. With increasing sophistication, it has become necessary to re-examine this assumption critically.

In general terms, three categories of experiment can be listed, which make rather different demands on the steps of histological preparation.

(a) *Precursor incorporation studies*

In this group of experiments, a radioactive precursor is presented to the tissue to determine the sites at which it is synthesised into the molecule under study. The autoradiograph is expected to show only the sites of synthesis into a larger molecule, and the histological processing to perform a selective extraction, removing all the unincorporated precursor.

The retention of radioactivity in the tissue after administration of labelled aminoacids was examined by Vanha-Perttula and Grimley[1], amongst others. They compared fixation with either formaldehyde or glutaraldehyde, followed by buffer washing, post-fixation in osmium tetroxide and dehydration, with the standard biochemical step of precipitation in trichloroacetic acid (TCA), at a time after the administration of the labelled aminoacid when very little of the radioactivity would be expected as free aminoacid. When [^{3}H]leucine was used as a precursor, the TCA precipitate contained all but 13.9% of the total radioactivity. Formalin fixation and subsequent processing retained all but 14.5%, while glutaraldehyde permitted the loss of only 3.5%. Quantitatively, it looks as if formaldehyde and the subsequent processing leave the equivalent of a TCA precipitate in the tissue section, though several questions remain unsolved. The chemical form of the radioactivity that was lost was not established: it may have been free amino acid, or represented incorporation into other molecules such as polypeptides or proteins of low molecular weight. Whatever its form, it is clear that the fixative glutaraldehyde retained much of this material in the tissue. Peters and Ashley[2] have demonstrated that glutaraldehyde can link free amino acids to the fixed and precipitated protein, giving spurious results for incorporation into proteins: this effect will be particularly serious at very short times after the administration of the amino acid. A similar effect has been found, though to a lesser extent, for formaldehyde[1].

Mitchell[3], studying the retention of a labelled protein in tissues, found very variable losses in fixation and embedding from one tissue to another. Losses in processing were particularly severe in neonatal animals, with formalin fixation. The failure of formalin to retain protein quantitatively in tissues has also been described by Merriam[4].

With the nucleic acids, it is reasonable to assume that DNA will be retained by fixation in formaldehyde or glutaraldehyde: the latter fixative can link free thymidine to precipitated material, though this effect is small by comparison to that seen with aminoacids[1]. With precursors of RNA, the position is far less clear. Sirlin and Leoning[5] have shown that 4S RNA is effectively retained after fixation with formaldehyde or Carnoy's, and work reported by Edström[6] on isolated neurones shows that Carnoy's fixative can quantitatively retain RNA. Schneider and Maurer[7] and Schneider and Schneider[8] have reported loss of RNA after formalin fixation and the figures obtained by Vanha-Perttula and Grimley[1] with [³H]uridine show 22% of the total radioactivity in their system extracted by glutaraldehyde and the subsequent buffer washes, and 24% loss with formaldehyde. It would seem that Carnoy's fixative is the one of choice for RNA retention, but further work is needed to define the precise patterns of retention of this group of macromolecules after different fixation procedures. Monneron and Moulé[9] have studied the retention of soluble precursors of RNA by fixation, which was quite considerable after osmium tetroxide as a primary fixative. Routines of fixation with glutaraldehyde or formaldehyde, extensive buffer washes over two days at 0°C, and post-fixation in osmium tetroxide were free from this artefact. Extended washing after fixation may have its own problems, however. Rowinski, Adamski and Sawicki[10] have reported a puzzling but extensive reduction in grain counts over cells labelled with [³H]thymidine after prolonged washing in either distilled or tapwater. No similar effect was seen with [¹⁴C]thymidine.

There is little work available on the quantitative retention of polysaccharides after fixation. With mannose as a precursor, losses of about 20% have been found with glutaraldehyde and formaldehyde[1]: binding of the precursor does not seem to be a problem in this one example.

Cope and Williams[11,12] have studied the preservation of choline and ethanolamine phosphatides in tissues for electron microscope autoradiography, and Stein and Stein[13] have written an admirable review on autoradiographic methods for lipids, both for light and electron microscopic autoradiographs. Both groups favour glutaraldehyde fixation with post-fixation in osmium tetroxide, but it is clear that variable and often extensive losses of radioactivity from the various classes of lipids in tissues occur in the final stages of dehydration and embedding. Stein and Stein[13] comment on the possible use of Aquon, a water-miscible derivative of Epon, to avoid this loss. Lipid losses after embedding in GACH or Durcopan have been studied by Ward and Gloster[14].

(b) *Distribution studies of drugs and hormones*

The second group of autoradiographic experiments in this context involves the injection of labelled drugs or hormones into the biological system, and the use of autoradiography to identify the sites at which the labelled material is "bound". Here the assumption is sometimes made that histological processing will remove all the "free" reagent. This situation has seldom been examined with the care that is required, and little evidence is available on the retention of active agents at their sites of binding, or on possible spurious incorporation of unbound radioactivity by fixation. Nor is the assumption always justified that sites of binding represent sites of action of drugs and hormones.

Amongst the steroid hormones, oestradiol binds tightly to a protein present in the cytosol of target cells but, as Stumpf and Roth[15] have shown, the hormone–protein complex cannot be quantitatively demonstrated except by autoradiographic techniques designed for freely diffusible compounds. Similarly, the drug methotrexate, which binds very tightly to folate reductase, requires techniques for diffusible compounds for its demonstration[16]. On the other hand, the quantitative demonstration of the covalently bound enzyme inhibitor DFP has been carried out on fixed and embedded material[17]. Three factors seem to be involved. The first is the nature of the binding to the tissue component. Covalent binding usually survives processing: non-covalent bonds seldom survive. However high the affinity of the labelled compound for its binding site, the transfer of the tissue through relatively large volumes of aqueous solutions in the early stages of fixation will remove label. Secondly, fixation may produce stereochemical changes on the binding site, with loss of the labelled molecule. Finally, the binding site itself may not be completely precipitated by fixation.

(c) *Ions and other diffusible molecules*

Studies of the distribution of labelled ions and small, freely diffusible molecules are clearly impossible with conventional histological techniques. Methods have been described for the preliminary precipitation of particular ions in tissue prior to fixation (e.g. Halbhuber and Geyer[18]), but these have their own sources of error which must be considered.

It may be possible to modify the conventional histological procedure in some way to minimise the loss of radioactivity. The use of fat solvents may be avoided by cutting frozen sections, or by embedding in a water-soluble wax[19], for example. Alternatively, techniques for the autoradiography of freely diffusible materials can be used, as described in Chapter 8. These may quite simply be combined with extractions with specific solvents, if removal of unbound radioactivity is required.

In selecting a method of fixation for tissue for autoradiography, the possibility of interaction between tissue and emulsion must not be over-looked. Fixatives, such as Zenker's, which contain salts of mercury or lead should be avoided, as they tend to cause positive chemography[20]. If the excellent cytological detail given by fixation in Zenker's is essential, it is possible to treat the sections to obtain a reasonable emulsion response[21]. Those fixatives containing picric acid, such as Bouin's may be used provided the section is treated with ammoniacal alcohol or a solution of lithium carbonate to remove all traces of yellow colouration before autoradiography. Even with formaldehyde and glutaraldehyde, extensive washing should follow fixation: otherwise desensitisation of the emulsion may result from traces of free fixative in the tissue (see Fig. 39).

HISTOLOGICAL SECTIONS

The majority of autoradiographs are based on histological sections of embedded tissue. It is surprising how little attention has been paid to the difficulties of producing sections of known thickness. In a recent series of experiments, I was shaken to find a variance in grain counts over sections labelled with carbon-14 so great that I was unable to detect differences between experimental groups that had been obvious on parallel experi-ments with liquid scintillation counting. A block of paraffin wax was uniformly impregnated with radioactive testosterone, and serial sections cut on our microtome under normal conditions were dissolved and counted individually in the liquid scintillation counter. The variability in section thickness was such that the standard deviation of the counts from adjacent sections was nearly 30% of the mean figure. Tests of several combinations of microtome and technician in surrounding laboratories showed a range of values about this figure. By investing in an expensive and large microtome, and paying considerable attention to the details of knife sharpening and sectioning, we can now obtain sections at a nominal 4 μm in which the standard deviation of a series of thickness measurements is not more than 7.5% of the mean value.

Sections cut at a nominal 1 or 2 μm from material embedded in plastic is even more difficult to obtain at a uniform thickness. With blocks of uniformly labelled methacrylate, we have had series in which the standard deviation has been as high as 70% of the mean thickness. We have also found wide differences in mean thickness between one series and another, particularly when different microtomes and operators are involved.

Probably the most detailed examination of sectioning is presented by Hallén[22], who developed an ingenious method of measuring section thickness. He concluded that "even if the sectioning is performed under ideal conditions the variation of thickness between and within the section is remarkably large". In many experiments this variation will impose a corresponding uncertainty on the grain counts obtained from the sections. Some care and attention to the technique of sectioning may reduce this source of variability, but cannot eliminate it. Either the experimental design must ensure that grain counts from each group are based on a relatively large number of sections, or some method must be employed to select for autoradiography sections from a narrow range of thicknesses. Reference to the section in Chapter 5 on self-absorption will show that, with carbon-14, sulphur-35 or isotopes of higher energy, variations in source thickness around a mean of 5 μm will introduce corresponding changes in grain count; while at 1 μm sources labelled with tritium will show variation in grain count related to thickness.

Unfortunately section thickness is not often measured. With a uniformly labelled block, as described above, the mean volume of each section can be estimated, provided that the section is complete. This method gives no information about variations in thickness due to compression of the section, and variations within each section cannot be measured. For plastic sections up to 2 μm thick, mounted on a glass slide, interference measurements in reflected light provide a quick and simple way of finding the thickness in absolute terms. Sections in paraffin wax can also be measured in the same way prior to dewaxing, provided they do not exceed about 6 μm, but their upper profile after dewaxing is so irregular that this type of measurement, or indeed any estimate at all, becomes very difficult indeed. Hallén's apparatus[22], while not commercially available, is not too complicated, and provides accurate estimates up to 20 μm, or even up to 200 μm with reduced accuracy.

Estimates of section thickness at the electron microscope level are usually made on the basis of the interference colour of the section, so that the variability of thickness is considerably less for those sections selected for autoradiography. Even in this case, Williams[23] has drawn attention to the variations that can occur between different observers and in different conditions of observation. In a laboratory where the apparatus for measuring section thickness by interference in reflected light is available and frequently used, it is possible to improve on the figures presented by Williams[23] (Salpeter, personal communication). Williams has also drawn attention to variations in thickness within sections[23], which could introduce errors into the analysis of grain distributions at the electron microscope level.

While the problems of achieving reproducible section thickness have received some attention over the years, little has been written about shrinkage of tissues during histological processing. This is perhaps surprising in a technique which usually expresses its results in terms of silver grains per unit area of section. Experience would suggest that shrinkage can be quite variable from one processing routine to another, and can also vary with the tissue.

Strict self-absorption corrections require a detailed knowledge of the density of tissue components after fixation, embedding, and sectioning. This is a problem that has been skated around in the literature, and little information is available apart from estimates in very general terms.

The step of mounting the section on the slide requires a little care, as the gelatin layer on the subbed slide differs somewhat in its characteristics from the usual adhesives such as glycerine and albumin. When mounting the section on the slide, it is a good idea to allow the gelatin subbing to hydrate fully before drying down again in contact with the section. It is possible to pick sections up from a water surface and to dry the slide again so quickly that there is virtually no adhesion between gelatin and section, and dipping in liquid emulsion can either introduce emulsion into spaces between section and slide, or float the section off altogether. I find it best to place the subbed slide, with a few drops of distilled water on it, on a level hotplate, and to float the wax section on the surface of the water until it has flattened out. This gives the gelatin an opportunity to hydrate, and it will stick firmly to the section on removing the water.

Paraffin wax sections require dewaxing before applying the emulsion. Scrupulous care should be taken over this process: traces of wax left on the slide will produce very uneven coating with liquid emulsion, and prevent adhesion if stripping film is used. Clean solutions, changed frequently if many slides are being processed at once, will ensure that the surface of the slide wets evenly with distilled water when dewaxing is complete.

STAINING THE AUTORADIOGRAPH

Correlations between the emulsion response and the underlying specimen are only possible if structural detail can be easily recognised. This usually requires some form of staining of the section at the light microscope level. Phase contrast can be used very successfully to view autoradiographs, particularly if the tissue is familiar to the microscopist, but most workers still prefer staining, which improves the range of detail that can be recognised and provides a closer correlation to normal histology.

There are two possibilities in staining: the first is to introduce the stain before applying the emulsion layer; the second to stain after photographic processing is complete.

(a) *Prestaining*

This choice has some advantages, and many limitations. Staining before autoradiography is identical to the normal routine of histology — the same reactive groups are present in the section, which should look just like a standard histological specimen after staining. The gelatin of the emulsion layer will not be stained, and in general it is easier to get precise and vivid staining than if the emulsion layer is already present. Some reactive groups are lost in photographic processing, particularly enzymes, which often cannot be demonstrated histochemically after autoradiography.

Unfortunately, prestaining has to be approached with caution. It introduces another step between the collection of the tissue from the animal and the application of the emulsion. It must be demonstrated that this step does not remove significant radioactivity from the specimen, or introduce a source of chemography, which will affect the performance of the emulsion layer later. The loss of radioactivity on prestaining is well documented with the Feulgen reaction for DNA[24,25]: it seems to be maximal at the washing stage after acid hydrolysis[25]. Bryant[26] has described a curious phenomenon on hydrolysis, in which radioactivity appears to be displaced from onion tip nuclei in the S-phase and simultaneously increased over metaphase cells. The periodic acid–Schiff reaction also can cause trouble, though this has been attributed to Schiff's reagent rather than to the acid hydrolysis[27]: if the hydrolysis is carried out before autoradiography and the Schiff staining afterwards, grain counts are not reduced.

There are many examples in the literature of stains applied before autoradiography affecting the emulsion layer. To quote only two examples, prestaining with Celestin blue causes severe positive chemography[28], as do attempts to stain plastic embedded sections with toluidine blue.

One further problem with prestaining is that the photographic processing often alters or removes the stain. This effect, which is quite significant with the haematoxylins, for instance, can be reduced in severity by the selection of processing conditions which do not involve wide changes in pH. Amidol will act as a suitable developing agent when buffered to a pH only slightly on the alkaline side of neutrality: a distilled water stopbath can be followed by fixation in sodium thiosulphate, using the same buffer system that was employed for the developer. Recipes for suitable solutions will be found on p. 218.

Prestaining is sometimes accompanied by the use of an impermeable membrane or layer, designed to prevent both the stained tissue affecting the emulsion and the stain itself being lost in photographic processing. Such membranes are discussed in detail later (p. 146). The layer of carbon evaporated on to electron microscope sections serves the same purposes. Applied in a slightly thicker layer, carbon may also be used over semithin plastic sections, stained for light microscopy before autoradiography[29].

In summary, staining the section before autoradiography is acceptable if control experiments show that grain counts over stained and unstained sections are identical, and there is neither positive nor negative chemography. In these circumstances, if removal of stain in processing can be avoided, the clarity and precision of staining will usually be better than with post-stained material.

(b) *Post-staining*

If the section is stained through the photographic emulsion after processing is complete, there is clearly no problem about the removal of radioactivity, or possible chemography, nor will the stain be affected by processing. This approach has two potential headaches, the first the removal of silver grains or even the emulsion itself from the specimen, the second the difficulty of adequate staining after processing has modified the reactive groups present in the specimen and covered it with a layer of gelatin.

The removal of developed silver grains in staining is often related to the pH of the solutions used. Photographic fixation in sodium thiosulphate leaves each developed grain surrounded with a shell of adsorbed thiosulphate ions (p. 31) which cannot be removed even by long washing times in water. On transfer to an acid solution these ions can erode and remove the silver grains completely. I have known this happen after staining with several of the haematoxylins, on differentiation with acid alcohol. It has been described on acid hydrolysis prior to Schiff staining[27]. This effect can be minimised by iodide treatment of the emulsion prior to fixation[30], but it is probably best avoided by controlling the pH of all staining solutions, so that the acid conditions likely to produce this attack on the developed grains do not occur.

Removal of the emulsion itself can occur with staining methods that require heat or very alkaline solutions. The staining of plastic sections with toluidine blue, which usually involves heating for a short time to 60°C, is one example. Deliberate removal of the gelatin by a lead stain of high pH has been suggested by Revel and Hay[31] for autoradiographs for electron microscopy. Although this improves the final contrast, the strong possibility of loss or movement of developed grains during digestion of the gelatin has been a deterrent to the use of this technique.

The processed emulsion layer may provide quite a barrier to precise staining. Reagents have to diffuse through the gelatin to reach the specimen, and the gelatin may retain traces of thiosulphate, affecting the reagents or providing an environment of an inappropriate pH. It may make staining less variable if autoradiographs are soaked in a buffer at the optimum pH for the staining reaction before the reagents are applied. The gelatin frequently takes up the stain, which is not surprising since it is a protein, like much of the histological specimen. Stains for specific reactive groups in the tissue are often used in an attempt to colour the section without affecting the gelatin; methyl green–pyronin for nucleic acids, for instance, has been used by Ficq[32] through layers of emulsion 50 μm or more thick. The very non-specific counterstains of conventional histology, such as eosin or chromotrope, are the worst for colouring the gelatin. Grossly over-staining the specimen, followed by a rapid removal of excess stain, will tend to bring down the levels of staining in the emulsion before the section is affected.

Some reactive groups are lost from the specimen in autoradiography and photographic processing, and histochemical reactions will at times be much weaker than in normal material. It is all too easy to assume that a particular reaction will be impossible through the emulsion, and this assumption should always be examined in a critical and optimistic spirit. Peroxidase activity can be demonstrated by post-staining[35], and even fairly complex sequences such as the reactivation of organophosphate-inhibited acetylcholinesterase by pyridine-2-aldoxime, followed by Karnovsky's technique for the demonstration of sites of cholinesterase activity — can be successfully carried out, as has been shown by Rogers, Darżynkiewicz, Ostrowski, Barnard and Salpeter[34] (Fig. 43).

In general, post-staining will be found preferable to prestaining. Before a method is used on experimental material, it should be demonstrated that grain densities over stained and unstained specimens are identical.

A few general points remain to be made about the selection of stains. If the autoradiograph is to be viewed by transmitted light, a light colour, such as pink or yellow or light green, may be preferable to dark blues or purples which can produce darkly stained granules in the tissue rather similar to silver grains. If reflected light is to be used, giving a dark-field effect for viewing, methods that produce a light-scattering precipitate in the tissue should be avoided, and also stains such as eosin, which give a considerable fluorescent glare by this method of illumination.

Several lists of stains which have proved compatible with autoradiography are available[20,35–37]. In many cases, several slightly differing recipes are available in the histological literature for a single staining method, and local differences in the sources of dyes and reagents may also occur. It is sensible to check that a chosen technique does not interact with the autoradiograph in the conditions within one's own laboratory before adopting it.

145

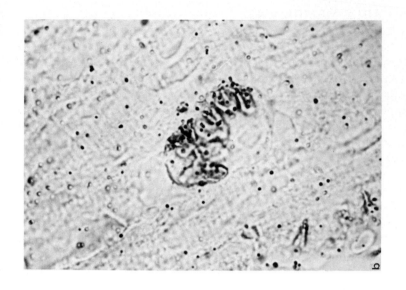

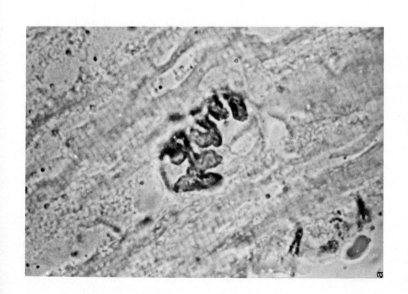

Fig. 43. An endplate of mouse diaphragm, which was labelled in vitro with [³H]DFP, sectioned on a cryostat and autoradiographed with Ilford K2 emulsion. After processing, the DFP was removed with pyridine-2-aldoxime and the reactivated acetylcholinesterase demonstrated histochemically. (×800). (Material prepared by Dr. Z. Dařynkiewicz)

IMPERMEABLE MEMBRANES

It may be essential to stain the specimen before autoradiography by a method that causes chemography, or with a stain that is altered by photographic processing. It may be impossible to remove a source of chemography from a section. In such cases a thin, impermeable membrane can be applied over the specimen before putting on the emulsion. Such a membrane should be very thin, to reduce to a minimum the separation of source from emulsion; it should also be reasonably strong, impermeable, and chemically inert.

Chapman-Andresen[38] investigated a number of possible materials, and described a nylon membrane which had suitable characteristics. More recently, Sawicki and Darżynkiewicz[39] described a very satisfactory membrane of polyvinyl chloride (Fig. 44). Unfortunately, the starting materials for both these techniques have been discontinued by their manufacturers. A suitable alternative has been found by Keyser and Wijffels[40], a copolymer that includes vinylchloride and vinylidenechloride, called Ixan SGA, manufactured by Solvay S.A., of Brussels.

To prepare the Ixan solution, use the following method.

Dissolve 9.7 g Ixan SGA in 30 g butyl acetate. This may take up to 48 h. When fully dissolved, add 57.9 g trichloroethylene with constant stirring, 0.8 g cyclohexanone, and 1.6 g dibutylphthalate. The latter acts as a plasticiser for the membrane, which can be made firmer and more brittle by reducing its concentration, or more pliable by increasing it. This stock solution keeps indefinitely, provided evaporation is prevented.

To prepare membranes, let one drop of the solution fall on to a clean water surface in a dish about 9 in. by 6 in. The drop will spread on the water, and the solvents evaporate, leaving a membrane which contracts a little in the next half-minute or so. This membrane can then be picked up on the specimen in much the same way as stripping film (Fig. 87). A membrane showing a red-green interference colour of the first order will be about 105 nm thick, and should not reduce the grain count over a tritiated source by more than about 20%. The efficacy of such a membrane in preventing chemography and retaining stain in the section was examined by Keyser and Wijffels for a number of different stains[40].

A membrane such as this is hydrophobic, and emulsion layers will not adhere to it as well as to a subbed slide. Thin layers of liquid emulsion can be kept in place on membrane-covered slides if they are kept horizontal through photographic fixation and washing, and if both these steps are carried out very gently. The use of a hardening stopbath (p. 29) might also reduce the chances of losing the emulsion. With stripping film loss of emulsion is even more likely, and it may be necessary to sub the slide with

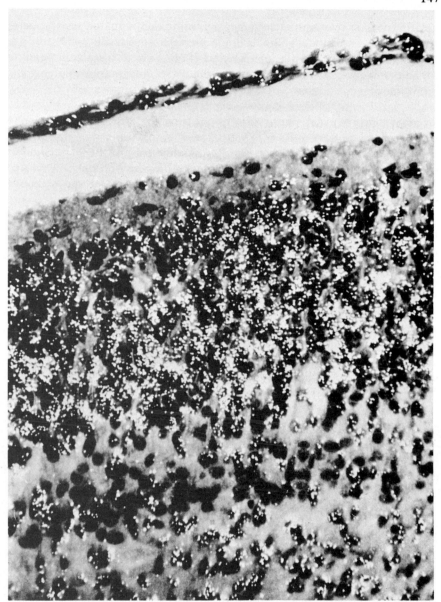

Fig. 44. Photomicrograph of the cerebral cortex of a rat at the 18th day of foetal life, following the injection of [³H]thymidine 2 days previously. The section was stained with Harris' haematoxylin and covered with a PVC membrane before autoradiography with Ilford G5 emulsion. Photograph by combined transmitted bright-field and incident dark-field illumination, using a Leitz × 22 Ultropak objective. (× 300). (Material produced in collaboration with Dr. M. Berry)

148

gelatin (p. 135) over the membrane, accepting the increased separation of source from emulsion, if this technique is to be used.

Once the membrane has stabilised, it is not easy to dissolve it again in organic solvents, so that sections should be stained before applying the membrane.

HISTOLOGICAL METHODS FOR PARTICULAR APPLICATIONS

In whole-body autoradiography, the section should be stained after exposure. This seldom presents any difficulty as the section is separated from the X-ray film before processing. The section, on the sellotape backing, can be stained in Mayer's haematoxylin and counterstained with eosin before being mounted on a glass support under a coverslip[41]. Many alternative stains can be used. Ullberg[41] has often used Goldner's collagen stain. The glass-mounted sections are relatively heavy, and large coverslips are expensive. An alternative method is to spray section and sellotape backing with "Trycolac", producing a much cheaper and lighter specimen for examination and storage[42].

Reference has already been made to published lists of stains compatible with the autoradiography of paraffin-embedded material[20,35-37]. In addition, articles are available which describe the staining of autoradiographs of central nervous system[43], of chromosome squash preparations[44,45], and by enzyme histochemical reactions[46]. Many individual staining techniques have been investigated: these reports are often published in Stain Technology.

Semithin sections of plastic-embedded material may be difficult to stain adequately, without damaging the emulsion layer. Stevens[47] discusses this problem at some length. The usual stains, applied before autoradiography, cause serious chemography, unless separated from the emulsion by a relatively thick layer of carbon[29]: aldehyde–fuchsin seems to be an exception in this respect[48]. Post-staining runs the risk of damaging or losing the emulsion layer, since heat is generally necessary to get adequate penetration of stain into the section. Richter and King[49] have pointed out the value of hardening the emulsion layer in 5% formalin after photographic processing and before staining. Treatment of the sections with sodium hydroxide followed by periodic acid before applying the emulsion may improve the stainability of the sections after photographic processing[50]. Whatever method is used, careful control of the temperature and time of staining is essential if post-staining is to be successful. A quick warm-up in a Bunsen flame is not good enough. A list of stains compatible with autoradiography of plastic-embedded sections has been prepared by Grimley, Albrecht and Michelitch[51].

Hard tissues such as bones and teeth present special problems. Demineralisation may precede sectioning, provided the radioactivity of interest survives the process[52]. Alternatively, sections may be ground or sawn for autoradiography[53], or even cut on a microtome if from a small animal such as a mouse[54].

Fluorescence microscopy has been combined with autoradiography[55], and it has even been possible to combine immunohistochemical methods, such as the immunoglobulin–horseradish peroxidase technique, with the autoradiography of diffusible materials[56].

At the electron microscope level, prestaining with uranyl acetate and lead citrate in the routine way raises several difficulties. First, it is essential to cover the stained sections with a layer of evaporated carbon to prevent chemography. Even so, sections stained with lead often develop a number of granular deposits, chiefly in mitochondria, when in contact with an emulsion layer[57]. The stain is often faint, due to loss of contrast in the photographic processing. This whole problem is discussed by Salpeter and Bachmann[57], who suggest the possibility of dissolving away the celloidin membrane on which the section sits in the flat substrate technique, and restaining the section after photographic processing. An alternative Salpeter sometimes uses is to prestain in uranyl acetate alone for up to 3 h. With all these potential hazards, post-staining might seem to be the answer. Unfortunately, many laboratories have found that staining through the emulsion often produces heavy granular precipitates. It goes without saying that freshly made-up stain, carefully filtered and centrifuged, is essential, but even with such care precipitates often occur. The source of the trouble is the emulsion layer. After fixation and washing, it is usually at a pH inappropriate to the staining procedure. It may also contain thiosulphate ions adsorbed on to the developed silver grains[58] and soluble complexes of thiosulphate with silver. It may help to soak the sections and emulsion for 1–2 minutes in a buffer solution similar to that in which the stain is made up, between thorough washing and applying the stain. If stain deposits are still troublesome, try washing the emulsion layers thoroughly, then soaking briefly in a solution of potassium iodide (0.1 g/litre) in distilled water, followed by another wash in distilled water, a soak in buffer alone, and then the stain.

Prestaining with uranium salts introduces radioactivity into the section, though the amounts involved are likely to be extremely small in comparison to the experimental radioactivity. The effects of prestaining with heavy metals on the efficiency and resolution of electron microscope autoradiographs are discussed elsewhere (p. 70 and 93).

Negatively stained material has also been successfully autoradiographed at the electron microscope level[59].

PREPARING AUTORADIOGRAPHS FOR MICROSCOPY

Before viewing, autoradiographs are usually mounted under a coverglass to protect the emulsion layer from scratching. Processed stripping film is almost impermeable to many of the conventional mounting media, and attempts to prepare it under a coverglass may lead to a mass of tiny bubbles forming over the tissue section. Since it already carries a thick layer of gelatin over the emulsion, it is often stored and examined without a coverglass. If a coverglass is preferred, one remedy that usually prevents the formation of bubbles in the specimen is to dip the slides once in a 2% solution of polyvinyl alcohol during dehydration[60]. With plant material, Elvanol may not be successful, and Jona and Goren[61] have suggested using a 30% glycerol solution or a mixture of glycerol and gelatin as a mounting medium.

Track autoradiographs are easier to interpret if they are viewed with the emulsion in a swollen, hydrated state (p. 376): the same may be true of grain density autoradiographs if there is any danger of confusing silver grains with stained granules in the tissue section.

REFERENCES

1 T. Vanha-Perttula and P.M. Grimley, *J. Histochem. Cytochem.*, 18 (1970) 565.
2 T. Peters and C.A. Ashley, *J. Cell Biol.*, 33 (1967) 53.
3 J. Mitchell, *Austr. J. Exptl. Biol. Med. Sci.*, 44 (1966) 225.
4 R.W. Merriam, *J. Histochem. Cytochem.*, 6 (1958) 43.
5 J.L. Sirlin and U.E. Leoning, *Biochem. J.*, 109 (1968) 375.
6 J.E. Edström, *J. Neurochem.*, 1 (1956) 159.
7 G. Schneider and W. Maurer, *Acta Histochem.*, 15 (1963) 171.
8 G. Schneider and G. Schneider, *Proc. 2nd. Int. Congr. Histochem. Cytochem.* Frankfurt/Main, 1964, p. 169.
9 A. Monneron and Y. Moulé, *Exptl. Cell Res.*, 56 (1969) 179.
10 J. Rowinski, M. Adamski and W. Sawicki, *Histochemie*, 28 (1971) 250.
11 G.H. Cope and M.A. Williams, *J. Microscopy*, 90 (1969) 31.
12 G.H. Cope and M.A. Williams, *J. Microscopy*, 90 (1969) 47.
13 O. Stein and Y. Stein, *Adv. Lipid Res.*, 9 (1971) 1.
14 B.J. Ward and J.A. Gloster, *J. Microscopy*, 108 (1976) 41.
15 W E. Stumpf and L.J. Roth, *J. Histochem. Cytochem.*, 14 (1966) 274.
16 Z. Darżynkiewicz, A.W. Rogers, E.A. Barnard, D.-H. Wang and W.K. Werkheiser, *Science*, 151 (1966) 1528.
17 A.W. Rogers, Z. Darżynkiewicz, E.A. Barnard and M.M. Salpeter, *Nature*, 210 (1966) 1003.
18 K.-J. Halbhuber and G. Geyer, *Acta Histochem.*, 31 (1968) 222.
19 H.F. Steadman, *Section Cutting in Microscopy*, Blackwell, Oxford, 1960.
20 G.A. Boyd, *Autoradiography in Biology and Medicine*, Academic Press, New York, 1955.
21 B.M. Kopriwa and C. Huckins, *Histochemie*, 32 (1972) 231.

22 O. Hallén, *Acta Anat.*, Suppl. 25 (1956).
23 M.A. Williams, *Adv. Opt. Electron Microscop.*, 3 (1969) 219.
24 R. Baserga and K. Nemeroff, *Stain Technol.*, 37 (1962) 21.
25 W. Lang and W. Maurer, *Exptl. Cell Res.*, 39 (1965) 1.
26 T.R. Bryant, *Exptl. Cell Res.*, 56 (1969) 127.
27 W. Sawicki and J. Rowinski, *Histochemie*, 19 (1969) 288.
28 E.M. Deuchar, *Stain Technol.*, 37 (1962) 324.
29 J.W. Sechrist and R.H. Upson, *Stain Technol.*, 49 (1974) 297.
30 G.W.W. Stevens and P. Block, *J. Phot. Sci.*, 7 (1959) 111.
31 J.P. Revel and E.D. Hay, *Exptl. Cell Res.*, 25 (1961) 474.
32 A. Ficq, in J. Brachett and A.E. Mirsky (Eds.), *The Cell*, Vol. 1, Academic Press, New York, 1959.
33 R.A. Popp, W.D. Gude and D.M. Popp, *Stain Technol.*, 37 (1962) 243.
34 A.W. Rogers, Z. Darżynkiewicz, K. Ostrowski, E.A. Barnard and M.M. Salpeter, *J. Cell Biol.*, 41 (1969) 665.
35 J.M. Thurston and D.L. Joftes, *Stain Technol.*, 38 (1963) 231.
36 L.F. Bélanger, *Stain Technol.*, 36 (1961) 313.
37 C.P. Leblond, B.M. Kopriwa and B. Messier, in R. Wegmann (Ed.), *Histochemistry and Cytochemistry*, Pergamon, London, 1963.
38 C. Chapman-Andresen, *Compt. Rend. Trav. Lab. Carlsberg Ser. Chim.*, 28 (1953) 529.
39 W. Sawicki and Z. Darżynkiewicz, *Folia Histochem. Cytochem.*, 1 (1964) 283.
40 A. Keyser and C. Wijffels, *Acta Histochem.*, Suppl. 8 (1968) 359.
41 S. Ullberg, *Science Tools*, Special Issue (1977) 2.
42 D.E. Farebrother and N.C. Woods, *J. Microscopy*, 97 (1973) 373.
43 A. Hendrickson, L. Moe and B. Noble, *Stain Technol.*, 47 (1972) 283.
44 D.M. Prescott and M.A. Bender, in D.M. Prescott (Ed.), *Methods in Cell Physiology*, Vol. 1, Academic Press, New York, 1964.
45 L.K. Schneider and S.L. Pierce, *Stain Technol.*, 48 (1973) 69.
46 M. Oehmichen and R. Saebisch, *Acta Histochem.*, 41 (1971) 353.
47 A.R. Stevens, in D.M. Prescott (Ed.), *Methods in Cell Physiology*, Vol. 2, Academic Press, New York, 1966.
48 R. Calvert and A. Pusterla, *Stain Technol.*, 49 (1974) 323.
49 C.B. Richter and C.S. King, *Stain Technol.*, 47 (1972) 268.
50 A. Hendrickson, S. Kunz and D.E. Kelly, *Stain Technol.*, 43 (1968) 175.
51 P.M. Grimley, J.M. Albrecht and H.J. Michelitch, *Stain Technol.*, 40 (1965) 357.
52 W. Beertsen and G.J.M. Tonino, *Arch. Oral Biol.*, 20 (1975) 189.
53 C.M. McQueen, I.B. Monk, P.W. Horton and D.A. Smith, *Cell Tiss. Res.*, 10 (1972) 23.
54 B. Hindringer and G. Büsing, *Histochemie*, 26 (1971) 333.
55 D. Masuoka and G.-F. Placidi, *J. Histochem. Cytochem.*, 16 (1968) 659.
56 D.A. Keefer, W.E. Stumpf, P. Petrusz and M. Sar., *Am. J. Anat.*, 142 (1975) 129.
57 M.M. Salpeter and L. Bachmann, in M.A. Hayat (Ed.), *Principles and Techniques of Electron Microscopy*, Vol. 2, Van Nostrand Reinhold, New York, 1972.
58 G.I.P. Levenson and C.J. Sharpe, *J. Phot. Sci.*, 4 (1956) 89.
59 N.M. Maraldi, G. Biagini, P. Simoni and R. Laschi, *Histochemie*, 35 (1973) 67.
60 M.J. Schlesinger, H. Levi and R. Weyant, *Rev. Sci. Instr.*, 27 (1956) 969.
61 R. Jona and R. Goren, *Stain Technol.*, 46 (1971) 156.

CHAPTER 8

The Autoradiography of Radioisotopes in Diffusible State

The methods for preparing sections of tissue for autoradiography derived from the conventional histological process of embedding in paraffin wax or a resin such as araldite were discussed in Chapter 7. For certain types of experiment, particularly for study of the incorporation of a labelled precursor into a macromolecule, these histological techniques have much to recommend them. The histological fixation may precipitate and retain the macromolecule, while the range of solvents used in fixation and dehydration prior to embedding may wash out of the tissue any precursor that has not been synthesised into macromolecules.

But there are many situations in which the process of section preparation removes radioactivity that is of interest. The degree of loss varies widely: it may be only a few percent of the total with some proteins, or it may go as far as complete removal with labelled ions. The mechanism of loss also varies: it may be limited to diffusion in aqueous solutions, or it may be due to the solubility of, for instance, lipids in non-aqueous solvents such as ethanol and xylene. Since the extent and mechanism of loss of radioactivity show this enormous variation from one experiment to another, it is not surprising that the literature contains a bewildering range of techniques that have been applied at one time or another to the localisation of radioactivity that is diffusible, when studied by conventional methods of histology and autoradiography. In some cases, relatively simple modifications of the process of embedding the tissue have proved sufficient: in others, nothing short of a method with the complete absence of solvents of any sort has been satisfactory; in many other cases, unfortunately, no clear evidence is offered on the suitability or otherwise of the technique used.

In consequence, the literature on the autoradiography of diffusible materials forms a dense jungle, bewildering and contradictory. No attempt will be made here to review it. Instead, I shall try to indicate the principles that should guide one in the selection of a technique, and give in some detail descriptions of a few methods that have found fairly general

153

154

acceptance. In many ways this is the most difficult area in autoradiography, but it is also one of the most interesting to the biologist, opening up possibilities for measuring the distributions of drugs and hormones in tissues, the concentrations of sugars and aminoacids in cells, and even the concentrations and diffusion rates of electrolytes between various cell compartments.

There is a clear progression in the difficulty of autoradiographing diffusible materials with increasingly high magnification of viewing. With whole-body autoradiographs, the standard technique described in Chapter 14 is suitable for diffusible substances, and has even been successfully modified, for the localisation of volatile anaesthetics such as chloroform[1] and diethyl ether[2]. For viewing with the light microscope, a number of acceptable techniques have evolved, and most of this chapter will be concerned with this level of work. The autoradiography of diffusible materials at the electron microscope level is at the very limit of present techniques, and only a handful of micrographs have been published: it is still far from being an established technique.

The problems posed by diffusible materials, and the techniques for coping with them, are best considered under two headings — those concerned with the preparation of the specimen for autoradiography, and those to do with the exposure of the specimen to the emulsion.

THE PREPARATION OF SPECIMENS FOR AUTORADIOGRAPHY

The complexity of the process between collecting the specimen and the start of autoradiography varies considerably with the type of experiment. The specimen must be thin enough to view at the appropriate magnification. With a suspension of separate cells this may present very few problems, but most specimens are solid tissues which require sectioning before microscopy is possible. In order to cut suitably thin sections, the tissue must have a certain hardness and uniformity. This can be achieved either by dehydrating the tissue and impregnating it uniformly with paraffin wax, or with a plastic of some sort, or the necessary consistency can be reached by freezing the tissue, effectively embedding it in ice. Embedding in wax or plastic provides conditions for sectioning which are better than those given by freezing. The uniformity of section thickness is better, the techniques are simpler, and the final product, the section, looks more acceptable to the histologist. So this type of section should be used wherever possible. Unfortunately, the range of solutions needed includes an aqueous fixative, dehydration in increasing concentrations of alcohol, and replacing the alcohol with a solvent such as xylene or benzene before impregnation with the embedding agent. The method which gives the best sections involves the greatest risk of removing the very radioactivity which is to be autoradiographed. The first question facing one is the following.

DO CONVENTIONAL HISTOLOGICAL METHODS REMOVE THE RADIOACTIVITY UNDER STUDY?

This can be tested in several ways. Labelled tissue fragments can be taken through all the solvents used in fixing and embedding, and the solutions analysed afterwards. If all the radioactive material is to be retained in the tissue, the presence of any detectable radioactivity in any of the solvents is enough to show that some modification of the process is necessary. If the extraction of radioactivity in one form is permissible, but retention in another is required, analysis of the solvents may be needed to establish the chemical form of the extracted radioactivity: extraction of inorganic iodide and retention of an iodinated compound, for instance. An alternative procedure is to measure the radioactivity remaining in the tissue fragment at each stage.

In general, these are rather insensitive controls. The volume of tissue is often small relative to that of the processing solutions, so that a total loss of as much as 10% of the radioactivity through the entire process may be difficult to detect. While a uniform reduction of 10% in the radioactivity of all tissue elements might not invalidate the experiment, complete removal from one tissue compartment, such as the extracellular fluid, would produce a totally different autoradiograph. This type of testing also fails to show whether there has been any redistribution of radioactivity within the tissue during processing.

A more time-consuming but more reliable control is to compare autoradiographs prepared by conventional means with those made from identical material treated by one of the techniques which prevent any loss or displacement during processing — the techniques which will be discussed in full later in this chapter (pp. 165–168). In many instances, if such a technique has to be set up at all, it may be quicker to use it for the whole experiment in the first place, rather than as a control to validate some other approach. In other cases, such as attempts to study the distribution of inorganic ions like sodium and iodide, it will be obvious that conventional histological methods should be replaced by frozen sectioning, and there is no point bothering with experiments into the rate of loss in different solvents.

Good examples of this type of investigation into the suitability of conventional histological methods for particular experiments are the papers by Attramadal[3] and by Darrah, Hedley-Whyte and Hedley-Whyte[4]. It is interesting that the latter authors found a similar distribution of radioactivity in the lung after administering [³H]cholesterol by a number of techniques which removed very different amounts of radioactivity from the tissue.

If there is an indication that radioactivity of interest to the experiment is being lost in tissue processing, two lines of action can be taken. The steps of tissue processing can be examined carefully to see if modifications to the method can cure the problem. Alternatively, a technique based on frozen sections can be adopted.

MODIFICATIONS TO CONVENTIONAL PROCESSING

There are many reasons why these should be preferred to frozen sections, provided that they meet the requirements of retention of radioactivity. The techniques are simpler, the histology is better, the section thickness is more reproducible and can be more accurately controlled and measured: the geometry of section and emulsion is more predictable with plastic embedding, and there is also the possibility of extending studies to the electron microscope level.

It may be sufficient to alter the fixation so as to retain radioactivity through processing. Williams[5] has illustrated how variable the retention of lipids can be with different fixation techniques. Loss of fat-soluble components may be prevented by using embedding media that are water-miscible, thus avoiding dehydration in alcohols and clearing in xylene or some similar reagent. Steedman[6] describes some of the water-miscible waxes available for light microscopy. A number of polymers exist for this purpose, including Aquon[7], Durcopan and GACH[8].

The most widely used modification of the methods for embedding tissues is to freeze-dry the tissue fragments, and then impregnate them directly with the embedding agent. Branton and Jacobson[9] described such a method with paraffin wax as the embedding medium. Wilske and Ross[10] tried the same thing with Epon as embedding agent (Fig. 45), and their method was further developed by Nadler, Benard, Fitzsimmons and Leblond[11]. A detailed study of the application of this approach to the autoradiography of steroids has been presented by Attramadal[12,13]. Since then, Stirling and Kinter and their collaborators[14,15] have developed this approach further: both they and Frederik[16-18] have succeeded not only in producing light microscopic autoradiographs of high quality but also in progressing to electron microscopy.

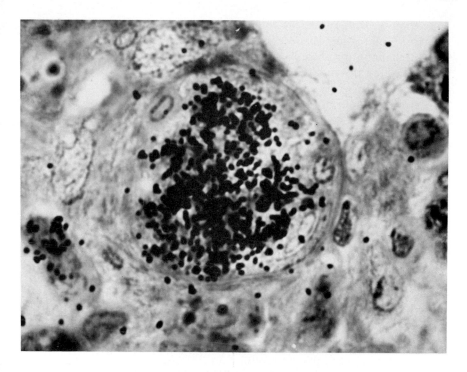

Fig. 45. Photomicrograph of a small artery in the lung of a rat, 5 minutes after the administration of [³H]aspirin. The tissue was frozen in isopentane cooled in liquid nitrogen, freeze-dried and infiltrated in vacuo with Epon. The section, 1 μm thick, was autoradiographed with Kodak NTB-2 emulsion. Note the localisation of this highly soluble drug to the lumen of the vessel. (× 1600). (From Wilske and Ross, 1965)

Nadler, Benard, Fitzsimmons and Leblond[11] showed an almost complete loss of inorganic iodide from the thyroid gland during fixation in Bouin's fluid and embedding in paraffin wax in the conventional way; with freeze-drying, fixation in osmium vapour and impregnation with Epon, this loss was reduced to about 5%. Similar figures are presented by Attramadal[3] for labelled steroids. The loss of radioactivity on fixation, dehydration and embedding in Epon was 75–80% for the uterus. Osmium tetroxide fixation produced very little loss of activity by comparison with formaldehyde, while glutaraldehyde perfusion with osmium post-fixation retained practically all the activity: in spite of this promising beginning to the process, losses on dehydration were very high, producing comparable total extractions whatever fixative was used. The losses from liver were lower, in the range 45–50%. With freeze-drying, vapour fixation and impregnation with Epon, the total loss from the tissue was about 4% (ref. 12). It is interesting to note that collecting the sections on water and floating them out produced a significant further loss of radioactivity — 10% in the case of thyroidal iodide[11], and 2.5% with labelled steroids in the uterus[12].

In Kinter's laboratory, the loss of radioactivity from sections while in contact with water was even more serious when either [³H]galactose or [³H]phlorizin was used[14]. Diffusion artefacts could be recognised in the autoradiographs, due not only to translocation of radioactivity while collecting and floating out the sections, but also to movement during dipping in liquid emulsion. These artefacts were ascribed to water-permeable channels in the araldite sections, and to loss from the surface of the section. Considerable improvement was obtained by incorporating a silicone fluid in the araldite: this was thought to close off the water-permeable channels in the sections. Surface loss to the water on floating out remained high at 25% of the total radioactivity, but did not increase above this level on prolonged floating. The autoradiographs showed much better resolution and no obvious diffusion artefacts when silicone was used. Even this surface loss can be reduced if the sections are cut on a dry glass knife[14]. Frederik's method[16] is based largely on the work of Stirling and Kinter[14], but with modifications to the methods of freeze-drying and, for electron microscopy, of sectioning.

Several interesting features emerge from these experiments. First, very substantial reductions in the extraction of radioactivity can often be achieved by freeze-drying and vapour fixation instead of fixation and dehydration in solution, but there is almost always some loss of radioactivity. Next, the severity of this loss varies, as would be expected, with the chemical nature of the labelled material and with the precise processing method; in addition, it varies considerably from tissue to tissue under identical conditions of treatment. Thirdly, the retention of radioactivity through to the embedded block is not the end of one's worries since losses in cutting and mounting the sections may be surprisingly high.

Considering these findings, it is only possible to place reliance on the autoradiographs if one assumes that the radioactivity lost comes randomly and equally from all tissue compartments, and that it is not accompanied by significant translocation of the radioactivity left behind in the section. These assumptions are not easy to prove. However, Fig. 46 shows an autoradiograph prepared by Kinter of [³H]glucose in salamander plasma. On freezing, which was deliberately a slow process, large solute-free ice crystals have been produced bounded by "eutectic lines" of precipitated proteins and other solutes. Grain counting has shown that 85% of the silver grains lie within 1 μm of the centres of these lines, giving very respectable

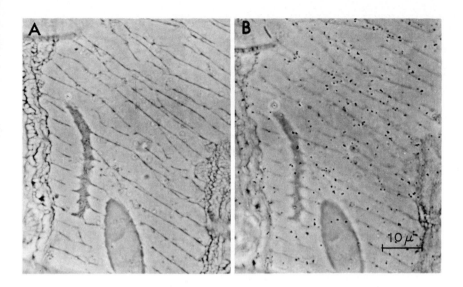

Fig. 46. A blood vessel of Necturus, which was slowly frozen to produce large ice crystals separated by eutectic lines, which contained all the solutes, including [³H]glucose. The tissue was then freeze-dried, and embedded in Epon and autoradiographed. (A), Focussed on the section; (B), with the silver grains in focus. Of the silver grains over the plasma, 85% lay within 1 μm of the centre of an eutectic line. (× 1400). (From Stirling, Schneider, Wong and Kinter, *J. Clin. Invest.*, 51 (1972) 438)

resolution and a clear indication that in this model system no significant translocation of the surviving radioactivity has taken place.

Freeze-drying and impregnation with an embedding agent are not universally applicable. Kinter informs me that the techniques which have given such useful information on the distribution of labelled sugars in his laboratory are far from satisfactory for autoradiographs of labelled sodium. In each case, then, the onus of proof is on the experimenter, who must show how far his procedure retains the radioactivity under study in the tissue, and must consider possible translocation of label. Each detailed procedure, each labelled compound and each tissue may give different results. The critical steps seem to be the impregnation with embedding agent, the collection and floating out of sections and the coating with emulsion. In spite of these obstacles, when the technique is carefully established it can give very valuable results (Fig. 47).

THE FREEZING OF TISSUE FRAGMENTS

The freezing of specimens and their handling while frozen are common to those techniques that rely on freeze-drying and embedding in plastic and the alternative ones that involve sectioning the frozen block directly. The chief problem for the autoradiographer is the formation of ice crystals within the specimen. If these grow to any considerable size, they distort the microstructure giving a poor or even unrecognisable section. In addition, the growth of an ice crystal in a cell results in all the solutes being pushed to the rim of material surrounding the crystal, distorting its distribution pattern (Fig. 46).

Ice crystal formation is particularly liable to appear on freezing the tissue. The aim should be to keep the crystals so small that they do not seriously interfere with microscopy, and this is best achieved by freezing very rapidly. Whatever the method used to cool the tissue, heat loss from the centre of the block will be damagingly slow if the block itself is large, since heat transfer through tissue is not rapid. No point in a block should be more than 0.5 mm from the surface. At a temperature of $-120°C$, ice crystal growth ceases, so this temperature or lower should be achieved as rapidly as possible. Frederik[19] has shown that cooling in Freon 22 at the temperature of liquid nitrogen gives very rapid cooling, more rapid than the use of isopentane in liquid nitrogen. So a very small fragment of tissue should be cooled in a reasonably large volume of Freon 22, held at the temperature of liquid nitrogen.

Once satisfactorily frozen, the tissue should ideally be kept at less than $-80°C$ in subsequent handling. Nei[20] has documented the growth of ice crystals in frozen specimens when the temperature is allowed to rise above this figure. This poses a problem for the autoradiographer. Attachment of a specimen to a microtome chuck or other support before freezing greatly increases the mass of material to be frozen, and hence the freezing time. On the other hand, it is very difficult to attach a frozen specimen to a support later without a rise in temperature. In previous editions, I described a method for attaching a frozen tissue fragment to a microtome chuck without letting it thaw. I now believe this to be the poorer of the two methods, and recommend freezing the tissue on to its support if the frozen block is to be directly sectioned.

Although the initial freezing will be slower with the tissue attached to the chuck, the later probability of a significant rise in temperature is avoided.

One advantage of the techniques based on impregnating the freeze-dried block with Epon or araldite is that the tiny tissue fragment does not come above $-50°C$ again until the ice has been removed by sublimation.

Ice has an appreciable vapour pressure even at $-70°C$, and tissue fragments and sections will freeze-dry surprisingly rapidly. If frozen

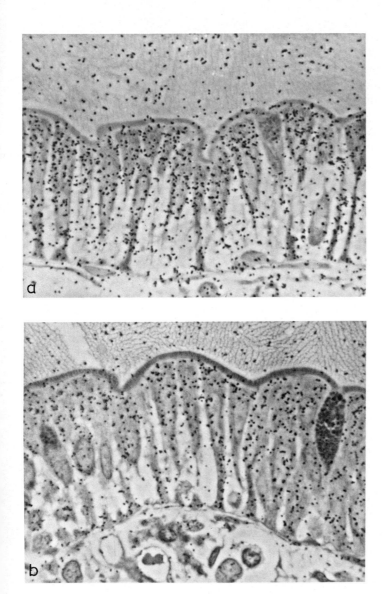

Fig. 47. Autoradiographs of biopsies of human intestinal mucosa, incubated in a medium containing [³H]galactose. The fragments were then freeze-dried, embedded in Epon and sectioned at 2 μm. (a), Grain densities considerably higher than are found over the incubating medium occur over the cytoplasm of columnar absorptive cells, but not over their nuclei or over a goblet cell. (b), After similar incubation in the presence of phlorizin, grain densities over columnar cells were only 50% of those seen in the absence of this glucoside. (× 800). (From Stirling, Schneider, Wong and Kinter, *J. Clin. Invest.*, 51 (1972) 438)

sections are to be cut from the tissue directly, freeze-drying of the block amounts to removal of the embedding medium, and produces a block that does not have sufficient hardness for sectioning. Blocks can be stored for many weeks under liquid nitrogen and still be sectioned satisfactorily, but it is quite impossible to cut sections from blocks that have been kept in the deep freeze or, worse, in the refrigerator after freezing.

Cryoprotectants, such as glycerol or dimethyl sulphoxide, greatly reduce ice crystal formation on freezing. They have seldom been used in autoradiography as their effects on the distributions of labelled compounds are uncertain. They have been known to cause changes in the ultrastructure of cells[21].

Freezing at high pressure may also reduce ice crystal formation[22], but the apparatus needed will deter many biologists.

METHOD FOR FREEZE-DRYING AND PLASTIC EMBEDDING

Step-by-step accounts of the method are published, with slight differences, by several authors[10–12,14,16]. The following short account is based on the process as used in Kinter's laboratory.

Small fragments of the tissue to be sectioned are placed on pieces of aluminium foil and rapidly frozen in propane or Freon 22 cooled in liquid nitrogen to about $-184°C$. The size of fragment is critical to adequate preservation of cytological detail: 1 mm cube is about the upper limit. The foil and tissue, after draining off excess propane, are then transferred to a freeze-dryer cooled to liquid nitrogen temperature and the apparatus is pumped down to 10^{-5} torr (10^{-5} mm Hg). After a total of 50 hours, the temperature is raised to $-70°C$, and thence by stages to room temperature over a further 40–48 hours.

The tissue fragments, which are very fragile by this time, are transferred to a desiccator containing phosphorus pentoxide and 0.1–0.5 g osmium tetroxide; after evacuating the air, the tissue is left for 12 hours to fix in the osmium vapour. Next, the tissue is placed in the side-arm of a Thunberg tube, which has about 0.5 ml of embedding medium in the lower arm. This is pumped down to 0.2 torr for an hour, and then evacuated a second time before allowing the tissue to fall into the embedding medium. After warming the tube to 60°C it is evacuated for a third time, sealed, and kept at 60°C for 12 hours. The tissue is transferred to a capsule containing degassed medium with catalyst added, and cured for 2 hours at 0.2 torr at room temperature, followed by 36 hours at 48°C.

The embedding medium consists of 54 volumes Araldite 502 (Ciba), 45 volumes dodecenyl succinic anhydride and 1 volume of silicone fluid 200

(Dow-Corning): the latter is dispersed in the other constituents by vigorous shaking for 10–12 minutes. The catalyst is 2 volumes of benzyl-dimethylamine. The sections can be cut in the normal way on an ultra-microtome: alternatively, a dry glass knife can be used with the back of the knife (the surface extending from the cutting edge away from the block face) coated with Teflon. Using Wantz T-fix (Du Pont), spray the back of the knife briefly, and heat the knife on a hotplate at 70°C for 5 minutes. Sections collected dry may be flattened with chloroform, either as a drop of fluid or as vapour.

Frederik and Klepper's technique[16] is similar in general, though they use a specially designed freeze-drier. They attach sections cut on a dry knife to microscope slides by means of a thin layer of the adhesive, Entalan (Merck). Ultra-thin sections for electron microscopy were cut on an LKB Cryokit, with the specimen at − 70°C and the knife at − 120°C.

METHODS BASED ON FROZEN SECTIONS

We have seen that methods based on the use of conventional embedding media, while capable of giving interesting results in particular cases, fail to retain all the radioactivity in tissue fragments in many instances. A number of techniques have been described which attempt to preserve all the radioactivity, whatever its chemical form, through the process of sectioning by the simple expedient of avoiding all contact between the tissue and solvents of any sort. These methods in general have the great advantage that loss of radioactivity from the tissue may be confidently excluded. Technically they are more difficult than producing sections from an embedded block, and the reproducibility of section thickness and quality of histological picture will usually compare unfavourably with the latter. The idea common to these methods is that the tissue should be rapidly frozen, and kept in the frozen state through sectioning and autoradiography so that translocation or diffusion of radioactive material is kept to a minimum. Alternatively, the frozen sections may be freeze-dried before exposure, permitting their return to room temperature under controlled conditions of humidity.

The frozen section approach has been widely used. Cryostat sections are difficult to cut by comparison to embedded tissues, as mentioned above; in addition to this, the retention of the tissue and section in the frozen state for relatively long times may favour the growth of ice crystals, with disruption of cellular structure and the movement of labelled solutes. Sectioning a frozen block of tissue carries the risk of thawing at the line of pressure of the knife edge. The frozen section contains all the reactive groups that were present in that volume of tissue in vivo, and is much more likely to interact chemically with the emulsion than a comparable section in Epon or araldite.

In spite of these difficulties, some of which will be discussed in more detail later on, techniques based on frozen sections have given a lot of information, and are the best we have at present for the autoradiography of labelled compounds that are lost in embedding.

OBTAINING FROZEN SECTIONS

The freezing of tissues and cutting of frozen sections are discussed in a number of articles[23,24]. Tissue blocks may be cut at almost any temperature from $-20°C$ downwards: the lower the temperature the thinner the section that can be conveniently cut. A thin frozen section can so easily thaw, with spread of labelled solutes in a totally uncontrolled fashion, that a frozen cabinet with reasonable working space around the microtome (a cryostat) is essential for this type of work. There are many commercially available cryostats, which are not all suitable to this particular application. It is crucially important that the very thin frozen section should not thaw, even transiently, so that any design of cryostat that permits warm air to reach the region of the knife should be avoided. It is a fairly simple job to modify many commercial designs to convert the cabinet to a closed working area with glove ports[25]; some manufacturers will provide this modification on request.

The better the design of the microtome in the cryostat, the easier it will be to cut reproducible and thin sections: a rotary microtome is clearly preferable. There is often a gain in reproducibility of sectioning from motorising the microtome: a slow cutting stroke is often required, and this is difficult to reproduce by hand without hesitations and changes of speed. The quality and angle of the knife are crucially important. I have found that machine-sharpened knives are a great help: the bevel angle remains constant from one knife to the next. Once the optimum knife angle has been determined, knives sharpened by machine can be changed without any alteration in sectioning characteristics. The temperature of sectioning should be $-20°C$ or less to keep the section frozen. At $-20°C$, thicker sections in the range of 10–20 μm can be cut reasonably well: at -25 to $-30°C$, it becomes easier to cut below 5 μm. It should be possible to cut at 3 or 2 μm and collect practically all the sections without undue difficulty. Stumpf and Roth[26] have shown that sections down to 1 μm can be cut at temperatures down to $-90°C$, and even occasional ones at 0.5 μm; for routine autoradiography at the light microscope level these very low temperatures are not essential, provided care is taken to prevent temperature fluctuations around the knife.

It has been suggested that the tissue must thaw at the knife edge, like ice under a skate, with consequent redistribution of radioactivity. This seems undeniable. However, the zone of thawing is probably very narrow and re-freezing must occur almost instantaneously on the surface of the knife. Certainly many series of autoradiographs of radioiodide produced in my laboratory[27] have given good localisation with no evidence of a smearing of radioactivity from regions of high grain density to adjacent areas (Fig. 48). As an added precaution against this effect, it may help to cool the knife to a considerably lower temperature than the rest of the cryostat cabinet: this can be done by packing small chips of solid carbon dioxide against it.

A completely enclosed cryostat may contain air that is so dry that a build-up of static electricity on constant use may make the sections very difficult to handle. I have seen them hopping about between glass and

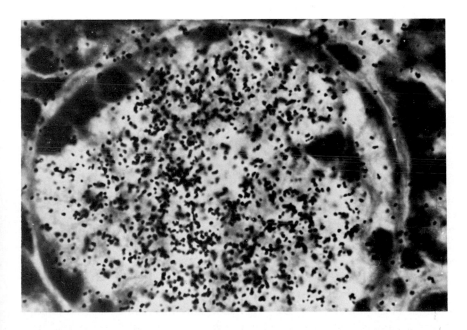

Fig. 48. An autoradiograph of the submaxillary salivary gland of the mouse after an injection of [^{125}I]iodide. The tissue was sectioned at 3 μm in a cryostat and autoradiographed by the technique of Stumpf and Roth. Note the excellent resolution, with heavy grain density over the cytoplasm of cells of the convoluted granular tubule, but not over the nuclei, and a clear edge to the high grain density following the basal cell membrane. ($\times$ 400). (From Rogers and Brown-Grant, 1971)

metal surfaces inside the cryostat like a team of performing fleas. A small ion gun, of the type used to remove static from records and tapes, can bring order back to the process of sectioning. The guide plate, which is necessary in cryostat work to prevent the section from rolling up on the knife edge, can contribute greatly to this build-up of static electricity if, as is usual, it is made of transparent plastic. I have often replaced such a plate with one made from glass, with considerable advantage.

THE AUTORADIOGRAPHY OF SECTIONS CONTAINING DIFFUSIBLE RADIOISOTOPES

Let us assume that we have obtained a tissue section without any loss or displacement of the radioactive material within it. What problems arise in bringing this section in contact with an emulsion layer and subsequently processing the emulsion?

(a) *The autoradiography of embedded sections*

It is tempting to treat these as if they were sections of fixed radioactive material, and to forget that loss or movement of labelled material may still occur.

If the tissue has been embedded in paraffin wax, perhaps after freeze-drying[9], the temptation to dewax the section in xylene and to rehydrate them in a graded series of alcohols must be resisted: if the radioactivity will survive this extraction process, it can hardly be called "diffusible". Wax sections are best exposed by laying the ribbon of sections on a preformed emulsion layer, and gently flattening them with pressure. After exposure, the sections will have to be dewaxed before development, or removed from the emulsion altogether, to permit reagents to reach the emulsion beneath the sections. This can be done by dewaxing in xylene and careful washing, but sections are often lost at this stage. All in all, autoradiography is difficult with sections like this, and this method of embedding is best avoided.

Sections of tissue embedded in araldite or Epon are easier to autoradio-graph, as there is no need to remove the embedding plastic. Several methods are available. The sections may be mounted on slides and dipped in liquid emulsion or covered with stripping film in the conventional way (see pp. 366 and 342)[10]. But, as mentiond above, loss of radioactivity from the section can occur at this stage with many diffusible materials[11,12], and the experimenter must satisfy himself and others that such loss is not occurring before he can use such direct and simple methods.

Sections can be mounted on one slide and exposed by placing the slide face-to-face with another that has been coated with emulsion by dipping or by stripping film. This quick and simple method has the disadvantage that section and autoradiograph must be separated before photographic processing and viewing, with a consequent loss of resolution. Attempts to leave the section on the emulsion when separating the slides only make it difficult for the processing solutions to reach the emulsion under the section.

Perhaps the most satisfactory approach is to form a thin emulsion layer on a wire loop, using the Caro technique (p. 406), and to place this when dry over the section[28], mounted on some support such as a microscope slide. Reasonably reproducible geometry can be obtained without the section coming into contact with liquid emulsion or with water.

(b) *The autoradiography of frozen sections*

The handling of these very delicate sections is complicated by the fact that thawing produces gross redistribution of diffusible substances. The utmost care is needed if physical damage to the section is to be avoided, and thawing from warm instruments or air or even pressure prevented. There are two basic choices. Either the frozen section can be placed in contact with a preformed emulsion layer, which should be at $-15°C$ or lower, or the section can be freeze-dried. After this latter step it may be brought to room temperature under conditions of controlled humidity and placed in contact with a preformed emulsion layer.

The first and more direct method was described by Appleton[29]. He used slides coated with stripping film, applied so that the emulsion layer faced away from the glass; the techniques can just as well be followed after dipping the slides in liquid emulsion. The emulsion-coated slides are cooled, and gently touched against the cryostat section as it lies on the knife immediately after cutting. This method requires the cryostat to be in the darkroom, and really implies sectioning under safelighting. It is quite possible to section under full lighting and switch to safelighting to pick up the sections, but I find this alternation of light and dark more difficult to get used to than working throughout under safelighting. Cryostat cutting is never easy, but if care is taken to adjust the microtome so that satisfactory sections are being cut before switching off the main light, and if the safelight is positioned so that it is reflected by the knife surface to the operator, it is a reasonable technique. One advantage offered by the Ilford emulsions is their tolerance of a rather brighter level of safelighting than can be used with Eastman Kodak NTB or Kodak AR-10 emulsions (p. 297).

Appleton recommended keeping the emulsion-coated slides at -5 to $-10°C$, and stated that a temperature differential of several degrees between emulsion and knife was essential for good contact between section and emulsion. This has not been my experience, and I feel the margin of safety to prevent thawing of the section is too small if emulsion at $-5°C$ is used. If the emulsion-coated slides are kept in the cryostat cabinet at the temperature of the microtome itself, the dangers of thawing are less, and we have had no difficulty in picking up the section from the knife. The emulsion only needs to be touched gently against the section: why the section prefers the emulsion to the knife is not clear, but it is fortunate that it does. Pressing the slide against the section usually makes it stick firmly to the knife, and risks thawing the section by pressure. If the section does thaw, it will leave a frost mark on the knife which can easily be seen, and enables one to reject the section.

The Appleton technique has the advantage of simplicity. The frozen section is directly placed in contact with a frozen emulsion layer. It is reasonably easy to produce ribbons of sections on the emulsion, and serial sections are quite possible. It has been criticised on the ground that the section probably thaws, at least transiently, at the instant of picking up from the knife. This can certainly happen, particularly if the slide and emulsion are not cold enough, or the design of the cryostat makes it necessary to introduce warm air and a warm hand in order to pick up the section. With slides at the temperature of the cryostat interior and an enclosed cryostat design with glove ports, thawing at this stage should be simply avoided.

Contact between section and emulsion may not be uniformly close with the Appleton technique. Some sections seem to curl away from the emulsion during exposure. An alternative, called the "modified sandwich technique", has been suggested by Kinter and Wilson[30]. Here, the cryostat sections are picked off the knife with a mounted needle or bristle and gently flattened on a slide with a very smooth surface such as Saran or Teflon. This slide is brought face-to-face with a slide coated with emulsion, and kept in contact throughout exposure. All these steps, of course, take place at temperatures well below freezing. At the end of exposure the slides are separated, and the section remains in contact with the emulsion. An excellent and detailed review of this method has been written by Horowitz[31].

The second approach with freeze-dried frozen sections was described by Stumpf and Roth[32]. The frozen sections are collected from the knife with a mounted bristle or fine forceps, and placed in a small container for transfer to a freeze-drier. A simple and convenient system for freeze-drying is the cryosorption pump[33]. It appears from data presented by Stumpf and Roth that a vacuum of about 10^{-3} torr at $-70°C$ for a period of 12 hours is sufficient to dry tissue sections to constant weight. We usually leave sections in the cryosorption apparatus overnight. It is a good idea to fit a vacuum gauge to the apparatus, as the undetected presence of a leak may result in the sections thawing and the material being wasted on allowing them to come to room temperature. Stumpf and Roth have shown that the freeze-dried sections can take up moisture again from the ambient air on bringing them to room temperature and opening the pump. The weight increases by about 1% of the dry weight for every 10% relative humidity in the air. At high humidities this may well produce movement of labelled material. It is as well to use dry air (R.H. < 30%) to break the vacuum and to store the sections in a desiccator until they are ready for mounting on the emulsion.

To place the freeze-dried section on the emulsion layer, slides are first coated with emulsion, either by dipping or with stripping film, and carefully dried. Then, in safelighting, the sections are placed on a very smooth surface, such as Teflon, slightly smaller in area than the emulsion-coated slides. This Teflon support is brought to the edge of the bench so that it overhangs slightly, and an emulsion-coated slide is placed emulsion down on the support. Slide and support are picked up together, gently pressed between finger and thumb, and the support allowed to fall away. The sections remain on the surface of the emulsion. A detailed account of the full method has recently been given by Stumpf[34].

Wedeen[35] has evolved a slightly different method. Cutting at −60°C in a cryostat, he places a small piece of Saran plastic sheet, on which an adhesive has been smeared, on the surface of the frozen block. After sectioning at 3–5 μm, the Saran with section attached is left in the cryostat overnight to freeze-dry. It is then warmed to room temperature in dry air, and pressed against a preformed emulsion layer for exposure. Before development the Saran is loosened and removed in xylene, leaving the section in contact with the emulsion.

THE EXPOSURE AND PROCESSING OF AUTORADIOGRAPHS OF FROZEN SECTIONS

With the Appleton technique it is essential to expose at temperatures low enough to prevent thawing; with the Stumpf and Roth technique it is not essential, but still preferable to do so. The presence of a drying agent is a good idea. The temperature of exposure will be discussed later, in connection with the prevention of chemography (p. 171).

Frozen sections are no appreciable barrier to the diffusion of reagents into the underlying emulsion. The main problem is to keep them in contact with the emulsion. With the Appleton technique the slides should be brought out into the darkroom while still cold. Condensation will form on the slides, which should be left to warm up to room temperature and to dry out. This wetting and drying helps to fix the sections to the emulsion. Stumpf[34] achieves the same end by breathing on the sections several times and letting them dry off. Even with this precaution, sections may sometimes float off the emulsion. If this is going to happen it is most likely to occur in photographic fixation or the subsequent wash. Coating emulsion and section with a thin layer of gelatin after development, but before fixation, prevents this.

The sequence in which we routinely process our autoradiographs is as follows:

(i) The slides are brought out of the deep freeze in which they were exposed, placed on the bench, and allowed to warm up and dry in a gentle current of cool air.

(ii) The slides are then placed in 4% formaldehyde in phosphate buffer (pH 7.4) for 15 minutes at room temperature to fix the sections.

(iii) After rinsing in 3 changes of distilled water for 1 minute each, the slides go into developer and stopbath.

(iv) After the stopbath, the slides are dipped once in 0.5% gelatin in distilled water, and allowed to dry in air.

(v) Photographic fixation follows, with gentle washing in tapwater to finish.

The end product is usually quite adequate for histology and photography (Figs. 48 and 49). If reflected light is to be used to view the final autoradiograph or as a basis for photometric grain counting, it is essential to have the emulsion on top of the stained section and immediately under the coverslip. This can be done by picking up the section from the cryostat knife on a coverslip coated with emulsion, rather than a slide.

Unfortunately, a coverglass has a very small heat capacity, and it is difficult to control its temperature accurately during the handling involved in picking up the section, so that the section may sometimes thaw, with a consequent redistribution of radioactivity. It is possible to get the best of both worlds by fixing coverglasses to slides with a drop of histological mounting medium, and then coating them with emulsion. The slide provides the required heat capacity, and makes handling simpler and, after staining, the coverglass can be loosened from the slide in xylene, and can be mounted emulsion side downwards, on a clean slide for microscopy.

CHEMOGRAPHY

We have seen (p. 116) that even after fixation, an extractive process and embedding, sections can sometimes affect an emulsion layer chemically.

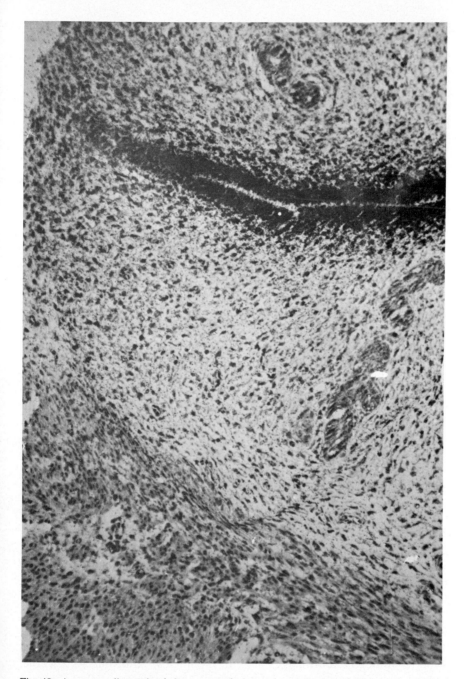

Fig. 49. An autoradiograph of the uterus of an ovariectomised rat 24 hours after a single injection of progesterone. The rat was given [^{125}I]iodide 2 hours before sacrifice. A cryostat section of the uterus was autoradiographed by a modified Appleton technique. Note the high grain density over the luminal epithelium (× 130). (From Brown-Grant and Rogers, 1972)

This is even more likely when all the ions and reactive groups that were present in vivo are next to the emulsion through exposure. Some tissues are worse than others in this respect: we have found that brain, salivary gland and pancreas are particularly bad. It must be realised that chemography is quite likely with frozen sections, and that its influence on the final grain densities can be anatomical in distribution (Fig. 34). No autoradiograph of frozen sections can be interpreted without controls to exclude chemography.

For this artefact to occur, material from the section must diffuse into the emulsion and react there. From first principles, the removal of water (or rather, ice) from section and emulsion should reduce the chance of interaction, and exposure at very low temperatures should also help. These possibilities have been examined[25], and reductions in negative chemography were achieved by thorough drying of the emulsion layer before picking up the sections in the Appleton technique. Freeze-drying the section is even more effective, whether this is done in a deliberate and controlled fashion as in the Stumpf and Roth method, or by allowing the sections to stay for several hours in the cryostat before placing them on emulsion.

Exposure at the temperature of solid carbon dioxide ($-79°C$) gives less chemography than exposure at $-20°C$ (ref. 25). Appleton[36] has claimed that the sensitivity of Kodak AR-10 stripping film falls rapidly with exposure temperatures below $-20°C$, but it is important to note that his experiments were carried out with constant development conditions. Our experience has been that the efficiency of an autoradiograph exposed at $-79°C$ is very similar to that at $-20°C$ or $+4°C$, provided development is adjusted. It seems as if the latent images formed at lower temperatures are smaller or more diffuse: they are still there, however, and can be visualised with increased development.

Rogers and John[25] found an interesting effect on the severity of negative chemography from the choice of developer. An Amidol developer, used in conditions which gave reasonable development in the absence of negative chemography, produced autoradiographs of salivary gland in which loss of developed grains was very severe. By contrast, similar preparations developed with Ilford 1D-19 (or Kodak D-19b) were hardly affected at all. They suggested at the time that Amidol is only affecting latent images at the surface of the crystal, the very ones most likely to be affected by chemography; the other developer, with a considerable solvent action, could reach deep latent images also. In the nuclear emulsions as a whole, such a high proportion of the latent images are surface ones that this may not be the complete explanation. Whatever the reason for it, this effect exists, and if chemography is a problem in autoradiographs of frozen sections it may be wise to look at several developers. Gold latensification may help to visualise latent images that have been severely eroded by chemography (pp. 33–35).

If chemography does occur in spite of drying, low temperature exposure and a suitable choice of developer, it is very difficult to know what to do about it. It is not easy to place an impermeable layer, such as that described on p. 146, between section and emulsion without preventing access of developer and fixer to the emulsion, or of stain to the section. Nor is it easy to get the final sandwich together with good contact between the various layers without the frozen section thawing. A possible but clumsy technique is to drill a hole about 1 cm diameter in a thin piece of plastic and cover it with an impermeable membrane (p. 146). Liquid emulsion can be placed on one side of the membrane and the frozen section picked up by the Appleton method on the other (Fig. 50).

It seems likely that most of the interaction between frozen section and emulsion affects the surface layer of silver halide crystals only. It may be that coating the emulsion with a very thin layer of gelatin by dipping the slide once in a dilute solution, such as 0.5%, before putting the section on the emulsion, would reduce the severity of this artefact.

ASSESSMENT OF THE VARIOUS AVAILABLE TECHNIQUES

It is notoriously difficult to compare different techniques. A procedure that works admirably in one laboratory may not be satisfactory in another, for reasons that are very difficult to sort out. Also, some labelled substances are more difficult to autoradiograph than others in this general group of diffusible materials, and tissues vary in the way in which they bind

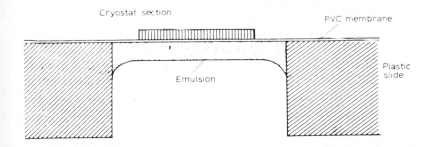

Fig. 50. A diagram to illustrate the position of cryostat section and emulsion relative to the PVC membrane in the suggested method for the autoradiography of soluble isotopes in the presence of chemographic artefacts.

drugs or hormones and in their ease of sectioning and liability to chemography.

The methods for autoradiographing diffusible materials that involve embedding can only be used after very careful examination of isotope loss through the entire process, including bringing the section in contact with the emulsion. The careful work of Attramadal[3,12,13] suggests that for steroid hormones and their target tissues this method can be made to work. But the hormones oestradiol and testosterone are tightly bound to receptor molecules in their target cells, molecules that are fairly large proteins which must restrict the mobility of the labelled hormones very considerably. There is no guarantee that the same method will be universally suitable. Kinter's work with labelled sugars[14] showed that useful results can be obtained with much smaller molecules that do not appear to be bound in the tissue in the same way, but loss of radioactivity does occur even with this technique, and this loss is so severe with labelled sodium that the method as used is impracticable. While embedding gives excellent histology, more reproducible section thickness, and the possibility of extension to the electron microscope level in very favourable cases, it is not universally applicable and always leaves some measure of doubt of its validity. In a direct comparison of an embedding technique with the freeze-drying of frozen sections, Stumpf and Roth[37] found the former method to give slightly different distributions of [^{3}H]oestradiol and [^{3}H]mesobilirubinogen, suggesting preferential loss of radioactivity from some tissue compartments.

The frozen section techniques are, in general, preferable. A direct comparison of the Appleton and the Stumpf and Roth techniques in my laboratory in two experiments has failed to show any difference in the distribution of radioactivity, or any very obvious advantage of the one method over the other. One experiment examined the distribution of iodine-125 as iodide in the salivary glands of mice[27] (Fig. 48). This required very short exposure times, often of 24–48 hours, and it was noticed that freeze-dried sections gave higher grain densities than the Appleton method; presumably self-absorption was less after drying. The other experiment studied the distribution of [^{3}H]progesterone and [^{3}H]megestrol acetate in the rat uterus, with exposure times of 2–6 weeks[38]: this time there was no demonstrable difference in efficiency, since the Appleton sections presumably freeze-dried during the first day or two of exposure.

Both methods, sensibly and carefully carried out, can give valid data, and one is left with a choice which may depend on the circumstances of the experiment. The Appleton method requires a cryostat in the darkroom, which may be impossible if the darkroom is small or badly ventilated. Chemography is reduced by freeze-drying, which should be used for particularly difficult tissues. The Appleton technique, suitably modified, is quick and simple, however, and can produce serial sections much more easily than the method of Stumpf and Roth, which requires each section in the series to be freeze-dried in a separate container of some sort.

With both methods, controls for chemography are essential. Statements will be found in the literature that some emulsions do not show chemography, or that freeze-drying the sections will prevent it altogether. These are not true.

THE RESOLUTION AND EFFICIENCY OF AUTORADIOGRAPHS OF DIFFUSIBLE SUBSTANCES

The resolution obtainable by methods for diffusible materials is determined by the factors discussed in Chapter 4; in addition, two other effects must be taken into account. The first is the possibility of diffusion of the radioactive compound in the tissue. This, whether it occurs in the interval between obtaining the specimen and freezing it, during transient thawing, or even in the frozen state during exposure, can only make the resolution worse by comparison with the best that can be done with the same autoradiographic system. Large molecules diffuse more slowly than small ones, and the resolution of frozen section autoradiographs may well vary with the molecule under study. The diffusion rate also depends critically on the medium through which the molecule is trying to move. A true solution of low viscosity will permit more rapid diffusion than cell cytoplasm, with its submicroscopic organisation. This may be clearly seen in different areas of the same autoradiograph. When iodine-125 in ionic form is examined in the salivary gland of the mouse, the cells of the convoluted granular tubules concentrate the iodide, giving autoradiographs which are often very precisely localised. In the lumen of the collecting ducts, the iodide often gives a clumpy distribution with poor resolution. So the influence of diffusion on the resolution of autoradiographs varies with the labelled molecule and with its immediate environment.

A second factor has been identified by Clarkson[39], working with frozen section autoradiographs of plant material. On freeze-drying, the vacuole in the centre of plant cells will dry down to a very thin layer of solute in contact with the emulsion, while the cellulose cell walls remain the original thickness of the section. While more obvious perhaps with plant material, this effect must occur with any frozen tissue section, giving geometry which varies from place to place in a very striking and non-random fashion. There have been no serious attempts to take this variation into account so far in the literature. Some of the problems involved in the autoradiography of plant tissues containing diffusible radioactivity are discussed by Sanderson[40] (Fig. 51).

Measurements of resolution in the literature suggest that it is of the order of twice that of fixed and precipitated material. Appleton[36] found a resolution of 9–13 μm with sodium-22, compared with expected values of 4–5 μm; with tritium in the form of thymidine, he claimed 2–4 μm, compared with an expected 1 μm. On similar definitions of resolution, Horowitz[31] found values of 7 μm for sodium-22, and 1 μm or less for tritium. Creese and Maclagan[41], working with [³H]decamethonium in muscle, claimed a resolution of 1–1.5 μm. For embedded material, Kinter's demonstration of the concentration of [³H]glucose in the eutectic bands of frozen plasma indicates that a resolution practically as good as that of non-diffusible material can be obtained (Fig. 46).

Efficiency should also be similar to that of conventional autoradiographs. Once again the variable geometry produced by freeze-drying described by Clarkson[39] must have quite an effect on the efficiency of recording from

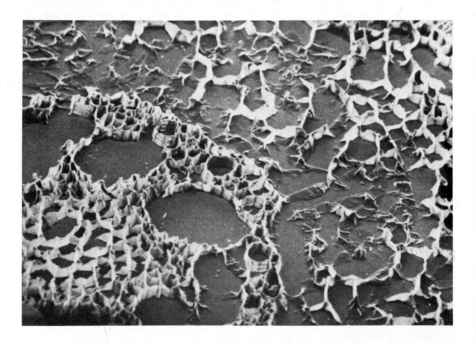

Fig. 51. A scanning electron micrograph of a frozen section, 20 μm thick, from maize root. The section has been mounted on Kodak AR-10 stripping-film and freeze-dried. Note the variations in thickness of the specimen, with cell walls remaining at about the original 20 μm while cell contents and conducting systems have dried down to a thin film in contact with the emulsion. (Material provided by D.T. Clarkson and A.W. Robards)

adjacent tissue compartments. The influence of this on grain counting has not so far been investigated in detail. The one great advantage of the "modified sandwich" technique[31] (p. 167) is that freeze-drying of the section during exposure is prevented. It would be interesting to compare the efficiencies in nucleus, cytoplasm and extracellular space of a tissue autoradiographed by this technique and by the Appleton method.

If care is taken to base the results on a sufficiently large series of sections to compensate for variation in thickness, autoradiographs based on frozen sections can give precise quantitative results. The technique can be combined with track autoradiography[42].

THE AUTORADIOGRAPHY OF DIFFUSIBLE MATERIAL IN CELL SUSPENSIONS

The previous sections have dealt with solid tissues that require sectioning before autoradiography and microscopy. How about specimens that are already thin enough, such as cell suspensions or bacterial preparations?

It is tempting to prepare an air-dried smear of suspended cells and then place it in contact with a preformed emulsion layer. But there are few diffusible substances that retain their in vivo position after this treatment. As drying proceeds, the fluid outside the cell becomes hypertonic and many intracellular materials will leak out. In fact, the production of a halo of silver grains around such cells was used by Miller, Stone and Prescott[43] to differentiate diffusible material from radioactivity firmly incorporated into macromolecules in the cell. A similar criticism can be applied to touch preparations from solid tissues[44].

One can of course rapidly freeze a small drop of the suspension and treat the block as if it were a solid tissue, sectioning it before autoradiography. Alternatively, a technique has been described by Darżynkiewicz and Komender[45] for spraying a cell suspension on to a Teflon slide, freezing it rapidly, freeze-drying, and then pressing it against a preformed emulsion layer. At the end of exposure the Teflon slide is separated from the emulsion leaving the cells on the emulsion. With this technique they were able to achieve a very reasonable resolution with sodium-22: grain counts fell to 10% of the levels over the cytoplasm at a distance of 10 μm from the cell membrane.

DIFFUSIBLE MATERIALS AND ELECTRON MICROSCOPY

The last few years have seen real progress made at this most difficult combination of techniques, and the publication of several acceptable

178

autoradiographs. As with light microscopy, the methods have followed two main lines — freeze-drying with infiltration by plastic, and the direct cutting of frozen sections. I still would not consider either of these to be techniques that are available for any laboratory to take up — to demonstrate the distribution of freely diffusible materials at this level is still rather like climbing Everest: although it has been done, it still requires an immense investment in effort with a fairly high chance of failure.

To start with radioactive substances that are not freely diffusible, but still require special techniques, Darrah, Hedley-Whyte and Hedley-Whyte[4] have examined the distribution of [3H]cholesterol in the lung by a number of different techniques at the light microscope level, and shown that it remains reproducible even with techniques that leach out a high percentage of the radioactivity. In other words, the loss on embedding fixed material is randomly affecting all the cells and compartments equally. It is therefore valid, in their experiment, to use a relatively conventional procedure to study the distribution of this compound at the ultrastructural level.

Boyenal and Droz[46] have adapted Bernhard's cytochemical techniques to the study of labelled lipids. Following glutaraldehyde fixation, frozen sections have been cut for electron microscopy.

For substances that are more strictly diffusible, however, these methods are of little relevance. As early as 1967, Stirling and Kinter[14] published autoradiographs of [3H]galactose taken in the electron microscope following their technique of freeze-drying and impregnation with araldite. More recently, Frederik, Van der Molen, Klepper and Galjaard[18] have used the same approach to examine [3H]testosterone in the testis (Fig. 52). By careful control of the processes of freezing and freeze-drying, Frederik was able to produce small blocks of tissue with a thin rim, 2–3 cells thick, at its surface in which the ultrastructure was well preserved and the ice crystal damage relatively insignificant. The sections of embedded material were cut on an LKB Cryokit on to a dry glass knife, transferred to a coated grid, covered with a layer of carbon and then with a monolayer of gelled L4 emulsion by the loop technique. Nagata[47] has also used this general approach with the Sakura emulsion, NR-H2.

There have been many attempts to cut frozen sections for electron microscope autoradiography. Recently, Baker and Appleton[48,49] have published micrographs showing autoradiographs of sodium-22 in kidney obtained in this way. The fresh frozen tissue was sectioned, and the sections mounted, frozen on to formvar-coated grids and freeze-dried. They were then brought up to ambient temperature in dried nitrogen, coated with carbon, and covered with a monolayer of L4 by the loop technique. By X-ray microanalysis, they have shown that there is no

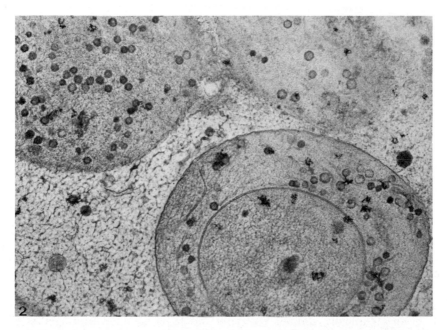

Fig. 52. An autoradiograph of rat testis, showing two round spermatids. After administration of [³H]testosterone, the tissue was frozen, freeze-dried, impregnated with Epon and sectioned on an LKB III ultramicrotome at low temperature. The section was mounted on a coated grid and a monolayer of Ilford L4 was applied by the loop technique, over a layer of evaporated carbon. Note the good preservation of ultrastructural detail. (× 7450). (From Frederik, van der Molen, Klepper and Galjaard, 1977)

significant redistribution of sodium during sectioning, and that the carbon coat prevents the moisture in the emulsion from affecting the sodium distribution. Nagata and Murata[50] have also obtained autoradiographs with a similar method, studying the distributions of tritiated precursors of nucleic acids and proteins.

The pictures obtained by freeze-drying and embedding are far better than the frozen sections, though the percentage of usable sections is probably very low by both techniques. However, there are doubts whether the embedding approach can give good localisation for labelled ions, whereas Baker and Appleton's work suggests that this is possible with the frozen section method.

The problems remain immense. Both approaches require very careful validation before the extent of translocation of labelled molecules during preparation is fully understood.

TECHNIQUES OF LABELLING AFTER SECTIONING

It is now well documented that sites which bind labelled steroids cannot be demonstrated by conventional histological methods, since the radioactivity is largely lost in the process. Uriel[51] has recently pointed out that the sites themselves do in fact survive routine preparative methods, and can be labelled on the dewaxed section successfully. In somewhat similar fashion, molecular hybridisation between DNA in the section and labelled RNA can also be achieved[52]. In this whole field of identifying specific sites in tissues by means of a labelled reagent molecule, it is possible that it will often be technically simpler to apply the label to fixed and sectioned material than to inject it into the animal, and be faced with the necessity of using special techniques to prevent the loss of radioactivity during specimen preparation.

REFERENCES

1 E.N. Cohen and N. Hood, *Anaesthesiology*, 30 (1969) 306.
2 E.N. Cohen and N. Hood, *Anaesthesiology*, 31 (1969) 61.
3 A. Attramadal, *Histochemie*, 19 (1969) 64.
4 H.K. Darrah, J. Hedley-Whyte and E.T. Hedley-Whyte, *J. Cell Biol.*, 49 (1971) 345.
5 M.A. Williams, *Adv. Opt. Electr. Microscop.*, 3 (1969) 219.
6 H.F. Steedman, *Section Cutting in Microscopy*, Blackwell, Oxford, 1960.
7 O. Stein and Y. Stein, *Adv. Lipid Res.*, 9 (1971) 1.
8 B.J. Ward and J.A. Gloster, *J. Microscopy*, 108 (1976) 41.
9 D. Branton and L. Jacobson, *Stain Technol.*, 37 (1962) 239.

181

10 K.R. Wilske and R. Ross, *J. Histochem. Cytochem.*, 13 (1965) 38.
11 N.J. Nadler, B. Benard, G. Fitzsimmons and C.P. Leblond, in L.J. Roth and W.E. Strumpf, (Eds.), *The Autoradiography of Diffusible Substances*, Academic Press, New York, 1969.
12 A. Attramadal, *Histochemie*, 19 (1969) 75.
13 A. Attramadal, *Histochemie*, 19 (1969) 110.
14 C.E. Stirling and W.B. Kinter, *J. Cell Biol.*, 35 (1967) 585.
15 K.J. Karnaky, L.B. Kinter, W.B. Kinter and C.E. Stirling, *J. Cell Biol.*, 70 (1976) 157.
16 P.M. Frederik and D. Klepper, *J. Microscopy*, 106 (1976) 209.
17 P.M. Frederik, D. Klepper, G.J. Van der Vusse and H.J. Van der Molen, *Mol. Cell Endocr.*, 5 (1976) 123.
18 P.M. Frederik, H.J. Van der Molen, D. Klepper and H. Galjaard, *J. Cell Sci.*, 26 (1977) 339.
19 P.M. Frederik, *Autoradiography of Diffusible Substances*, Thesis, Erasmus University, Rotterdam, 1977.
20 T. Nei, *J. Microscopy*, 99 (1973) 227.
21 H. Plattner, F. Miller and L. Bachmann, *J. Cell Sci.*, 13 (1973) 687.
22 U. Riehle and M. Hoechli, in E.L. Benedetti and P. Favard (Eds.), *Freeze Etching*, Soc. Francaise de Micr. Electronique, Paris, 1973.
23 L.J. Roth and W.E. Stumpf (Eds.), *The Autoradiography of Diffusible Substances*, Academic Press, New York, 1969.
24 E.L. Benedetti and P. Farard (Eds.), *Freeze Etching*, Soc. Francaise de Micr. Electronique, Paris, 1973.
25 A.W. Rogers and P.N. John, in L.J. Roth and W.E. Stumpf (Eds.), *The Autoradiography of Diffusible Substances*, Academic Press, New York, 1969.
26 W.E. Stumpf and L.J. Roth, *Nature*, 205 (1965) 712.
27 A.W. Rogers and K. Brown-Grant, *J. Anat.*, 109 (1971) 51.
28 O.L. Miller, G.E. Stone and D.M. Prescott, *J. Cell Biol.*, 23 (1964) 654.
29 T.C. Appleton, *J. Roy. Microscop. Soc.*, 83 (1964) 277.
30 W.B. Kinter and T.J. Wilson, *J. Cell Biol.*, 25 (1965) 19.
31 S.B. Horowitz, in D.M. Prescott (Ed.), *Methods in Cell Biology, Vol. 8*, Academic Press, New York, 1974.
32 W.E. Stumpf and L.J. Roth, *Stain Technol.*, 39 (1964) 219.
33 W.E. Stumpf and L.J. Roth, *J. Histochem. Cytochem.*, 15 (1967) 243.
34 W.E. Stumpf, in D.M. Prescott (Ed.), *Methods in Cell Biology, Vol. 13*, Academic Press, New York, 1976.
35 R.P. Wedeen, *Progr. Nucl. Med.*, 2 (1972) 147.
36 T.C. Appleton, *J. Histochem. Cytochem.*, 14 (1966) 414.
37 W.E. Stumpf and L.J. Roth, *J. Histochem. Cytochem.*, 14 (1966) 274.
38 P.N. John and A.W. Rogers, *J. Endocr.*, 53 (1972) 375.
39 D.T. Clarkson, personal communication.
40 J. Sanderson, *J. Microscopy*, 96 (1972) 245.
41 R. Creese and J. Maclagan, *J. Physiol.*, 210 (1970) 363.
42 A.W. Rogers, G.H. Thomas and K.M. Yates, *Exptl. Cell Res.*, 40 (1965) 668.
43 O.L. Miller, G.E. Stone and D.M. Prescott, in D.M. Prescott (Ed.), *Methods in Cell Physiology, Vol. 1*, Academic Press, New York, 1964.
44 W.E. Stumpf, *Acta Endocr.*, Suppl. 153 (1971) 205.
45 Z. Darżynkiewicz and J. Komender, *J. Histochem. Cytochem.*, 15 (1967) 605.
46 J. Boyenal and B. Droz, *J. Microscop. Biol. Cell.*, 27 (1976) 129.
47 T. Nagata, *Histochemie*, 18 (1969) 241.

182

48 J.R.J. Baker and T.C. Appleton, *J. Microscopy*, 108 (1976) 307.
49 T.C. Appleton, *J. Microscopy*, 100 (1974) 49.
50 T. Nagata and F. Murata, *Proc. 5th Internat. Congr. Histochem. Cytochem.*, Bucharest, 1976.
51 J. Uriel, in D.M. Prescott (Ed.), *Methods in Cell Biology, Vol. 10*, Academic Press, New York, 1975.
52 M. Buongiorne-Nardelli and F. Amaldi, *Nature*, 225 (1970) 946.
53 K. Brown-Grant and A.W. Rogers, *J. Endocr.*, 53 (1972) 355.

CHAPTER 9

The Microscopy and Photomicrography of Autoradiographs

The principal problems arise with autoradiographs for viewing with the light microscope. With whole-body autoradiographs the specimen and X-ray film are normally separated after exposure, and each photographed at the same magnification. The film can be transferred to a negative, and prints made from this, in which the radioactive areas will appear black. Alternatively, many laboratories use the X-ray film itself as a negative, printing black pictures in which the radioactive areas are white. The great depth of focus of the electron microscope makes it relatively easy to take micrographs with silver grains on. The only problem here may be obtaining a field with sufficient grains on it. It may help to expose a few specimens for a long time to a double layer of emulsion.

At the light microscope level, however, the developed grains are often tiny by comparison with the stained structures in the specimen, and it may be very difficult to see them, let alone record their distribution photographically, at low magnifications. At high magnifications, the grains will usually be visible, but the relatively small depth of focus of the objective will make it difficult to photograph grains and specimen without one or the other being out of focus.

The solution to these problems lies in the use of dark-field illumination, either alone or in combination with transmitted bright-field lighting. This chapter will discuss the effect of using dark-field methods at the light microscopic level on the whole process of preparing an autoradiograph.

CHOICE OF ILLUMINATION

Ideally, the presence of the emulsion layer should not interfere in any way with the microscopy of the specimen. The choice of staining method should not be limited, nor should the silver grains be so large or so closely packed as to hide important detail in the underlying tissue. At the same

184

time, it should be possible to examine the distribution of silver grains at low magnification over large areas of the autoradiograph, in order to assess the overall patterns of labelling and its relation to background. This ideal situation is not easy to meet with transmitted light alone. If the tissue has been stained with haematoxylin, for instance, giving blue-black chromatin and nucleoli, and cytoplasmic staining as well in some cells, it may be extremely difficult at low magnifications to distinguish the silver grains from dark granules in the specimen. Fig. 53 shows a low power view of the developing cerebral cortex of the rat after an injection of tritiated thymidine. It is extremely difficult to see the labelled cells, particularly in those parts of the section where the nuclei are very densely packed. One might get quite a mistaken impression about the distribution of silver grains, influenced by their relatively clear visibility in areas where the staining is less heavy.

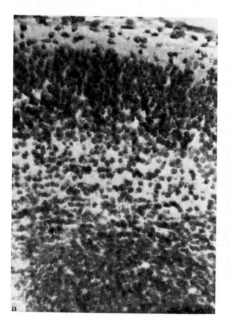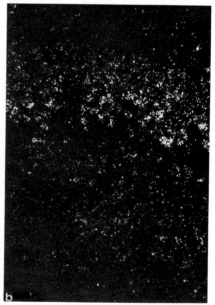

Fig. 53. Photomicrographs of the same field of an autoradiograph, taken by (a) transmitted and (b) incident dark-field illumination. The material is the cerebral cortex of a foetal rat, injected with [³H]thymidine on the 16th day of pregnancy and sacrificed 4 days later. In (a), the silver grains are easily seen across the middle of the field, where the density of stained cells is low. In (b), these silver grains are clearly insignificant relative to the heavily labelled band of cells just above. Autoradiograph prepared with Ilford G5 emulsion, applied over a PVC membrane; section stained with Harris' haematoxylin before autoradiography. (× 176). (Material prepared in collaboration with Dr. M. Berry)

If the silver grains are made very large, by selecting an emulsion with large crystal diameter and developing it fairly strongly, they may become easier to see at low magnifications. Under oil immersion, however, they will be enormous, obscuring detail in the specimen, and incidentally reducing both the resolution and the efficiency by comparison with a fine-grained emulsion.

In studying the correlation of labelling with the various stages of meiotic prophase in the developing ovary, for example, where the fine structure of the chromosomes is of crucial importance, and their condensed or beaded appearance with haematoxylin stains produces dark specks the approximate size of silver grains, very real problems in interpretation can arise.

It is possible to resolve many of these problems of observation and photography by making use of dark-field illumination, either separately or in conjunction with transmitted light. The principles involved, the choice of equipment, and the techniques of photography will be discussed in some detail, since the methods that will be used in microscopy have a considerable influence on the selection of the autoradiographic technique that is most suitable for a given experiment, and since only a partial description is available in the literature[1].

DARK-GROUND ILLUMINATION

Ideally, a method of microscopy is needed that distinguishes clearly between silver grains and stained material in the specimen. Developed grains are irregular, ribbon-like knots of metallic silver. They do not absorb light that falls on them, but reflect and scatter it. This is why they appear dark by transmitted light, just like stained objects. The latter appear dark through a different mechanism, however: they absorb light, rather than scattering or reflecting it. If an autoradiograph is illuminated by a beam of light so arranged that it does not enter the objective lens, silver grains will scatter light from its original path into the optical axis of the microscope, resulting in a light signal from each grain. Stained objects in the tissue will still absorb light of the appropriate wavelength: this will not give any image that can be seen through the objective.

In short, dark-ground illumination can produce an unmistakeable signal from silver grains, distinguishing them from material in the specimen that absorbs light. Viewed in this lighting system, the silver grains will be bright specks on a dark background. Very few components of vertebrate tissue will scatter light in this way. Curiously enough, pigment granules often appear bright in dark-field illumination, and may be very difficult to distinguish from silver grains. Naturally, if the technique of staining has

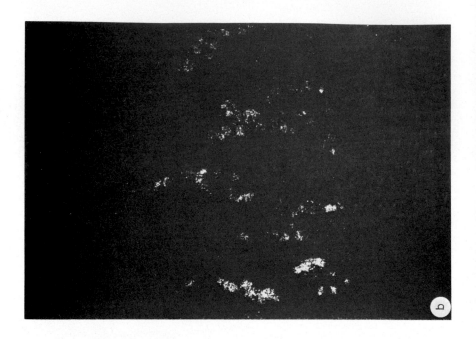

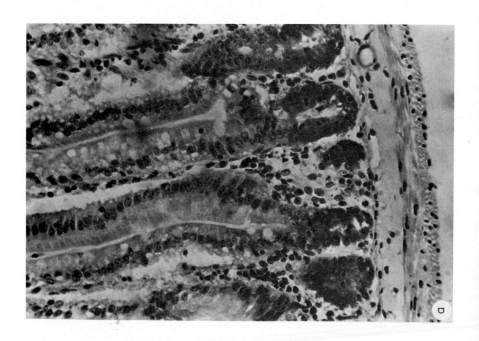

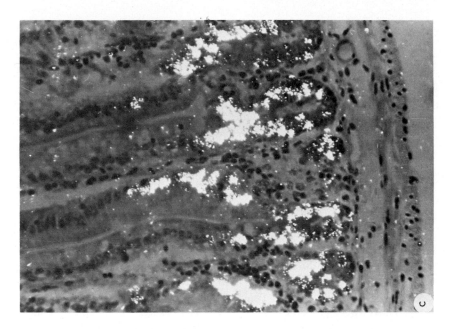

Fig. 54. Photomicrographs of a section through the small intestine of the bushbaby (*Galago demidoffi*), 24 hours after the injection of tritiated thymidine. (a), Section viewed by transmitted light alone: silver grains invisible. (b), Same field seen by incident dark-field lighting: silver grains clearly seen, but section invisible. (c), Same field by balanced transmitted and incident lighting. Autoradiograph with Ilford L4 emulsion. Leitz Ultropak × 22 objective. Stained with Harris' haematoxylin. (× 135). (From *Leitz Mitt. Wiss. u. Techn.*, 3 (1964) 43)

introduced metallic deposits, or anything that scatters light rather than absorbing it, these sites will also be bright against the generally dark background of the tissue.

In general, dark-field lighting does differentiate silver grains from tissue section. A further characteristic of this lighting system is that objects too small to be resolved in transmitted light by a given optical system may be seen clearly if they can be converted into sources of light on a dark background. In other words, silver grains at or even just below the limits of visibility by transmitted light can be seen, using dark-field illumination. In addition, this conversion of silver grains into light signals greatly simplifies the problems of automating the drudgery of grain counting. The use of dark-field lighting for photometric grain counting will be considered in Chapter 10 (p. 210).

With dark-field illumination, then, it is possible to recognise even very small grains against a stained background. As can be seen in Fig. 54, the stained material, when viewed by transmitted light alone, can look like a normal histological preparation. The use of dark-field illumination alone gives a clear idea of the distribution of silver grains, without any interference from the stained section. With both systems of lighting in use simultaneously, the relation of grains to stained material is obvious.

Optical systems for dark-ground illumination

There are two main methods of illuminating the specimen to obtain a dark-field effect. The system in general use in bacteriology and cytology employs sub-stage illumination with a special condenser, which produces a cone of light converging on the specimen from below. The metallurgical illuminators, on the other hand, being designed for use with solid, opaque materials, direct the light on to the specimen from above, i.e. through or around the objective lens itself. Both sub-stage and metallurgical methods can give the required effect with silver grains, but the latter is far more convenient. It can be used not only in rapid alternation with direct transmitted light, but even simultaneously. However rapid the switchover from dark- to bright-field illumination, it is not easy to carry over the mental picture of grain distribution accurately, and the simultaneous use of both methods of lighting is a great advantage.

With illumination from above the specimen, there are again two choices of optical system available. The first projects a converging cone of light on to the specimen from around the objective lens, in a manner rather similar to the sub-stage dark-field system. In the second, or vertical, system, the illuminating beam is actually directed down the optical axis and through the objective itself. Fig. 55 illustrates these two systems diagrammatically.

Dealing first with the convergent cone system, many manufacturers produce suitable equipment, such as the Ultropak illuminator and lenses of E. Leitz, G.m.b.H*. Since the equipment was produced for a very different optical task, not all the lenses available are suitable for autoradiographs. At very low magnification, lenses like the Ultropak X6.5 (N.A.O.18) can give a satisfactory dark-field effect, but they have a considerable depth of focus, and scratches and imperfections on the upper surface of the coverglass will scatter light and interfere with the dark-field picture. This can be avoided by using an immersion cone with the objective. At higher magnifications, the X22 (N.A.O.45) objective is very useful, and many of the illustrations in this book have been taken with it; at this level, no immersion cone is needed as the depth of focus is less. High dry objectives, such as the X50 (N.A.O.65) should be avoided, since with increasing numerical aperture much more light is scattered from the upper surface of the coverglass, so that the background against which the silver grains are viewed is quite bright and hazy, instead of being black. With oil immersion objectives such as the X75 and X100 (N.A.1.0) an excellent dark-field picture can again be seen. It may be difficult to obtain these two objectives since their manufacture has been discontinued: the X60 oil immersion objective, still available, can be used successfully in the same way.

These four Ultropak objectives X6.5, X22, X75 and X100 provide a very satisfactory series, with acceptable performance when used with transmitted light alone, as well as a good dark-field picture. This is the system of choice for observing autoradiographs, and also for photomicrography.

A surprising amount of light is reflected back through the specimen by the sub-stage condenser when the Ultropak illumination alone is being used. In many cases this scattered light is enough to outline structural detail in the specimen, and there is no need to supplement it with transmitted lighting. If a really dark field is wanted, the top lens of the sub-stage condenser can be flipped out of the optical axis, or the sub-stage condenser itself racked down. Either manoeuvre will abolish the reflections from the upper surface of the condenser.

The second type of system for giving dark-field lighting from above the specimen is the vertical illuminator, in which the light is directed down on to the specimen through the objective lens itself. This is very different in use. I have never been able to get a satisfactory dark-field effect except with oil-immersion objectives of high numerical aperture. This reduces the usefulness of this particular system for autoradiography. However, with an oil-immersion objective one can obtain a finely focussed beam of light which gives a higher ratio of light reflected per silver grain to light scattered

---

*Now a subsidiary of Wild Instruments.

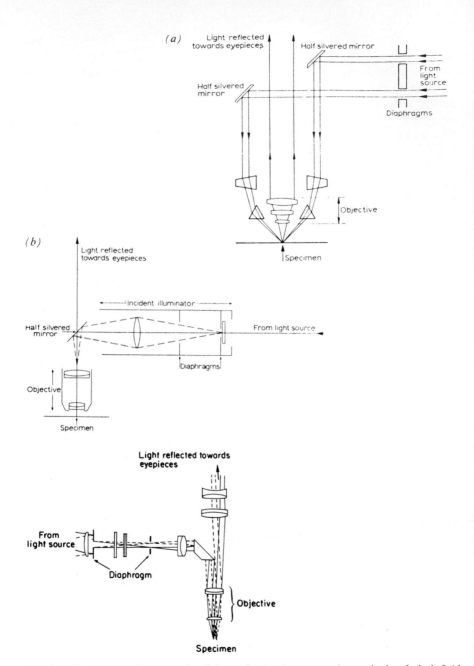

Fig. 55. Diagrams to illustrate the light paths in the two main methods of dark-field illumination from above the specimen. (a), Typified by the Leitz Ultropak system, in which a hollow cone of light converges on the specimen from around the objective. (b), The vertical-incident system, in which the illuminating beam travels through the objective itself to reach the specimen. This method can use a half-silvered mirror in the light path, or a prism, to direct the illuminating beam into the objective. (By courtesy of Wild Instruments)

from the various optical interfaces than is possible with the Ultropak type of lens. It is also possible to limit the area of emulsion illuminated to the precise area from which observations are required, by using the field diaphragm of the illuminator. This, as will be discussed in Chapter 10, has practical advantages in the design of photometric devices for grain counting. In this particular application, therefore, the vertical illuminator is the system of choice. For routine viewing and for photomicrography, the convergent cone type of illuminator is preferable.

Let us summarise the advantages to be gained from using the Ultropak lenses for dark-field viewing of grain density autoradiographs. Provided staining methods are avoided which produce a light-scattering precipitate in the specimen, it is possible to view the silver grains as bright specks on a dark background. This makes it feasible to keep the size of the developed grains relatively small, so that fine detail in a heavily labelled cell can still be recognised. At the same time, small, darkly stained granules in the specimen can be confidently identified, without confusing them with silver grains.

Even in autoradiographs in which the grains appear small under high magnification, it is still possible to observe the overall distribution of silver grains over a stained specimen at low magnifications, using dark-field illumination (cf. Figs. 3a and b).

THE PHOTOMICROGRAPHY OF GRAIN DENSITY AUTORADIOGRAPHS

If dark-field illumination with Ultropak lenses is useful in viewing autoradiographs, it is invaluable in their photomicrography.

As has been shown in Fig. 54, it is possible to photograph the section at relatively low magnification, and produce a print as good as with histological material alone. It is possible to photograph the silver grains without the section being visible. Finally, it is possible to produce a composite photograph, in which the silver grains, appearing as bright specks, can be clearly seen against the stained section. At low magnifications, the depth of focus of the objective is sufficient for both section and autoradiograph to appear sharp on the same exposure. This type of picture is best taken with transmitted and dark-field illuminating systems both on together. It will be found that the bright-field lighting has to be kept to a very low intensity, or it will flood out the dark-field effect, and the silver grains will be very difficult to see. The best method is to adjust the two lighting systems under direct visual control, and then take the photomicrograph when the best balance has been obtained.

With the increasing use of exposure meters and automatic cameras, this balancing process may not be easy. If one reduces the intensity of transmitted light in order to emphasise the silver grains, the camera will compensate by increasing the exposure time. It may be necessary to deceive the camera by providing it with false information. Most of the photomicrographs in this book were taken on Kodak High Contrast Copy Film, for which an ASA rating of 6 normally gives good results. If one sets an automatic camera for an ASA value of 12, this effectively shortens the exposure, resulting in a darker print. This is ideal for a photomicrograph at low magnification with both lighting systems balanced, as the bright silver grains show up well against the darker specimen.

At high magnifications, the biggest problem in obtaining clear photomicrographs is to focus section and silver grains sharply on the same negative. The solution to this is to employ a double exposure technique, superimposing a dark-field picture of the silver grains, critically focussed, on a transmitted light picture of the specimen. Once again, the duration of the two exposures is extremely important, in order to see sufficient detail on the specimen without fading out the light signal from the silver grains.

If one considers first the dark-field picture of silver grains, the exposure should give as black a background as possible, with the grains contrasting brightly. The exposure time that achieves this is really independent of the number of silver grains present in the field of view: if one grain is correctly exposed, so will be one hundred. But all meters and automatic cameras relate the light output of the field to the exposure time, so that if the field with one hundred grains in it is correctly exposed, the one with only one grain will be exposed for a very much longer time, resulting in a background that is not quite so black.

The best solution is to standardise the conditions of illumination and of processing the negative as far as possible, and then to try various exposure times on a field with relatively few grains in it. When this gives a negative with clear black specks on a background that is really white, that exposure time should then be used for all subsequent pictures under those conditions of microscopy and of autoradiography. As a rough guide, the dark-field exposure I use for oil-immersion photomicrographs is 2 minutes; by contrast, the automatic camera that is available may indicate exposure times from 1 up to over 20 minutes for similar material, depending on the number of grains in the field. These long exposure times for dark-field work require complete freedom from vibration. The microscope should preferably be on a slate bench. Dust on the coverglass, in the immersion oil, or on the emulsion, may prove embarrassingly obvious in dark-field photographs, and dust or scratches on the negative will be very difficult to ignore when over 90% of the finished print is intended to be uniformly black.

For a double exposure photomicrograph, after the dark-field picture is taken, the film, of course, must be restrained from winding on. The dark-field illumination is then replaced by transmitted, bright-field lighting. The stained section is carefully focussed and the second exposure made. For this exposure, the same comments apply as for the low power photomicrographs. The light intensity must not be such as to make the bright silver grains invisible on the finished photograph. I find that half the exposure that would normally be required for a simple photomicrograph of the histological specimen in transmitted light gives about the right picture. This can be achieved with automatic cameras by setting them for an ASA value twice that of the film.

Fig. 56 illustrates the advantages of this double exposure method. When photographed in transmitted light alone, focussing on the silver grains has put the tissue out of focus; even so, many of the grains are invisible against the stained background. The double exposure method has produced a much clearer picture of the tissue, with increased clarity of recognition of the silver grains.

Colour pictures and transparencies may be obtained by precisely the same methods that have been outlined above for monochrome. The trick of shortening the transmitted light exposure of a double-exposure picture to half its normal length will obviously result in some distortion of colour in the final picture. But, in an autoradiograph, the precise reproduction of colour in the specimen is normally of secondary importance to the clear demonstration of the distribution of silver grains. In colour pictures it may be a help to make the silver grains some brightly contrasting colour. This can easily be done by putting a colour filter in the incident illuminator. This trick may be very helpful if many of the grains are over unstained areas of autoradiograph, and do not show up very clearly as bright specks.

THE IMPLICATIONS OF DARK-FIELD VIEWING FOR AUTORADIOGRAPHIC TECHNIQUES

The use of dark-field illumination for viewing nuclear emulsions is not new. Walmsley and Makower[2], in 1914, published pictures of α tracks taken by this method. It has not achieved much popularity with autoradiographers, however, in spite of its obvious advantages for the design of photometric grain counters[3-5]. The reason for this is suggested in the discussion following Gullberg's paper[3].

If an autoradiograph prepared for viewing by transmitted light alone is examined by dark-field illumination, many tiny silver grains are often visible, constituting an embarrassingly high and hitherto unsuspected background.

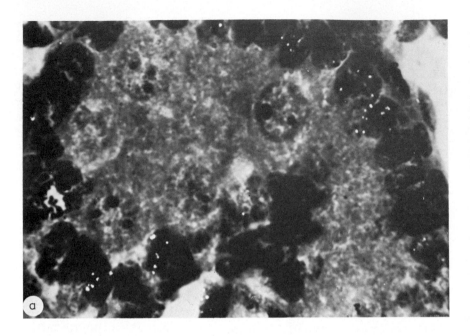

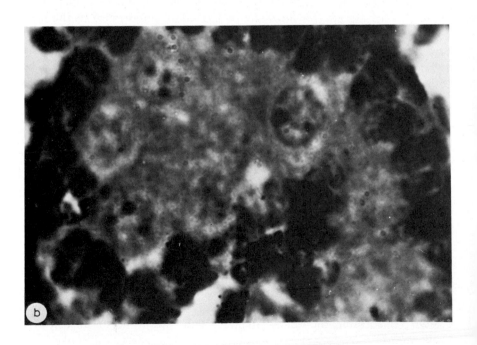

The reason for this high background of tiny grains becomes clearer if one looks again at Fig. 3 (p. 22). Here, identical autoradiographs were developed under constant conditions for increasing lengths of time. When viewed with dark-field illumination, the second picture is probably the best, with nice, clear grains and relatively low background. In the third picture, the small background grains characteristic of overdevelopment have begun to appear. Yet, seen by transmitted light, the silver grains are still too small for comfortable viewing in the second picture, and only just sufficiently developed in the third. Certainly, with transmitted light, there is no hint of gathering hordes of background grains in the latter photograph.

Objects too small to be seen with transmitted light may become visible in dark-field conditions. It follows that the optimum development time for a series of autoradiographs will be shorter if dark-field methods are to be used than if they will be viewed by transmitted light alone. Development must be stopped short before these tiny background grains begin to appear. Fig. 57 shows that, with correct development, the background levels with dark-field viewing can be very low indeed.

If, therefore, dark-field methods are to be used to view the finished autoradiograph, the development schedule must be appropriate. In this sense, it is obviously important to match the autoradiographic technique to the system of microscopy available.

But I feel that the use of dark-field methods can and should have a greater impact on autoradiographic techniques than this. We have already seen that reducing the dimensions of the silver halide crystals in the emulsion layers of grain density autoradiographs can improve both the resolution and the efficiency considerably (p. 71 and p. 94). Dark-field microscopy enables one to take full advantage of this fact, for the small developed grains can still be adequately recognised. At the same time, it becomes possible to study fine cytological detail, even in a heavily labelled cell, for the grains are so small that they cover relatively little of the cell. Fig. 57 illustrates this point. This is an autoradiograph of the ovary of *Galago demidoffi* after the injection of tritiated thymidine. The central cell

---

Fig. 56. Section of the testis of a newborn rat, 24 hours after the injection of tritiated thymidine. The section was stained with Harris' haematoxylin. (a), A photomicrograph of the silver grains by incident dark-field illumination superimposed on a transmitted-light picture of the section by the double-exposure technique described in the text. The small bright grains can be clearly seen against the darkly stained nuclei, and both are sharply in focus. (b), The same field photographed by transmitted light alone, with the silver grains critically in focus. The very small grains are invisible against the darkly stained section, which is out of focus. Autoradiograph prepared with Ilford K2 emulsion. Leitz Ultropak × 100 objective, with mirror condenser. (× 660). (From *Leitz Mitt. Wiss. u. Techn.*, 3 (1964) 43)

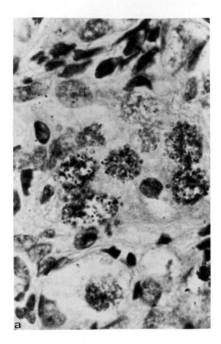

Fig. 57. Photomicrographs of the same field of an autoradiograph taken by (a) transmitted, and (b) incident dark-field illumination. The material is a section of the ovary of the bushbaby (*Galago demidoffi*), 24 hours after the injection of [³H]thymidine. The autoradiograph was prepared with Ilford K2 emulsion, and the grain size kept small by use of a short development time. In spite of the high grain densities over these 5 labelled oogonia, they can be seen to be in mitotic prophase. (× 720). (Material prepared in collaboration with Dr. J. Ioannou)

is an oogonium in the prophase of mitosis. Even though there are many grains over the nucleus, the fine thread-like chromosomes can still be distinguished.

There are clear advantages to be gained from the intelligent use of dark-field methods of viewing, together with an emulsion with crystals of small diameter, such as the Ilford K or L series.

PHOTOMICROGRAPHY IN TRANSMITTED LIGHT

In spite of my obvious preference for dark-field methods of photomicrography, it is still quite possible to make a satisfactory record of many autoradiographs by the use of conventional optical systems and transmitted light.

Variations on this method have been described by several authors, to try and avoid some of the difficulties inherent in the technique. An effect rather similar to dark-field incident lighting combined with transmitted light can be obtained with phase contrast, if the tissue is accurately focussed rather than the silver grains: the latter appear as an overfocussed phase image, rather brighter than the underlying tissue[6].

The problem of the visibility of the underlying specimen is acute in squash preparations of chromosomes[7]. Several methods have been proposed to improve the chances of identifying the chromosomes below the silver grains. After photography of the autoradiograph, the silver grains may be removed with iodine and potassium iodide, followed by hypo treatment. After further staining, the same areas of specimen may be found, and the chromosomes alone photographed[8]. Alternatively, the Feulgen-stained chromosomes may be photographed before autoradiography, and then again afterwards[9].

THE VIEWING AND PHOTOGRAPHY OF TRACK AUTORADIOGRAPHS

Dark-field techniques are difficult to use successfully with thick emulsions. So much of the incident light is scattered in the upper layers of the emulsion that silver grains more deeply placed do not appear bright, and the background against which they are viewed is not black. Occasionally, it may be possible to use dark-field methods with track autoradiographs of carbon-14 or sulphur-35 where the emulsion layer is about 20 μm thick (Fig. 58). Similarly, α tracks can be recorded satisfactorily in a relatively thin emulsion layer, and dark-field techniques can be used to photograph them (Figs. 8 and 9).

With emulsion layers thicker than about 20 μm, however, it is virtually impossible to get a reasonable dark-field picture. Track autoradiographs are therefore usually viewed by transmitted light alone. This makes desirable the use of an emulsion with a fairly large crystal size. Since most of the quantitative data on the characteristics of β particle tracks in nuclear emulsion have been obtained from Ilford G5, this is probably the most convenient choice.

Track autoradiographs remain extremely difficult to photograph. Fairly high magnifications are needed to see the silver grains clearly, and the tracks climb and dip in the emulsion in such a way that only short lengths of track can usually be seen in any one focal plane. This is particularly true if the emulsion has been re-swollen prior to microscopy (p. 376). It often pays to prepare special slides for microphotography. These should be exposed

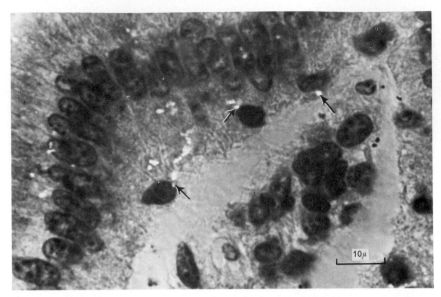

Fig. 58. A section of the small intestine of a mouse, killed one hour after the injection of [³⁵S]sulphate. The autoradiograph was prepared with a 25 μm layer of Ilford G5 emulsion. β-Tracks can be seen entering the emulsion at the nuclear membranes of three adjacent epithelial lymphocytes (see arrows). Section stained with Harris' haematoxylin and photographed with a double-exposure technique, using a Leitz Ultropak × 100 objective. (× 1300). (From Darlington and Rogers, *J. Anat.*, 100 (1966) 813)

for longer than is usual, in order to collect a relatively high track density around the source. Instead of re-swelling the emulsion in glycerol, it should be dehydrated prior to mounting under a coverglass in the normal way. In this way, the chances of finding several tracks in the same focal plane as the source will be increased. Even then, only relatively short lengths of track can usually be included in the photograph (Fig. 59).

For track autoradiographs that have been re-swollen for microscopy to thicknesses of 60 μm or more, special objectives with a long working distance are usually needed, such as the Leitz KSX53 and KSX100 oil-immersion objectives. Most manufacturers produce objectives of this type.

Fig. 59. A photomicrograph of a darkly stained endplate on a single muscle fibre, obtained by microdissection from the sternomastoid muscle of a mouse. The endplate was treated with [³²P]DFP, and autoradiographed with a 60 μm layer of Ilford G5 emulsion. Two β-tracks can be seen entering the emulsion from the endplate. Leitz KS × 100 objective. (× 1300). (From Rogers et al., *Nature*, 210 (1966) 1003)

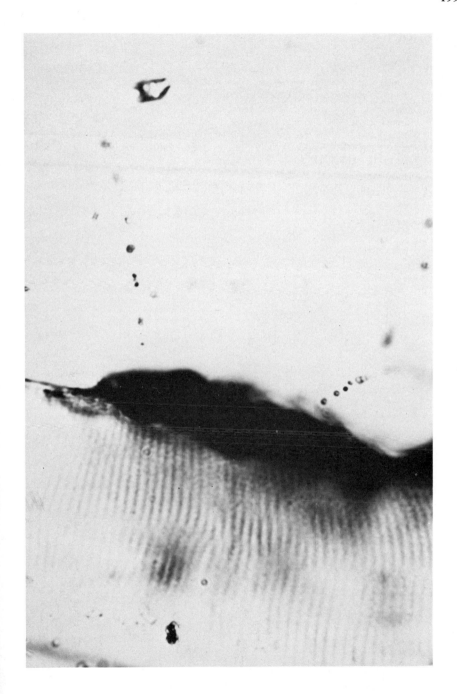

REFERENCES

1 A.W. Rogers, *Sci. Tech. Inform.*, 1 (1965) 62.
2 H.P. Walmsley and W. Makower, *Proc. Phys. Soc. (London)*, 26 (1914) 261.
3 J.E. Gullberg, *Exptl. Cell Res.*, Suppl. 4 (1959) 222.
4 P.P. Dendy, *Phys. Biol. Med.*, 5 (1960) 133.
5 A.W. Rogers, *Exptl. Cell Res.*, 24 (1961) 228.
6 D.E. Schlegel and G.A. de Zoeten, *J. Cell Biol.*, 33 (1967) 728.
7 D.M. Prescott and M.A. Bender, in D.M. Prescott (Ed.), *Methods in Cell Physiology, Vol. 1*, Academic Press, New York, 1964.
8 A. Fröland, *Stain Technol.*, 40 (1965) 41.
9 M. Callebaut and P. Demalsy, *Stain Technol.*, 42 (1967) 227.

CHAPTER 10

The Collection of Data from Autoradiographs

Most scientific experiments, whatever their initial aims, end up with some unfortunate sitting down to measure something. Autoradiography is no exception. In this chapter we will examine the various properties of nuclear emulsions that can conveniently be measured, and how they are related to the radiation dose to which the emulsion layer has been exposed. The methods available at the present time for collecting data from autoradiographs will then be discussed in detail. The problems of selecting appropriate areas for measurement and of interpretation of the data will be considered in Chapter 11.

THE RESPONSE OF NUCLEAR EMULSIONS TO RADIATION

When a volume of nuclear emulsion is exposed to β particles, a number of silver halide crystals form latent images which, on subsequent development, are converted into metallic silver grains. This basic response of the emulsion to radiation bears a defined relationship to the number of β particles entering it. The emulsion response can be looked at in many ways: the number of silver grains produced, the mass of developed silver per unit volume, and the transparency or blackening of the emulsion are all examples of measurable parameters that will change as a result of exposure to radiation. Which parameters can be conveniently measured, and how are they related to radiation dose?

(a) *Grain number*
This is probably the most obvious and basic change in an emulsion exposed to radiation, and grain counting has for years been the most widely used method of assessing the radiation dose.

The relationship between radiation dose and number of grains per unit volume of emulsion has been investigated by Goldstein and Williams[1] and by Dörmer[2], amongst others. It is not a direct and linear relationship, but a logarithmic one. As the number of β particles entering the emulsion increases, the increase in grain number becomes less and less. Two factors are responsible for this. The first is that as the radiation dose increases the probability of silver halide crystals being hit more than once also increases, and a crystal hit three times by separate β particles can only produce one developed grain. The second factor is that the recognition and counting of silver grains become very difficult at high radiation doses, adjacent grains overlapping and even fusing during development.

If one plots the grain density produced in a given autoradiographic system against the density of radioactive disintegrations in the source, the first part of this logarithmic curve can be approximated to a straight line. In other words, up to a limiting grain density, grain density can be considered directly proportional to radioactivity. This limit is reached when the probability of multiple hits on single crystals becomes unacceptably high, or the developed grains are so closely packed that counting becomes impossible. Both of these factors are related to the size of the undeveloped crystals in the emulsion: the smaller the crystal diameter, the higher the grain densities that will still be proportional to radioactivity.

We have seen (p. 97) that multiple hits on crystals can be ignored up to the point where 10% of the available crystals have already been hit. For practical purposes, only the monolayer of crystals nearest the source is "available" to the β particles of tritium. Although for isotopes of higher energy crystals in many layers above the first will be "available", at grain densities higher than one-tenth of the crystals in a packed monolayer, the chances of grains overlapping or fusing during development become so high that it is advisable to make this the upper limit of density for grain counting. So, with Ilford L4 emulsion, where a packed monolayer contains between 45 and 50 crystals per square micron, grain number can be considered proportional to radiation dose up to about 5 grains per square micron. With Ilford G5, the corresponding limit would be 1.5 grains per square micron.

There have been attempts to calculate or measure "coincidence coefficients" for grain density, which would enable one to correct for the non-linearity of grain number and radiation dose[3]. In practice, however, the problems of reproducible grain counting by direct visualisation in the microscope become so great at these high grain densities that it is better to choose some other emulsion response to measure.

As a means of measuring the density of β particles entering the emulsion, then, the grain density may be considered as directly proportional up to limits corresponding to one-tenth the density of undeveloped crystals in a packed monolayer of the emulsion. In this range counting should not present great problems. Obviously, counting of individual grains can only be done at sufficiently high magnifications for the grains to be recognised. Each counting area will necessarily be rather small with the nuclear emulsions, and large areas can only be examined by repeated sampling. This is not a severe restriction in most histological experiments where sources of cellular dimensions are being studied, since the grain densities often vary abruptly within distances of 10–20 μm, and small counting areas positioned over many cells of the selected type have to be examined in any case to sample a cell population. Where relatively extended uniform sources are being autoradiographed, however, the necessity to work at high magnification and make repeated samples may be avoided by choosing some other parameter of the emulsion response to measure.

(b) *Track number*

The grain density mentioned above may be proportional to the number of β particles entering the emulsion, but it is often extremely difficult to determine the number of β particles responsible for a given number of grains. Many factors can influence the efficiency of an autoradiograph (see Chapter 5), and it may not be possible to calculate the absolute efficiency of recording, or to prepare standard sources which correspond accurately enough with the experimental ones.

If an isotope of maximum energy equal to or higher than carbon-14 is used, a sufficient percentage of the β particles entering a thick emulsion layer will give rise to recognisable tracks for these to be made the basis of data collection.

There will be few applications for which track autoradiography offers clear advantages, and these are discussed in more detail in Chapter 17. The recognition and counting of tracks is slow relative to grain counting, and it becomes very difficult and inaccurate at high track densities. In fact, the range of radiation doses in which track counting is reliable is much shorter than for grain counting. The limits for carbon-14 and sulphur-35 are reached when 10 tracks or so start at one surface of the emulsion within an area of 500 square microns: for the longer, straighter tracks of phosphorus-32, the corresponding figure is about 15 tracks per 500 square microns.

Within these limits, track counting is a precise and reproducible method of measuring the number of β particles entering the emulsion. As with grain counting, the counting areas will be fairly small as the emulsion must be viewed at a high enough magnification to recognise individual grains.

(c) *The reflectance of the emulsion layer*

We have seen in Chapter 9 that if an emulsion layer is viewed by dark-field incident light the silver grains reflect and scatter the light, appearing as bright specks on a black background (Fig. 53). Obviously this method of illumination can be used as a basis for recognising and counting the individual silver grains, and everything that was said above about grain number applies directly to this procedure. But it is also possible to measure the total light reflected by a given area of emulsion and to relate this to the radiation dose there[4]. In fact, the reflectance of a thin layer of a given emulsion is directly proportional to the grain density, as has been demonstrated by a number of authors[1,2,4]. So, as with measurements of grain density, the reflectance is a logarithmic function of the radiation dose, but may be approximated to a linear response over the first part of the curve[1]. The direct correspondence of reflectance to grain density has been shown to extend up to 7 grains per square micron for Kodak AR-10 (ref. 2), which is higher by far than the upper limit of linearity between grain density and radiation dose, and higher also than the reproducible upper limit for visual grain counting.

Reflectance measurements are at their best with oil immersion objectives and, as with the counting of silver grains, would seem to be most useful for fairly small measuring areas[2].

(d) *Optical density*

The presence of developed grains in a thin layer of emulsion reduces its transparency. From the early days of photography, measurements of some function of the transparency of film have been related to the exposure of the emulsion[6]. The transparency (T) is defined as the ratio between the light transmitted by the film and the light incident on the film; this ratio is only valid when the area of film measured has been uniformly exposed to light and uniformly processed, and is large relative to the size of the silver grains. Hurter and Driffield[7] defined the density as $\log(1/T)$, and demonstrated that it was proportional to the mass of developed silver per unit area for a given emulsion. Plotted against the logarithm of the exposure to light, density produces a curve (known as the H and D curve) which contains a long central linear portion[6]. Over a specific range, then, the density of a film is directly proportional to the logarithm of the light exposure.

The response of nuclear emulsions to β particles differs from that of photographic emulsions to light. This is due to the fact that a single hit by a β particle will have a high probability of producing a developable latent image, whereas multiple hits by light photons are needed to achieve the same probability. With β particles the density is proportional to the exposure at low densities[1,8], rather than to the logarithm of the exposure. This proportionality extends to considerably higher radiation doses than the approximately linear relationship between grain number and radiation dose[1]. It follows that density measurements on the emulsion layer provide satisfactory data over a much wider range of radiation doses.

Density measurements are quick and simple to make. They impose certain restrictions on the autoradiographic material, however. The measurements must be made at a magnification low enough to avoid resolving individual silver grains. The areas of emulsion chosen for measurement must be uniformly irradiated, so that sources that vary abruptly in radioactivity over very small distances should be autoradiographed in a way that averages out the radiation dose to the film over areas that are big enough for sensible measurements. Finally, the specimen must either be transparent (i.e. unstained), or separated from the emulsion layer during measurement.

In general, density is a useful parameter to choose when fairly large, homogeneous areas of emulsion are to be compared. It tends to be a difficult measurement to apply to sources of cellular dimensions, where the problems of relating source to emulsion are much more critical, and where the variations of radioactivity between cell and extracellular space, nucleus and cytoplasm, have been resolved by the autoradiograph.

(e) *Other possible parameters of the emulsion response*

Back in the last century, it was realised that the mass of developed silver per unit volume of emulsion could be related to the exposure of the film[7]. This parameter has not been utilised on more than an experimental scale for measuring the radiation dose to the emulsion. I have heard of a laboratory in which conversion of the developed silver to iodide, using iodine-131 in controlled conditions, was followed by the measurement of radioactivity in selected volumes of emulsion with a Geiger counter. The adsorption of labelled thiosulphate to the surface of freshly developed silver grains has been used in investigations into the photographic process itself[9]. In scanning electron microscopy, the measurement of the mass of silver present per unit area of autoradiograph by X-ray microanalysis represents a further development of this theme[10].

When one comes to consider television image analysis systems and their place in the collection of data from autoradiographs, other parameters of the silver grains in the emulsion will have to be considered; but this rather specialised topic will be deferred to p. 223.

MEASUREMENTS OF EMULSION RESPONSE

It is obvious that several responses of the emulsion to radiation can be observed. What methods are available for measuring these responses, and what factors might lead one to choose one method rather than another for a particular experiment?

VISUAL GRAIN COUNTING

Basically, this is a very simple process. The observer records the number of developed grains visible in the volumes of emulsion selected for study. Two problems require consideration. The first is one of deciding what is to be accepted as a silver grain. The second concerns practical planning, so that the best possible conditions are provided for the observer.

"The definition of a silver grain" may sound unreasonably academic. It is usually easy to reject specks of dust, or granular deposits from the histological stain on the basis of their appearance. The precise definition of a grain is nevertheless important.

At the end of its track (see p. 52), a β particle loses energy more rapidly than it did earlier, and the silver grains it produces lie closer together and tend to be larger. These grains may even make contact in the course of development (see, for instance, Fig. 11), and it may be very difficult to decide whether a given blob of silver represents one large grain, or several smaller ones fused together. In fact, blob counting has replaced grain counting in some of the analyses of particle tracks by physicists. The β particles of tritium are an example of this rapid rate of energy loss by particles of very low energy, and it is a common observation that the grains produced by tritium are usually larger than the background grains on the same autoradiograph. One observer may feel each blob represents one grain; another may estimate that, since the larger blobs are twice as big as the average background grain, they probably represent two grains fused during development. There is no simple way of deciding which interpretation is correct. If counts by different observers are to be compared, it is best to agree on one or other definition before counting starts.

There is no reason why dark-field conditions should not be used for grain counting, provided the development routine has been appropriate (see Fig. 57). In such a case, every grain that can be seen should be counted. It is impossible to discriminate successfully against the smaller grains, and only count the larger ones, as the distribution in grain size is continuous, and this type of decision introduces an unacceptably high element of subjectivity.

In deciding whether to plan for dark- or bright-field conditions, it should be remembered that the former give much lower light intensities than are usual in a well-lit laboratory. During the period of accommodation to darkness, the eye's threshold of recognition for silver grains must be expected to alter, and any interruption in counting will produce another period of accommodation, if the lighting in the room is very bright. If constant, subdued ambient lighting is not available, it may well be better to count grains by bright-field illumination.

The conditions for visual grain counting

With most observers the reproducibility of their grain counts begins to fall off after the first hour, though the precise time relations of this effect obviously vary with the experience of the microscopist. Beginners may feel actual nausea, particularly if the work involves much scanning of the slide, with frequent movements of the microscope stage. Even with the most experienced grain counter, the accuracy of the counts will decrease if they are attempted for more than a few hours each day.

Perhaps the most tiring aspect of grain counting is the necessity to hold the head in a constant position relative to the microscope. Even when sleeping we are always shifting around, and the constraint imposed by the need to keep the eyes level and still at the eyepieces of the microscope is physically wearing. A comfortable and relaxed position is quite essential. Many systems have been suggested to free the observer from this constraint while grain counting. Micou and Goldstein[11] thought it worthwhile to publish an account of a projection system for throwing the field of view on to a screen. Closed-circuit television has been used in the same way. Sometimes, photomicrographs are taken of each field, and the grains counted on the developed print[12]. All these methods extend the period of counting before fatigue begins to affect the results, and make it easier to employ personnel who have no training or experience in microscopy.

Whatever system of counting is chosen, it takes a certain time to adjust to the conditions of work. Freedom from interruption will give more reproducible counts, so that a quiet room with subdued lighting is obviously better than a busy and well-lit laboratory. Extending this principle further, two people can usually work faster and more accurately than one. The microscopist concentrates on the specimen itself, and the assistant records the counts. The need to write down each count before going on to the next can break the rhythm of work significantly. In one laboratory where physicists are carrying out particle track analysis, each microscope has a tape-recorder beside it, so that measurements and comments can be recorded without the observer taking his eyes from the microscope.

Decision-making should be reduced to a minimum while counting. The appropriate volumes of emulsion to be scanned should be determined in advance and translated into practical instructions before counting starts. "All cells of type A on a line from X to Y", or "All cells in the field of view at such-and-such settings on the microscope stage" should be counted, and these instructions strictly adhered to. Quite apart from the time spent in deciding whether to count grains over this cell or that one in the absence of such precise instructions, the subjective element introduced opens the way to the choice of "typical cells" to count, and makes the results as a whole suspect.

Similarly, the shape and size of the volume of emulsion to be examined around each cell should be predetermined. If this area is large or contains many silver grains, it should be subdivided by a grid in the eye-piece into smaller squares, and each square counted in turn, as with blood cell counting in a haemocytometer. Small tally counters are very useful if the numbers to be counted are higher than about 20.

Once the influence of external conditions on the accuracy and reproducibility of visual grain counts is realised, it is possible to plan the procedure in such a way that reliable results can be obtained fairly rapidly. Perhaps the most important single step is to arrange that grain counting is done in relatively short periods, with breaks for other work. In this way, all the counts can be obtained during the period of 1–2 hours when the observer is fresh.

TRACK COUNTING

This is inevitably a slower and more difficult task than grain counting. Tracks are three-dimensional, and each field chosen for examination must be scanned at many focal levels. The tracks themselves must be followed for some distance, to try and determine the direction of travel of the β particles, and to record their points of entry into the emulsion as precisely as possible.

Track counting requires a certain amount of experience, and it is not a good idea to turn over the task of track counting to an untrained person, whereas grain counting can be delegated with the minimum of explanation to anyone who is competent to use a microscope. When this has been said, it must also be admitted that many autoradiographers are unnecessarily frightened by the difficulties of track recognition. On several occasions I have asked students to count tracks on material prepared with carbon-14, sulphur-35 or phosphorus-32; after half-an-hour of explanation and demonstration, their counts have been reasonably accurate and reproducible.

The characteristics of β particle tracks are dealt with in detail in Chapter 3 (p. 52). Briefly, their chief characteristic is their variability. The distance between successive grains, the changes in direction, and the grain size vary in a disconcerting way, so that any statement about them only reflects the statistically probable behaviour of an idealised particle. High energy β particles, such as those from about 400 keV and upwards, tend to run in approximately straight or gently curving lines for considerable distances. The grains lie, on the average, about 2 μm apart in Ilford G5 emulsion[13], though there will be gaps of 4–5 μm and other regions where several grains lie more closely spaced. The majority of grains will be rather small under normal conditions of development. Below about 75 keV, whether we are dealing with a low energy particle or the terminal part of the track of one of higher energy, the grains lie, on the average, 1 μm apart. The chances of a gap of more than 3 μm are quite low, and the grains in the terminal few microns of track may even fuse together. The grains themselves are larger than those usually found at higher energies. The track usually changes direction many times in the last 25 μm, whereas abrupt changes of course, due to deflection of the β particle by an atomic nucleus, are generally much more widely spaced at higher energies (Fig. 11).

It is usual to define a β particle track as 4 or more silver grains in a row. It is clearly possible to arrange any 4 background grains in some sort of linear array. A track of only 4 grains, however, represents the terminal 10–20 keV from a β particle, and these grains one would expect to be large, and very closely spaced. If the ends of undoubted β particle tracks are used as a basis for comparison, it will be found, in any reasonably prepared material, that the chance of observing 4 background grains of comparable size and similar spacing is very low indeed.

δ Tracks often prove confusing. These are the tracks left by electrons that have been ejected from orbit by the passage of a particle through their parent atom. Fig. 17b shows such a track taking origin from the long, straight track of a high energy cosmic ray. The only difference between a δ track and that of a β particle is the mechanism of production of the particle itself. If a track that resembles that of a typical β particle starts from another track, it should be considered a δ track, and not a β particle. If the parent track is that of a β particle, the two tracks fork, with an angle of approximately 90 degrees between the two branches, both of which have typical terminal portions, with large grains closely spaced (Fig. 12).

It may be possible to recognise and reject background tracks caused by β particles that did not arise from the source being studied. If the specimen is mounted on a glass slide, with a thick layer of emulsion over it, it is evident that tracks entering the emulsion from its upper surface must be considered as due to background particles. Background tracks may also be recognised if their length or grain number is greater than expected from the known characteristics of the isotope being used. Unless the emulsion layer has been accurately re-swollen after processing to its thickness during exposure (p. 376), the track length is not such a reliable guide to the initial energy of the particle as the number of grains in it. The equations relating initial energy to track length and grain number are given on pp. 53–55 and are illustrated graphically in Figs. 13–15. To give an example, the mean number of grains produced by a β particle of initial energy 155 keV in Ilford G5 emulsion is 74. There is a statistical scatter of grain numbers in each track around this mean figure, and tracks of up to 90 grains may be encountered from carbon-14 (E_{max}, 155 keV), but the probability of finding a track with more than 90 grains in it from this isotope is extremely low. If, therefore, one finds a track of 105 grains in an autoradiograph of carbon-14, it is reasonable to conclude that it is a background track.

Track counting with β particles is just about impossible to automate, because of the wide variety of track lengths and patterns that can occur with one isotope. It is slower and requires more patience and experience than visual grain counting. It is not unduly difficult, however, and it can become very precise and reproducible.

The most important single step in the technique of preparing track autoradiographs for microscopy is probably the re-swelling of the shrunken emulsion layer immediately prior to mounting it under a coverglass (p. 376). There are often many scattered background grains at the upper surface of the emulsion, and these are separated clearly from the specimen beneath the emulsion, by the process of re-swelling. Tracks which cross in a hopeless tangle before re-swelling can often be seen to pass each other at an appreciable distance after this step. The task of recognising and recording the tracks of β particles is considerably eased by this simple procedure.

PHOTOMETRIC MEASUREMENTS IN REFLECTED LIGHT

Gullberg[14,15] was the first to devise a semi-automatic method of measuring grain densities using dark-field conditions to view the silver grains, but his was a scanning system, counting the individual bright specks produced by the silver grains. Rogers[4] described the first apparatus for measuring the light reflected by the silver grains, and showed that this gave an integrated value for the mean number of grains in the measuring field. Since then, several descriptions of the technique have come from Dörmer and his associates[2,3,16], it has been examined by Goldstein and Williams[1], and a discussion of its characteristics and limitations has been presented by Rogers[5].

Briefly, the autoradiograph is illuminated with a fine, narrow beam of light directed vertically downward on the emulsion through the objective lens itself. The light reflected by the metallic silver grains back through the objective is collected on the photocathode of a photomultiplier, and the current flowing through the photomultiplier generates a reading. Since the light on which the reading is based enters and leaves the upper surface of the emulsion without passing through the underlying specimen, a stained section can be used in contact with the emulsion without, in many cases, affecting the reading at all. A rather similar method described by Dendy[17] was based on substage dark-field illumination, but this has fewer applications, since the light has to traverse the specimen to reach the emulsion, introducing the possibility of variations in reading due to light absorption by the specimen.

We have seen above (p. 204) that the reflectance of an area of emulsion is directly proportional to the grain count, after subtraction of machine background. This method, then, gives a reading that can be directly related to the radiation dose received by the emulsion layer, provided the number of developed grains per square micron does not exceed one-tenth the number of undeveloped crystals per square micron of a densely packed monolayer. It is clear that measurements can be made satisfactorily up to far higher grain densities: Dörmer[2], working with carbon-14 and AR-10 stripping film, suggests that the photometer reading is proportional to the radiation dose up to 30 grains per 10 square microns, while reproducible measurements can be taken up to over 70 grains per 10 square microns. Goldstein and Williams[18] have shown that the grain density at which it becomes necessary to correct for double hits is actually higher if the reflectance of the emulsion is measured by polarised light. It seems that the presence of a high density of grains in the measuring area increases the percentage of light that has been depolarised, probably as a result of multiple scattering between adjacent grains, thus effectively correcting for the reduction in grain number through multiple hits.

The method gives its best results with oil immersion objectives in the range of magnifications × 60 to × 105, and with measuring areas of 10–200 square microns. It is not affected by grain densities immediately outside the measuring area, so the method can be used in autoradiographs where grain densities vary over structures of cellular dimensions.

The speed of operation is obviously determined by the time necessary to find the next object for measurement in the specimen. With biological specimens of reasonable complexity, 200 measurements per hour can be made without strain[5], representing a considerable mass of data in one working day. With suitably prepared material, this method of data collection offers many advantages over the much slower visual grain counting.

(a) *The apparatus needed for photometric measurements*
The most important single step in making reflectance measurements is to

obtain the best possible conditions of dark-field illumination, with bright, clear silver grains on a black background (Fig. 60c). The two types of illuminating system available have been discussed in Chapter 9, and illustrated in Fig. 55. The first, providing a convergent cone of light entering the emulsion from a condenser system surrounding the objective lens itself, gives a good dark-field picture, but has the great disadvantage that it is not possible to limit the area illuminated by means of a field diaphragm. Silver grains are not plane mirrors, but coiled masses of filamentous silver (Fig. 2), and they scatter and reflect light in every direction, behaving as point sources of light in this dark-field illumination. Adjacent silver grains can thus illuminate each other with light scattered in the plane of the emulsion layer, and also, since the emulsion after processing is never completely clear optically, produce a certain amount of background glare from the surrounding gelatin. With the convergent cone method of lighting, reflectance measurements from small areas of emulsion without a single silver grain will vary considerably with the number of developed grains immediately outside the measuring area.

This light scattered from outside the measuring area can be avoided if the emulsion outside the measuring area is not illuminated, and this can only be done with the vertical incident system of lighting, which permits control of the diameter of the illuminating beam with a field stop. This system, in which the light travels to the specimen through the objective lens itself, is essential if reflectance measurements are to be made from chosen small areas of emulsion within a specimen of varying radioactivity. Obviously, the convergent cone system can be used for special situations where the grain densities surrounding each measuring area are always uniform: smears of cells, widely separated by emulsion with background grain densities only, might be examined in this way.

There are clear advantages from using polarised light[18]. Not only is the scattered light from optical surfaces cut down, but the light scattered from pigment granules, which can be very troublesome, is also reduced. Many staining methods can be used which would otherwise give an annoying background measurement that varies with the nature of the underlying tissue.

Most manufacturers of microscopes provide vertical incident illuminators that are suitable. It is important to ensure that the field stop can close far enough to produce an illuminated area the same size as the measuring area that will be required. The Leitz Pol-Opak illuminator has the most convenient diaphragm system I have seen from this point of view, giving rectangular or circular areas of illumination down to 1 μm diameter.

The light source must be well stabilised, so that variations in light output do not introduce errors into the measurements of light reflected. This can

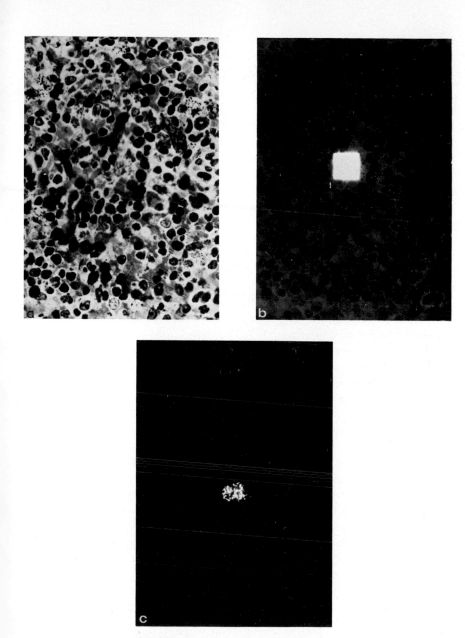

Fig. 60. Three photomicrographs of the same field of an autoradiograph of mouse spleen following an injection of [³H]uridine, to illustrate steps in the setting up of a photometric grain counter. (a), The section is viewed by transmitted light and the area selected for measurement is centered in the field. (b), The diaphragm defining the area from which light will be collected by the photocathode is matched in shape, size and position to the area illuminated by the incident beam. (c), Viewed by incident light only, an integrated measurement is taken of the light reflected by the measuring area. (× 820)

be cheaply and conveniently done by a constant voltage transformer and a stepped transformer providing current to the light source at reproducible levels. For measurements in reflected light, even with the simultaneous use of colour filters on the incident and reflected light paths, I have found a 30 W tungsten light source adequate. If polarised light is to be used[18] (p. 212), it is better to go to a 100 W source. Since the light output and colour temperature of the source will vary for some time after first switching on, it is better to allow the light source a warm-up period of 20–30 minutes before starting to make measurements.

The area for measurement is identified and centred by the microscopist, who may need transmitted light to view the stained section. The transmitted light must be cut off, and the reflected light picture focussed, before the reflected light is switched over to the photomultiplier. The microscope head should present all the light entering the objective lens either to the eyepieces or to the photomultiplier tube. Systems that give a partial separation of light, permitting observation of the specimen during the act of measurement, should be avoided. The light intensities reflected by a few silver grains are very low, and it is a mistake to reduce them still further by diverting a fraction of the light to the eyepieces, while light entering the system through the microscope eyepieces may affect the measurement.

It is possible to use the objective lens as a simple projection device, throwing the light from the silver grains on the photocathode without an intervening lens system[4]. It is also possible to image the light on the photocathode by a lens system similar to an eyepiece, and introduce a diaphragm to define the measuring area accurately. In the first case, the measuring area can only be defined by the field diaphragm on the illuminator; in other words, the area of emulsion illuminated becomes the measuring area. With the second, more complicated system, it is necessary to match the shape, size and position of the field diaphragm (defining the area of emulsion illuminated) with that of the diaphragm in front of the photomultiplier (which defines the area of emulsion from which light is accepted by the photocathode). The second system does reduce the scattered light reaching the photocathode, improving the signal-to-noise ratio. If there is any doubt about the uniformity of response of the photocathode from one point on its surface to another, it is advisable to place a diffusing screen between the second diaphragm and the photocathode.

Many microphotometers are commercially available which can do the job required by reflectance measurements. The one in use in my laboratory at present is a Photovolt model 520-M (Photovolt Corp., New York), which has about the right sensitivity. A full scale deflection on the most sensitive range corresponds to 5×10^{-6} foot-candle, or 0.01 microlumen.

This, then, completes the list of the basic equipment needed for reflectance measurements: in my case, I began with a Leitz Ortholux microscope with a Pol-Opak illuminator, and a 30 W tungsten bulb, which is stabilised with a constant voltage transformer. An FS trinocular microscope head carries an MPV microphotometer attachment with Photovolt 520-M photometer.

This basic equipment is not too expensive. I have since gone over to a 100 W stabilised bulb on the incident illuminator, and crossed polars. The output of the photometer is fed to a digital display and print-out system. Following the suggestions of Clarkson and Sanderson[19], a stage drive and pen-recorder are available, enabling traces of grain density to be taken across autoradiographs — a very useful method of presenting information.

It is possible to go further, driving the stage with an X–Y plotter and desk computer. It is interesting that Bisconte, Fulerand and Marty[20] found the simple, semi-automatic machine faster to operate than the more elaborate version with scanning stage and computer.

(b) *The characteristics of a photometer measuring light reflected by silver grains*

Goldstein and Williams[1] have estimated that silver grains reflect only about 3% of the light falling on them. The light intensities to be measured are low. It makes sense to instal the photometer in reasonably constant, low light levels, to reduce the chance of variations in ambient light affecting the readings. A suitably prepared autoradiograph (pp. 217–221) is placed on the microscope stage, and a heavily labelled area found. The transmitted light is switched off, the dark-field picture focussed, and the reflected light directed to the photomultiplier. By adjusting the current to the light source and the high voltage supply to the photomultiplier, conditions of measurement are found which give a satisfactory reading, with a sufficient difference between the heavily labelled area and one with no silver grains. At the same time, the field diaphragm is adjusted to the size required for the measuring area for that particular set of readings, and the diaphragm in front of the photomultiplier is adjusted to correspond to the field diaphragm in size, shape and position. When these adjustements are completed, the photometer is left for a further 10 minutes or so to stabilise. With every series of measurements, it is essential to have a reference standard which can be examined before, after and at intervals during the series of measurements, to check that variations in operating conditions are not occurring. I use an autoradiograph of a section of labelled methacrylate for this purpose, but any standard which gives reproducible results can be used. Dörmer[2] has described the use of a reference objective for this purpose.

216

As with visual grain counting, decisions should be made about the choice of areas for measurement before starting the series. With the autoradiographs available, and the instructions about finding the next area to hand, the microscopist finds and centres the appropriate areas, switching off the transmitted light to focus the silver grains before sending the reflected light to the photometer (Fig. 60). The vertically incident light should be left on throughout the whole series of measurements.

Dörmer[2] recommends operating with the illuminating aperture fully open. He has also shown that there is a linear relationship between the diameter of the measuring area and the background reading; having determined this relationship for any given set of autoradiographs, it becomes possible to vary the diameter to match the objects being measured within a series of readings, and to correct for the change.

In these conditions the photometric readings are directly proportional to the number of silver grains in the measuring area (Fig. 61), above a level, the machine background, which is mainly due to light scattered by various optical interfaces in the system, and by the gelatin itself. This machine background can be found by making several measurements from areas with no visible silver grains in them, or simply summed with the autoradiographic background by taking measurements from suitably chosen areas in the autoradiograph.

Not only is the mean photometric reading proportional to the mean visual grain count, but the use of the photometer does not add to the

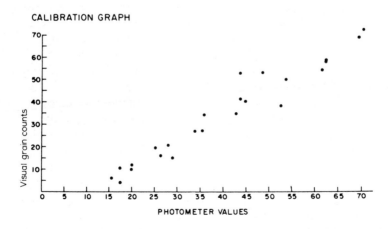

Fig. 61. A graph relating the number of developed grains in the measuring area, determined by visual counting, to the photometer reading from the same area. Above a given value, the photometric measurement is directly related to visual count over a considerable range of grain densities.

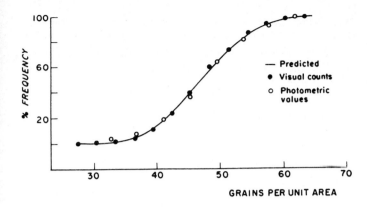

Fig. 62. The results of visual and photometric estimates of grain density from the layer of L4 emulsion over a section of uniformly labelled [³H]methyl methacrylate. A cumulative distribution curve built up from 200 counting areas over the section shows a close correlation with the curve predicted from the Poisson theory, whether visual or photometric grain counts are used. (From England and Rogers, 1970)

variance of measurements about the mean. England and Rogers[21] have shown that the visual and photometric grain measurements taken from over a uniformly labelled source are Poisson distributed (Fig. 62).

Large silver grains reflect more light than small ones. In any series of measurements involving large numbers of grains, this variation evens out. Grains deep in the emulsion layer may reflect significantly less light than grains at the surface, particularly if the gelatin is not clear, or has been stained. With emulsion layers 3–4 μm thick during exposure, the developed grains after dehydrating the processed emulsion ought to lie within 1 μm of the section, and there is seldom a serious problem from the depth of the grains in the emulsion.

The observed reading varies with the focus of the microscope, but does not vary very abruptly (Fig. 63). At about the level where the majority of silver grains are in focus visually, the reading reaches a maximum[2]. I have found readings to be very reproducible when taken with the grains in focus as judged visually: it is possible to alter the focus until a maximum reading is obtained, and record this, but it is slower to work this way and the readings have been no different to those obtained by visual focussing, in our experience.

(c) *Preparing autoradiographs for photometry in reflected light*

The specimen can scatter and reflect light, and every care should be

218

taken to reduce this source of error to a minimum. Clean working is essential in preparing the sections for autoradiography, and the processed emulsion should never be allowed to dry out or to acquire dirt on its surface. The processed emulsion can scatter light also. Dichroic fog, due to active developer present in the fixing solutions, must be avoided at all cost by an adequate stop bath and rinse. Fine particulate deposits in the emulsion can also result from decomposition of sodium thiosulphate in the fixing solution. Fresh fixing solutions should be used, and a definite improvement in the clarity of the processed emulsions may result from making up the 30% thiosulphate used for fixing in a buffer of the following composition: dissolve 2.2 g sodium sulphite ($7H_2O$) in 100 ml water, add 0.46 ml of a solution of sodium hydrogen sulphite (specific gravity 1.34) to a further 210 ml and mix the two solutions.

Many stains flouresce: eosin is a particularly bad example, giving a coloured glow from the specimen and emulsion in incident dark-field lighting. We have found light staining with Harris' haematoxylin to be compatible with reflectance measurements, and to give very little trouble. If stain fluorescence is a problem it may be reduced by illuminating and measuring in monochromatic light of a wavelength chosen to avoid the absorption and emission wavelengths of the stain involved. Measuring in polarised light may also help[18].

In most experiments there will be slight fluctuations in the machine background reading from slide to slide, and from place to place on a slide, but these should be so small relative to the readings from labelled areas that they can be ignored. Fig. 64 illustrates this point from a paper that relied heavily on photometric measurements[22].

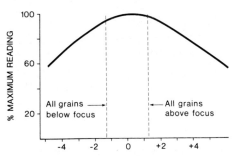

Fig. 63. The photometric reading from the same measuring area of an autoradiograph taken at different focal levels. The ^{14}C-labelled cell was exposed to Kodak AR-10 stripping-film. (Redrawn from Dörmer, 1972)

Treatment	Mean photometric reading			
	Lumen	Glands	Stroma	Muscle
Ethyl oleate	1.09	0.81	0.35	1.87
Progesterone	0.74	− 0.17	0.46	0.86
[³H]megestrol acetate	15.24	17.65	29.79	39.14

Fig. 64. Photometric grain counts taken from sections of rat uterus that were non-radioactive, and similar sections from rats injected with [³H]megestrol acetate. Note the low values from the non-radioactive specimens, showing that light scattered by the sections, which were stained with Harris' haematoxylin, contributed very little indeed to the readings.

(d) *The development of autoradiographs for photometry in dark-field lighting*

We have seen in Chapter 2 that the number of grains over a labelled source is critically dependent on the conditions of development. Fig. 4 illustrates the plateau in grain density that is found with increasing times of development. Unfortunately, when one measures the light reflected by the emulsion rather than the grain number, no such plateau appears (Fig. 65). With increasing development, the grains have grown larger or more reflectile. This makes comparisons between batches of autoradiographs difficult by photometric measurements, as the ordinary, rather sloppy method of developing autoradiographs on the bench at room temperature does not give adequate control of development, or sufficiently reproducible reflectivity per silver grain from one day to the next.

A specially designed developing tank is available from John Varney and Co., Blidworth, Mansfield, Notts., which greatly improves the reproducibility of processing. Temperature is closely controlled in all the solutions, and nitrogen burst agitation is used in the developer. With standard sections of labelled methacrylate autoradiographed and developed in this tank in separate experiments, the reproducibility of photometric measurement is excellent. If a really close correlation of photometric readings between different batches of autoradiographs is essential, the exposure of some standard preparation on each occasion should provide a basis for accurate comparison.

Naturally, the development conditions chosen should be matched to the requirements of viewing the autoradiograph in dark-field lighting (see p. 195 and Fig. 3). Overdevelopment causes the appearance of a second population of tiny silver grains through the emulsion, which can make measurements of reflectance quite meaningless.

Neely and Combs[23] have investigated a number of different developers

VISUAL COUNTS ON DEVELOPMENT TIME

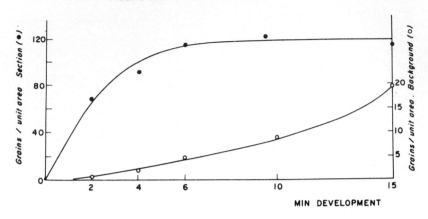

PHOTOMETER READING vs. DEVELOPMENT TIME

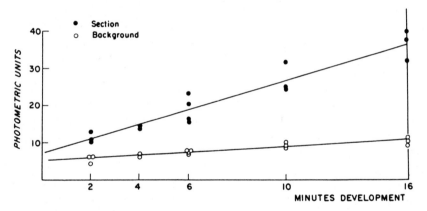

Fig. 65. The effects of increasing development on visual grain density (a), and on reflectance measurements in incident light (b). The specimens were sections of [³H]methyl methacrylate, autoradiographed by dipping in Ilford L4 emulsion. (a), Note the long plateau in grain number that is found with visual grain counts. (b), Photometric readings, however, show no such plateau. With Ilford ID-19 developer there appears to be a steady growth in size of the developed grains, and a consequential increase in their reflectivity, even though their number is constant. (From Rogers, 1972)

for their suitability for use with incident light photometry. Under optimal conditions of development there was a wide variation in the light output from standard areas of emulsion, depending on the developer used. Microdol consistently gave very low readings, barely distinguishable above background. Dektol and D-19 gave good signals from L4 emulsion, Dektol and Rodinal were the best for NTB-2 emulsion. Dörmer and Brinkmann[24] have found an Amidol developer (D-170) to give a light signal per silver grain that was more reproducible from one specimen to another than that with D-19; they therefore prefer it, though the light signal is smaller with D-170. Clearly the choice of developer should be made with the method of measurement of emulsion response in mind.

(e) *Summary of reflectance measurements as a means of data collection*

The choice of reflectance as a parameter of emulsion response introduces technical difficulties which are not present when grain number is used — scattered light from specimen and emulsion, and the variation in grain size from batch to batch of autoradiographs, for instance. These difficulties can be overcome in the preparation and microscopy of the specimen in nearly every case.

For very low grain densities, below 5 per measuring area, visual grain counting is as rapid as photometry in incident light, and more accurate, since at this level variation in grain size and in the reflectance of the specimen can swamp the variations in grain density. For very irregular sources it may be simpler to count visually, since the photometer's counting area can only be rectangular or circular, and it may be difficult to devise a large enough area of simple geometry to fit inside the source.

These are exclusions which affect only a very small percentage of experiments, however. Counting over nucleus and cytoplasm of many mammalian cell types is quite feasible. If a nuclear profile is about 7 μm in diameter, a counting area 6 μm in diameter placed within it can contain over 120 developed grains with Ilford L4 emulsion and still be on the linear part of the curve relating grain density to radioactivity. Given preparations with a grain density that is sufficiently high — higher by a factor of at least two than material designed for visual grain counting — photometric measurement provides reliable results at a speed far higher than is possible with visual grain counting. The apparatus is not excessively expensive, and the techniques have been worked out in some detail already. As a method of data collection from high resolution autoradiographs at the light microscope level, it has a great deal to recommend it.

MEASUREMENTS OF OPTICAL DENSITY OF THE EMULSION

We have seen (p. 205) that exposing a layer of nuclear emulsion to β radiation reduces its ability to transmit light, and the density of the layer is directly proportional to the radiation dose over a considerable range[1,8]. Since measurements of optical density require the light to travel through the emulsion, the specimen must either be completely transparent, or separated from the emulsion before measurement. For a valid estimate of radiation dose, the optical density must be measured over an area of emulsion uniformly exposed to radiation, at a magnification that does not resolve individual silver grains. In short, density measurements are best suited to measuring areas rather larger than are customary for visual or reflectance measurements.

(a) *Apparatus needed for density measurements*

Goldstein and Williams[1] have examined the Joyce-Loebl microdensitometer and the Vickers flying-spot microscope, and we have extensively used a microscope fitted with a Photovolt 520-M photometer in our laboratory. As with reflectance measurements, the basic apparatus is simple, and capable of almost infinite elaboration to meet particular needs.

Silver grains both reflect and scatter light. If the light transmitted by the film in a direction perpendicular to its surface only is collected and measured, the density value calculated from this is known as the specular density. If all the light emerging from the surface of the film, regardless of its direction relative to the film, is collected, the diffuse density may be calculated[6]. In many cases the measurement made in an autoradiograph will be a sort of hybrid between specular and diffuse: this does not affect the validity of the measurements, but it is important to use the identical conditions of microscopy if comparable results are to be obtained on some later occasion. Density measurements based on an objective of a given numerical aperture are not immediately comparable with those made with another objective of different aperture since, out of the total light transmitted diffusely by the film, the proportion collected will be different in each case.

The conditions of illumination of the emulsion are also important, and the apparent density will be different if a wide angle of illuminating beam is used rather than a narrow, finely collimated pencil of light. Density measurements also vary somewhat with the wavelength of light used.

Any apparatus for this type of work must have the following features. The light source must be stabilised and of reproducible colour. The measuring area must be evenly illuminated by a beam of reproducible geometry. The light entering the measuring system should be specular, diffuse, or some fully reproducible combination of the two. The photometer should be of appropriate sensitivity, and preferably be calibrated directly in density units.

(b) *The apparatus in use*

In any series of measurements three types of standard are advisable. The first, a completely transparent emulsion layer, is used to calibrate the zero setting on the photometer scale. The second is an emulsion fully blackened by β radiation, used to calibrate the infinite density setting. The third is a standard preparation of intermediate density for reference purposes. When the size and shape of the measuring area have been determined, usually by a diaphragm just in front of the photomultiplier tube of the photometer, these three standards are examined, and the photometer is calibrated ready for use.

It is a good idea to make the measuring area smaller than the uniformly blackened area of emulsion over the source. If the measurement includes an element of diffuse light, the region immediately around the measuring area can contribute significantly to the reading.

If the biological source has local variations in radioactivity within the chosen measuring area, these can often be averaged out in terms of blackening of the overlying emulsion by choosing an autoradiographic technique of low resolution. If, for instance, one is interested in the rate of protein synthesis within a given part of the brain in animals treated with various drugs, it is clear that there will be differences within that area between neurones and glia, between cell bodies and fibres, between nucleus and cytoplasm. It is possible to obtain uniform blackening of the emulsion with sections labelled with carbon-14 autoradiographed against X-ray film, in spite of these local variations, and this type of material will favour rapid measurement and give a valid integrated reading for a fairly large area, which may be very difficult to extract from a high resolution autoradiograph with tritium.

Optical density measurements are quick and simple to make, and the results are proportional to radiation dose up to higher limits than with grain density or reflectance measurements[1]. They are best used with fairly large measuring areas at least 50 μm in diameter. They are applicable to whole-body autoradiographs, for instance, or to any other macroscopic specimen. A small photometer suitable for whole-body autoradiographs has been recently described by Cross, Groves and Hesselbo[25]. At higher resolution, such measurements have been used in the study of aminoacid incorporation in the brain in unstained sections viewed by phase contrast microscopy[26], and in the study of sugar absorption by the intestine[27]. Like photometric measurements of reflectance, optical density measurements provide a rapid and accurate means of collecting date from autoradiographs, in suitably prepared material[28].

COMPUTER-BASED COLLECTION OF DATA

In recent years, a number of papers have described attempts to extract data from autoradiographs by some computer-directed means. These machines are the lineal descendants of those produced long ago by Tolles[29] and by Gullberg[14]. They are usually designed for light microscopic autoradiographs, and they usually count silver grains in transmitted light. Most are based on television scanning systems, though "Chloe" is a flying spot machine[30], capable of constructing iso-density pictures of black-and-white images.

Such automatic grain-counters have made great progress. They remain very expensive in comparison to the semi-automatic apparatus needed for incident-light photometry or measurements of optical density. They require a very sizeable commitment of time and effort in order to overcome some of the difficulties inherent in the collection of data from autoradiographs. Now that very sophisticated instruments, such as the Quantimet 720 produced by Metals Research Ltd. of Cambridge and the TAS produced by E. Leitz of Wetzlar, are available commercially, it is likely that the next few years will see greatly increased interest in the full automation of grain counting.

One major problem is that silver grains do not all occur in the same focal plane in most autoradiographs prepared for light microscopy. If paraffin wax sections are examined, the upper profile of the section is very uneven, so that the lowest layer of silver grains is already at several levels. The situation is better with smears of flattened cells, and better still with semi-thin sections of Epon-embedded material; but most autoradiographs for light microscopy are prepared with emulsion layers 2–4 μm thick during exposure, and even with a completely flat upper profile to the section grains will be found at different levels.

A television camera effectively scans an image in a number of parallel lines. In many instruments the scan line is broken up into a series of individual points, separated by small gaps in the line. Each picture point generates an electrical signal, whose size is proportional to the light intensity at that point. The whole picture is then described by a train of electrical signals, generated in a regular sequence. This train of impulses can then be processed electronically in various ways.

Such a system "sees" a silver grain in transmitted light as an area of low light intensity, with a clearly defined margin, and is capable of recognising it from structures in the underlying specimen by such parameters as its blackness and its size and shape. If the silver grain is slightly out of focus, it will appear to the television scanning system as a larger area that is not as dark as expected, having a more gradual transition from light to dark. In consequence, it may not be recognised as a grain.

Most of the current literature describes attempts to make instruments that can analyse existing autoradiographs. I feel a lot more effort should go into designing autoradiographs that are suitable for machines. It is relatively simple to produce a specimen with a reasonably flat upper profile by embedding in Epon or Araldite. Methods exist for making packed monolayers of liquid emulsions. In combination, the flat specimen and the monolayer can give a preparation with all the grains in focus at once.

The difficulties of recognising silver grains against the varying densities of a stained section are often considerable. Here, the use of incident light to view the preparation converts the grains into bright specks on a black background, and the specimen itself becomes invisible. Scans of the same area by transmitted light, to record the stained specimen, and incident light for the silver grains, can provide the basis for much clearer recognition of each component of the autoradiograph. With semi-thin araldite sections, thin emulsion layers and viewing in incident light, I have found it relatively simple to count silver grains accurately and very fast with the Quantimet 720. This, of course, is only the first step. The recognition of patterns of density in the specimen and the counting of grains in areas of emulsion related to particular cells or structures are further problems to be solved before fully automated grain counting has really arrived.

To review the attempts that have been reported so far, Prensky[31] gave an account of grain counting with the earlier Quantimet B. He worked with lightly stained material viewed by transmitted light, which meant that the grains were seen against backgrounds of varying density, making it impossible to set the recognition threshold of the Quantimet to one level for all fields. He developed an instantaneous feed-back system through a digital display which told him the grain count the machine was making, and then adjusted the recognition threshold to give a "reasonable" count. This interaction with the machine introduced a rather subjective element. It was nevertheless possible to train an operator to work fast and accurately with this system. As well as difficulties with grains not fully in focus, he drew attention to the problems of grains whose images touch. At high grain densities, the grain counts can actually fall as the probability of adjacent grains touching increases. As with visual counting, the definition of a silver grain is crucial (p. 206).

Mertz[32] also used a Quantimet, and found that particle size analysis could be employed very successfully to distinguish between background grains and the rather larger grains produced by tritium in his autoradiographs. (For a brief account of why tritium grains should be, on average, bigger, see p. 296.)

Lipkin, Lemkin and Carman[33] based their work on the Quantimet 720, interfaced to a PDP8e computer and fitted with a special control unit. Working with autoradiographs of stained blood smears, they first counted the silver grains with a colour filter in the transmitted light path to minimize the influence of the stained specimen. Scanning the emulsion at successive optical levels, they recorded the position of every grain. After a change of filters the field of view was then scanned a second time, and the positions of cell nuclei identified. The computer then checked the grain positions against the nuclear data, assigning each grain to nucleus or non-nucleus. By pressing one of a series of 16 switches, the operator could allocate each nucleus to one of 16 cell types, together with its grain count.

The Quantimet itself has progressed significantly since their machine was described, and it is likely that much of the recognition of cell types could now be satisfactorily done by the machine itself, using pattern analysis on the shapes and sizes of the nuclei and the presence or absence of associated granules in the cell.

This two-stage type of analysis is promising, counting the grains that are associated with a particular pattern in a specimen that is complex.

Wann, Price, Cowan and Agulnek[34] have described their own machine, based on a Plumbicon television camera and a PDP-12 computer. This has been extensively tested with paraffin wax sections of brain autoradiographed by dipping in NTB-2 emulsion. The field of view is observed in transmitted light with a colour filter to minimize the influence of the stained section on grain counts. Grain counts are repeated at two different focal levels for each field to try to cope with the problems of grains not in focus. The operator can interact with the machine, on the basis of viewing the section without the colour filter, to instruct it about the positions of various structures within the area scanned, and the computer then allocates the grains it counts between these structures on the basis of their position. At low grain densities the machine is slower than visual grain counting, largely due to the difficulties posed by variation in focal level, and also tends to overestimate the number of grains. At very high grain densities, the machine has difficulty separating closely spaced grains, and it underestimates the numbers present. Over a range of 30–3000 grains per 1000 square microns, however, its accuracy compares very well with visual grain counting and its speed is significantly higher.

Bisignani and Greenhouse[35,36] have described a device called MITRE, which represents rather a different approach. Instead of building up a complete picture of the field of view from a series of very closely spaced scan lines, they use scans 0.5 μm wide, separated by 10 or 20 μm, to sample the field. On the basis of the densities in transmitted light, the machine recognises nucleus, cytoplasm and silver grain. From a programme based on the analysis of known standard preparations, the machine is able to compute the sizes and number of nuclei and the number of silver grains within and outside the nuclear profiles, for instance, from these sampling scans. MITRE represents an attempt to get away from the very expensive hardware of the television scanning systems by the intelligent use of the computer facilities that are so widely available nowadays.

In summary, then, for those with an interest in electronic systems, there is enough work in progress to suggest that many tasks of data collection can be automated. Not enough work has been done yet on producing autoradiographs suitable for automated analysis, or on the best optical systems for giving the machine an unmistakeable signal from the silver grains. This approach is expensive in time as well as money. It does hold out the tempting possibility of fully automated data collection from autoradiographs.

The semi-automatic systems described earlier, particularly those employing photometry in incident light, are, in contrast, well worked out, relatively cheap and undoubtedly quicker than visual counting without requiring a large investment in time and effort to set the system up. They all, however, require the operator to select fields and make decisions, and are only capable of measuring the emulsion response within the selected field.

REFERENCES

1 D.J. Goldstein and M.A. Williams, *J. Microscopy*, 94 (1971) 215.
2 P. Dörmer, in U. Lüttge (Ed.), *Microautoradiography and Electron Probe Analysis*, Springer-Verlag, Berlin, 1972.
3 P. Dörmer, *Histochemie*, 8 (1967) 1.
4 A.W. Rogers, *Exptl. Cell Res.*, 24 (1961) 228.
5 A.W. Rogers, *J. Microscopy*, 96 (1972) 141.
6 G.C. Farnell, in C.E.K. Mees and T.H. James (Eds.), *The Theory of the Photographic Process*, 3rd ed., Macmillan, New York, 1966.
7 F. Hurter and V.C. Driffield, *J. Soc. Chem. Ind. (London)*, 9 (1890) 455.
8 J.F. Hamilton, in C.E.K. Mees and T.H. James (Eds.), *The Theory of the Photographic Process*, 3rd ed., Macmillan, New York, 1966.
9 G.W.W. Stevens and P. Block, *J. Phot. Sci.*, 7 (1959) 111.
10 G.M. Hodges and M.D. Muir, *J. Microscopy*, 104 (1975) 173.
11 J. Micou and L. Goldstein, *Stain Technol.*, 34 (1959) 347.
12 K. Ostrowski and W. Sawicki, *Exptl. Cell Res.*, 24 (1961) 625.
13 H. Levi, A.W. Rogers, M.W. Bentzon and A. Nielson, *Kgl. Danske Videnskab. Selskab. Mat.-Fys. Medd.*, 33 (1963) No. 11.
14 J.E. Gullberg, *Exptl. Cell Res.*, Suppl. 4 (1957) 222.
15 J.E. Gullberg, *Lab Invest.*, 8 (1959) 94.
16 P. Dörmer, W. Brinkmann, A. Stieber and W. Stich, *Klin. Wochenschr.*, 44 (1966) 477.
17 P.P. Dendy, *Phys. Biol. Med.*, 5 (1960) 131.
18 D.J. Goldstein and M.A. Williams, *Histochem. J.*, 6 (1974) 223.
19 D.T. Clarkson and J. Sanderson, *Proc. Roy. Microscop. Soc.*, 6 (1971) 136.
20 J. Bisconte, J. Fulerand and R. Marty, *C.R. Soc. Biol.*, 162 (1968) 2178.
21 J.M. England and A.W. Rogers, *J. Microscopy*, 92 (1970) 159.

228

22 P.N. John and A.W. Rogers, *J. Endocr.*, 53 (1972) 375.

23 J.E.N. Neely and J.W. Combs, *J. Histochem. Cytochem.*, 24 (1976) 1057.

24 P. Dörmer and W. Brinkmann, *Histochemie*, 29 (1972) 248.

25 S.A.M. Cross, A. D. Groves and T. Hesselbo, *Int. J. Appl. Radiation Isotopes*, 25 (1974) 381.

26 J. Altman, *J. Histochem. Cytochem.*, 11 (1963) 741.

27 W.B. Kinter and T.H. Wilson, *J. Cell Biol.*, 25 (1965) 19.

28 H.A. Fischer and G. Werner, *Histochemie*, 15 (1968) 84.

29 W.E. Tolles, *Lab. Invest.*, 8 (1959) 99.

30 E. Lloyd, J.H. Marshall, J.W. Butler and R.E. Rowland, *Nature*, 211 (1966) 661.

31 W. Prensky, *Exptl. Cell Res.*, 68 (1971) 388.

32 M. Mertz, *Histochemie*, 17 (1969) 128.

33 L.E. Lipkin, P. Lemkin and G. Carman, *J Histochem. Cytochem.*, 22 (1974) 755.

34 D.F. Wann, J.L. Price, W.M. Cowan and M.A. Agulnek, *Brain Res.*, 81 (1974) 31.

35 W.T. Bisignani and S.C. Greenhouse, *J. Theoret. Biol.*, 54 (1975) 121.

36 W.T. Bisignani and S.C. Greenhouse, *J. Histochem. Cytochem.*, 24 (1976) 152.

CHAPTER 11

The Analysis of Autoradiographs

The final step in any autoradiographic experiment is the analysis of the response of the emulsion to exposure to the specimen. Since autoradiography may be used in such a wide spectrum of situations, there is no standardised method of analysis. Instead, it is up to the scientist to test his hypotheses in the manner that seems best suited to the particular experiment. Having said that, certain types of analysis have evolved for particular broad categories of experiment, and familiarity with these may help to suggest a suitable approach to what is bound to remain a solution to an individual problem — one which tests the grasp of autoradiographic principles, often severely.

The analysis of an autoradiograph always involves measurement. The method may be crude, and only able to detect gross differences in radioactivity, such as those found between thyroid follicles and skeletal muscle in animals given radioactive iodide, which can be detected on visual examination of a photomicrograph. At the other extreme, the patient application of sophisticated analytical methods to electron microscopic autoradiographs may identify the labelled organelles and measure their specific radioactivities in situations in which visual examination alone tells one nothing. The more carefully controlled the techniques and the greater the investment in the time and effort of analysis, the more sensitive and precise the final measurement.

No attempt will be made here to recapitulate the ground covered in earlier chapters on the factors which govern the efficiency, resolution and background of autoradiographs. Instead, this chapter will discuss how to select which areas of emulsion to examine in particular categories of autoradiograph, and how to handle the resulting data statistically.

Any series of measurements has several characteristics by which it can be described. The mean is clearly such a descriptor. In comparing two sets of measurements, the mean alone is practically useless without some statement of the scatter of the individual measurements about the group mean.

This scatter has two main origins, the biological material itself and the processes of recording its radioactivity as a response of the nuclear emulsion. Let us start by examining the variables in the autoradiographic process itself which can contribute to this scatter, to see if they can be minimised by the way in which data is collected and analysed. Any reduction in the scatter of measurements about the mean value immediately improves the precision with which mean differences between exprimental groups can be identified.

FACTORS IN THE EMULSION CONTRIBUTING TO VARIABILITY

Many of these factors are discussed in Chapter 5. Most of them will be controlled by the decisions on experimental design taken before preparation of the autoradiographs. The size, packing and sensitivity of the crystals in the emulsion are in general out of the hands of the experimenter, who relies on the supplier for a reasonably standardised product.

Emulsion thickness will often be satisfactorily controlled, either by the use of X-ray film or stripping film, or by some physical parameter like the maximum range of tritium β particles in nuclear emulsion. Wherever liquid emulsions are applied to the specimen by the autoradiographer, however, the probability is high that the emulsion layer will vary in thickness from one place to another, with consequent variability in emulsion response.

The conditions of drying and exposure of the emulsion can critically affect its performance: an incompletely dry emulsion exposed in the presence of oxidising agents will have a very low efficiency. Even within one batch of slides prepared for light microscopic autoradiography, variations in drying can easily occur, with the slides furthest from the fan, or those in an exposure box with less drying agent, showing significantly lower efficiencies. On the same slide, areas where the emulsion is thicker than usual may fail to dry adequately.

Development is another cause of variability. Without a specially designed developing tank (see p. 219), it is difficult to get reproducible conditions of development, without which direct comparisons between autoradiographs developed at different times are suspect. Even in autoradiographs developed in the same solutions, I have seen gross differences in efficiency resulting apparently from putting too many slides through a limited volume of developer.

Finally, chemography can introduce very significant variations in emulsion response into an experiment (Figs. 34 and 40).

Now one would hope that variability in the final response of the emulsion from these causes had been minimised in the technical steps of preparing the autoradiograph, long before the stages of collecting and analysing data. Even with reasonably careful techniques, they will still contribute something to the variability of the final emulsion response, however. If such minimal contributions are randomly distributed between the experimental groups, the final effect on the analysis is likely to be considerably less than variability due to the biological specimens them- selves. If, however, the emulsion is always thicker over the section from the control animal, which has always been placed nearer the bottom of the slide, or the experimental slides are always developed after the control ones, these sources of variability can introduce a bias into the data which will only become more significant the more measurements are made and the better the statistical analysis. Randomising the slides during prepara- tion and exposure is one answer, in order to distribute the many causes of variability in emulsion response as fairly as possible between the experi- mental groups.

As far as the analysis of the autoradiographs is concerned, it is clearly important to base this on a reasonably large number of preparations. It is likely to be better to count from 100 cells on each of 10 slides in each group than to count from 1000 cells all on the same slide. The allocation of counting effort between animals, slides, sections and counting areas will be discussed again later (p. 258).

One step in the preparation of the autoradiographs simplifies the control of emulsion variables enormously. This is the exposure of areas of emulsion to radiation from a standard source. The response of the emulsion to this source can be used as a test of whether to accept or reject a group of preparations as technically reliable: it can also serve as a calibration for all the data taken from that group of preparations. Some authors describe exposing a section of labelled methacrylate on every slide[1,2]: some use an external source of radiation over part of the slide: some even introduce labelled material into the blocks from which sections are cut[3]. In a sense, the basic control against negative chemography (p. 302), the slide exposed to a uniform flash of light or of other radiation, is a rather crude calibration. As autoradiography is increasingly used for more and more sophisticated measurements of radioactivity, the need for accurate reference standards becomes more pressing.

Finally, all the causes of variability due to the emulsion and its handling, with the possible exception of chemography, are likely to be insignificant in volumes of emulsion that are very close together. Differences due to emulsion thickness, drying or development may be serious from one slide to another, but are minimal in counting areas only a few microns apart. If an experiment can be designed around paired observations made very close to each other, the analytical design will effectively eliminate variability from these causes[4]. The paired t-test is discussed on p. 256, and is a powerful tool in autoradiographic analysis.

FACTORS IN THE SPECIMEN CONTRIBUTING TO VARIABILITY

Separation of the radioactive source from the emulsion can significantly reduce efficiency (p. 89). The reduction is related to the thickness and the density of the separating layer. The most common situation in which this geometrical variable affects efficiency is when radioactive sources lie at different depths within the specimen. If the sources are randomly scattered at all levels through the specimen, grain counts over them may vary all the way from a maximum down to zero. Modak, Lever, Therwath and Uppulri[5] discuss this situation with cell nuclei labelled with tritiated thymidine, and present graphs indicating the percentage of nuclei within range of the emulsion at various nuclear diameters and section thicknesses. With smears of whole cells, the nuclei and, in particular, nucleoli will always be separated from the emulsion by layers that will vary from one cell to the next: this situation has been examined in detail by Perry, Errera, Hell and Durwald[6] and by Maurer and Primbsch[7]. Reducing this source of variability is largely a matter of technique in preparing the autoradiograph: working with thin, plastic-embedded sections, for instance, will largely remove the intervening layers.

In much the same way, variations in specimen thickness and density can affect afficiency (pp. 89–93). The difficulties of cutting sections of reproducible thickness have been discussed on p. 139. It may be possible to work at infinite thickness as the self-absorption curve (Fig. 31), or to measure the thickness of plastic-embedded sections before covering them with emulsion, accepting only those that lie within a specified range[8]. In many experiments, however, variation in section thickness will unavoidably contribute to the variability of measurements. Variations in density within the specimen can occur, for example, in whole-body autoradiography, where radioactivity in bone may be recorded at significantly lower efficiencies than in soft tissue; in light microscope studies on erythropoiesis, the accumulation of iron in the cytoplasm increases the cell density considerably, affecting the efficiency of measurement: at the electron microscope level, staining with heavy metals prior to autoradiography can influence the efficiency[9-11].

At the stage of analysing the completed autoradiograph there is not a lot one can do to improve the situation. Variations in section thickness can be reduced in significance by the use of paired observations taken from immediately adjacent areas of section. It may be possible to restrict data collection to sources that lie at the upper level of the section where variation in depth within the section is causing difficulties. If comparisons of radioactivity are needed between sources that differ in density, or which are separated from the emulsion by layers that are consistently different in thickness or density, there is often no alternative to using a correction factor, which will usually require to be independently determined.

Two further factors require some attention. If a group of sources containing similar amounts of radioactive material vary greatly in size, grain or track counts from a standard area of emulsion over them will also vary considerably. Similarly, if the sources vary widely in shape, it will be very difficult to select a counting area that records with similar efficiency from each of them. These two situations are so important to the analysis of autoradiographs that they will be considered in rather more detail.

(a) Differences in the sizes of sources

Let us look first at sources which differ in size, in which the disintegration rate per unit area is the same. An isolated circular source whose diameter is small relative to the range of the β particles emitted will produce a distribution of silver grains in the overlying emulsion (Fig. 66), many of the grains lying outside the edges of the source itself. If the number of silver grains immediately above the source is counted, the volume of emulsion scanned has a given efficiency with which it records β particles from the

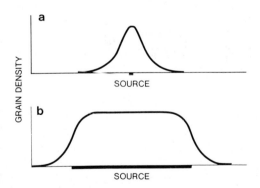

Fig. 66. Diagrams illustrating the grain densities seen on a line passing through the centres of two sources of different sizes and similar radioactivity per unit area: (a), A small source, giving a bell-shaped distribution of grain densities; (b), An extended source showing higher grain densities over the central part of the source caused by cross-fire effects.

source. It also records β particles originating from outside the source, though with lower efficiency, and the further away from the source that these β particles start their tracks, the lower the efficiency with which this particular volume of emulsion responds. If, therefore, one scans a similar volume of emulsion from the centre of a circular source of much greater diameter, but with the same concentration of isotope (Fig. 66b), the central part of that source, of diameter d, will give rise to exactly the same number of grains counted as the first source, whose total diameter was d. But, in addition, the rim of source around this central portion will also contribute grains to the volume of emulsion examined. In other words, the observed grain density will be higher over the larger source, even though both contain the same radioactivity per unit area. As the diameter of the source is further increased, the efficiency with which the central volume of emulsion records particles from the periphery of the source decreases, until, at their maximum range, it is zero. Above a certain source diameter, therefore, the small centrally placed volume of emulsion gives a record of the concentration of radioactivity that is independent of the size of the source. This critical diameter is governed mainly by the characteristics of the β particles emitted: the lower their energy, the smaller the diameter above which variations in the size of the source have no effect on the grain count over the central region of the source.

For the peripheral parts of a flat, circular source to have no effect at all on the emulsion directly over its centre, the diameter of the source should exceed that of the central measuring area by twice the track length of the highest energy particle emitted. Several factors combine to make this rather a counsel of perfection. Most isotopes emitting β particles have a relatively small percentage of particles at the high end of the energy spectrum (see Fig. 10). The track radius, or penetration, is always shorter than the point-to-point track length. The solid angle subtended to the central volume of emulsion by the periphery of the source is relatively small. In short, although the efficiency with which the central volume of emulsion records particles from the periphery does not fall to zero until the difference in diameters is twice the greatest track length, it approaches very close to zero at much smaller diameters. As a rough guide, the edges of the central volume of emulsion used for counting should be separated from the perimeter of the source by at least 2 HR (defined as the radius around a radioactive point source which contains half the silver grains produced by it). In these circumstances, grain counts over a series of sources should reflect their concentration of radioactivity, even if they differ in size.

These difficulties are illustrated in Fig. 67, in which Salpeter has calculated the percentage of the total grains produced in the emulsion that directly overlie sources of different radii. In the same table, it can be seen that

Radius of source (in HD units)	% grains over source	% grains over and within 1 HD	Source diameter for HD = 1600 Å
16	90	95	6 μm
4	60	70	1.5 μm
1	25	55	3500 Å
0.25	2.5	35	800 Å

Fig. 67. For solid radioactive discs of different diameters, the percentage of the silver grains produced that lie directly over the source, and that lie over and within a circle of 1 HD width outside the source, have been calculated. The source diameters in the last column are the measurements that would correspond in an electron microscope autoradiograph of HD 1600 Å. (Data provided by Dr. M.M. Salpeter)

increasing the area of emulsion scanned to include a rim 1 HD wide around the edges of the source, while improving the percentage of grains counted over the smallest source by a factor of $\times$ 14, still leaves a big difference between the percentage of grains counted over the smallest and largest sources.

If the total grains per source are to be used as a basis for comparing the radioactivities of sources of differing diameter, the grains should be counted in an area of emulsion overlying each source and extending for at least 2 HD on every side: if the size differences in the sources are very great, it may be necessary to extend this area of emulsion even further, to as much as 5 HD all round the source. The grain distributions around model sources from which this type of calculation can be derived are presented by Salpeter, Bachmann and Salpeter[12].

If grain densities (grains per unit area of emulsion) are to be the basis for comparing the radioactivities of sources which differ in diameter, the grain density in a central part of the source, separated by 2 HR from the edge of the source on all sides, can be found. The grain density from over the whole source is not a very good indication of the total radioactivity, as can be seen from Fig. 68. Here, for sources corresponding in size to those in Fig. 67, the densities were obtained by counting all the grains over the source and dividing by the source area. Here also are the densities obtained by counting all the grains within the emulsion over the source plus a rim of 1 HD around it, divided by the area of emulsion scanned. Although the percentage of all the grains emitted by the smallest source that is counted has been considerably increased by scanning the emulsion around as well as over the source, as seen in Fig. 67, the density of grains calculated in this way is in fact smaller, since the area of emulsion scanned has increased by

Radius of source (in HD units)	Grains over source/ source area	Grains within 1 HD/area within 1 HD	Grains within 1 HD/ source area	Source diameter for HD = 1600 Å
16	1.0	1.0	1.0	6 μm
4	0.7	0.5	0.7	1.5 μm
1	0.3	0.15	0.6	3500 Å
0.25	0.03	0.015	0.4	800 Å

Fig. 68. The relative grain densities over solid radioactive discs of different diameters are here presented, calculated in three ways: (a) grains over the source divided by area of source; (b) grains over and within 1 HD of the source, divided by the area lying within 1 HD of the source; (c) grains over and within 1 HD of the source, divided by the area of the source itself. (Data provided by Dr. M.M. Salpeter)

an even larger factor. A better "density" figure can be obtained by counting the grains in the larger area of emulsion, but relating this to the actual area of the source itself (Fig. 68).

Dörmer[13] has shown that the percentage of grains, from carbon-14 incorporated in the nucleus, found over the nuclear profile itself varies from less than 30% to over 60% with nuclei in the range 7–17 μm (Fig. 69). In his work he used a correction factor based on these observations in his analysis.

In summary, there are two approaches to the problem of sources that differ significantly in size. If the sources are large relative to the HD value of the autoradiograph, the emulsion response over central areas at least 2

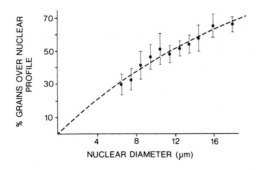

Fig. 69. The percentage of the observed grains lying directly over the nucleus in cells labelled with [14C]thymidine, plotted against nuclear diameter: autoradiograph prepared with smears of whole, isolated cells and AR-10 stripping-film. (Data from Dörmer, 1972)

HD within the edge of each source can be used as a basis for comparison. Alternatively, the emulsion over the source and outside it for a distance of at least 2 HD can be scanned if the sources are small relative to the HD. In either case, the observed grain counts can be related to the area of the source, if required. Note that the crucial measurement in each case is the HD for the autoradiographic system being used. It may be possible to simplify these problems of analysis by manipulating the HD values. By careful selection of the isotope and the conditions of autoradiography, an HD may be obtainable which minimises the effects of variation of source size on the emulsion response.

Both these procedures assume that the source of radioactivity is known, and that there is no appreciable contribution to the grain counts within the selected areas from other sources nearby.

(b) *Differences in the shapes of sources*

This is really an extension of the situation discussed in the previous section. If one compares a disc to an extended, linear source, keeping their total content of radioactivity, their surface area, and their isotope concentration equal, it is very difficult to select two equal volumes of emulsion that would contain the same number of grains. The grain density in the emulsion directly over the linear source will be lower than it is over the radioactive disc.

Once again, if the sources are large relative to the resolving distance, counts from a small centrally placed volume of emulsion as far as possible from the edges of the sources will give a measure of the concentration of radioactivity within them. If the sources are small relative to the resolving distance, it may be possible to ignore the differences in shape, and treat each as a point source.

In dealing with irregular sources, it may be possible to select volumes of emulsion over each source which are sufficiently similar geometrically to permit the assumption that the efficiency of measurement in each case is the same.

THE ANALYSIS OF AUTORADIOGRAPHS OF MACROSCOPIC SPECIMENS

These are usually free from many of the uncertainties that afflict light and electron microscopic specimens. The emulsion response is calibrated by the use of a radioactive ladder exposed with the specimen (Fig. 23). In the case of whole-body autoradiographs, a further reference is provided by the blood. It is usually possible to measure the emulsion response over blood in the ventricles of the heart, for instance, and to refer other activities to this; it is quite simple to withdraw a small volume of blood immediately before killing the animal for liquid scintillation counting.

Two difficulties do occur, however. The first is the very small source within the specimen, such as the contents of the biliary duct system. These tiny hot spots may provide blackening in the emulsion which is very difficult to quantitate. The blackening occupies a larger area of emulsion than the structure responsible for it, and the blackening falls off relatively sharply with increasing distance. Directly over the hot spot itself, emulsion saturation may even occur. The size of the area of emulsion from which a measurement is taken critically determines the observed degree of blackening. The situation really is very like that shown in Figs. 66a and b. It is usually preferable to give only a preliminary estimate of the radioactivity in such very small structures from the autoradiograph and to rely on the scintillation counting of small samples dissected out of the section for an accurate measurement.

The second source of difficulty in interpreting whole-body autoradiographs comes from the greater density of bone compared to blood and soft tissues. Increased self-absorption in heavily calcified tissue may result in significantly reduced blackening. Either a correction factor should be calculated for the isotope and the section thickness used, or reliance placed again on the final arbiter — scintillation counting of small samples punched out from the section.

THE ANALYSIS OF LIGHT MICROSCOPIC AUTORADIOGRAPHS OF WIDELY SEPARATED SOURCES

This situation is frequently encountered where smears are made from cell suspensions, or where sections are taken from tissues labelled with tritiated thymidine. In either case, the distances separating the labelled sources can be large relative to the distribution of silver grains around each one. Two main types of experiment will be considered here: the determination of the percentage of cells or nuclei that are radioactive, and the comparison of the degree of radioactivity present in two groups of sources.

The incorporation of tritiated thymidine into newly synthesised DNA occurs only during the S-phase of the cell cycle, giving a labelled sub-population. In very many experimental situations, the exact size of this sub-population is the basis for calculations of such parameters as the length of the cell cycle. It might seem a very simple decision to make, and indeed it can be in autoradiographs with very low background and high grain counts over labelled cells. A tradition has grown up in this sort of experiment that a threshold grain count is chosen by the scientist below which all the cells are non-radioactive, and above which they are all labelled. "Four or more grains per nucleus" is often chosen.

Now, background grain counts in areas of emulsion the approximate size of cell nuclei should be one or less. However, such counts are usually found to have a Poisson distribution[14] (see p. 252), and it is quite possible to find occasional volumes of emulsion with counts as high as 10 grains where the mean level is one grain. As discussed on p. 126, this variability in background grain counts can lead to "false positives" when a rigid threshold definition of labelling is used. Similarly, a uniformly labelled model source, such as a section of tritiated methacrylate, will have a Poisson distribution of grain counts about the mean value. Biological sources quite often have a wider distribution than Poisson, suggesting that it may not be reasonable always to assume uniform labelling[15-17]. Even with a Poisson distribution, the probability of finding sources with less than 4 grains over them when the mean grain count is 15 is finite, resulting in "false negatives".

Several papers have dealt with this problem[18-21]. Shackney[22] demonstrates clearly the errors that can creep into measurements of cell kinetics from the accumulation of false positives and negatives. Ideally one requires two distribution curves: that of background counts and that of counts of labelling plus background from over the population of sources suspected of including labelled members. It is then possible to calculate the percentage that are actually radioactive. The treatment given on p. 257 is based on the paper of England, Rogers and Miller[23].

The errors in interpretation that result from accepting a rigid threshold to define labelling vary considerably with the experimental design. They are greatest in the sort of experiment that involves the injection of labelled cells into an animal and their subsequent identification. Here, very large numbers of cells may be scanned by the microscopist in his search for labelled cells, and the improbable background counts of 4 or more grains have a greater chance of influencing the interpretation of results.

The comparison of the radioactivities of two populations of sources is relatively simple if they are closely similar in size, shape and density. The ideal situation involves comparison of the same sorts of source after different treatments provided, of course, the treatment has not induced a change in these three parameters. A volume of emulsion can be selected over each cell as a basis for measurement, and it may be assumed that the efficiency will remain constant between the experimental groups. Provided the most heavily labelled group is still below the level when double hits assume statistical significance (p. 97), the grain counts will directly reflect the radioactivities. If the difference between the most heavily labelled group and the most lightly labelled one is very great, it may be necessary to accept very high grain densities in the former and correct these for double hits.

The problems that arise when the groups of sources to be compared differ in size or shape or density have been discussed already (p. 232).

THE ANALYSIS OF LIGHT MICROSCOPIC AUTORADIOGRAPHS OF SOURCES WHOSE EFFECTS ON THE EMULSION OVERLAP

We now progress from sources which are separated from each other by a distance that is large relative to the HD of the autoradiograph to the next level of complexity. Here, the sources are close or maybe contiguous, and the emulsion over one source is registering radiation not only from the source itself, but also from surrounding sources. Let us assume that the emulsion response is adequately controlled, and that variations in the thickness, density, size and shape of the sources to be compared have been remembered and corrected for. How can one select suitable volumes of emulsion for analysis when there are crossfire effects between sources?

This type of problem is illustrated diagrammatically in Fig. 70.

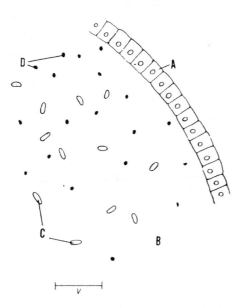

Fig. 70. Diagram illustrating crossfire problems in a biological specimen. (A), An epithelial surface, assumed to be heavily labelled. (B), Unlabelled connective tissue. C and D are nuclei of two distinct cell types, randomly distributed relative to A, whose radioactivity is to be compared. v is the resolution of the autoradiograph.

The areas A are known to be heavily labelled; those marked B are presumed to be inactive. The experiment is designed to determine the relative radioactivities of the small structures C and D. The HD is indicated for the autoradiographic system being used. It is clear that the volumes of emulsion overlying the Cs and Ds near to the areas A will receive very significant irradiation from their heavily labelled neighbours. The grain counts in them will reflect their distance from the nearest A far more accurately than they indicate the concentration of isotope in the underlying structures.

The simplest type of crossfire problem occurs when the two series of sources, C and D, are randomly scattered throughout the tissue section: in other words, both of them may be found at any distance from the highly radioactive A. In this case, it is possible to select volumes of emulsion to count on the basis of distance from A. For example, it may be decided to count grains over all the Cs and Ds that lie between 1 and 1.5 HD from the nearest A, and to count grains over a similar number of areas of the non-radioactive B. The B counts then represent the background in this position in the section, and the ratio C–B/D–B should reflect the relative radioactivities of C and D.

But there are many experiments in which this assumption of the random distribution of C and D relative to the highly radioactive A does not hold. In Fig. 71, for example, A is the content of the follicle, C is the epithelium around the follicle, D is an epithelium some distance away from the follicles, and the rest of the tissue is presumed to be unlabelled. The appropriate grain count for D can be fairly simply found, by counting over specified areas of D, and subtracting the counts over predetermined areas of inactive tissue far enough from A and D to give a reasonable control value, as at B. The epithelium of the follicle (C) represents a real difficulty. Fortunately, it is the distance from the nearest A on the tissue section that determines the crossfire effect in the autoradiograph, not the distance in vivo. While a section that passes through the epithelium at C at right angles to it will produce a single layer of cells around the follicle, a tangential section may give a much wider rim of epithelium several cells thick. It may be possible, therefore, to select areas of C in tangential sections that lie many microns from A, and to find areas of inactive tissue a similar distance from A in follicles sectioned perpendicularly to the epithelium (as at B_2). In this case, $C-B_2/D-B$ will give the relative activities of C and D. It may even be possible to identify small areas of C cut so fortunately that none of the radioactive follicular contents (A) appears in that particular part of the section (as at C_2).

There are no hard and fast rules for treating these experiments in which crossfire may play a significant role. The central principle is the selection of

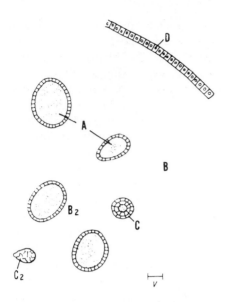

Fig. 71. Diagram illustrating crossfire problems in a biological specimen. D is an epithelial surface, below which a number of follicles lie in connective tissue. The content of the follicles (A) is highly radioactive: the connective tissue is assumed to be unlabelled (B). The radioactivity of the follicular cells (C) is to be compared with that in the epithelial cells (D). C_2 represents a tangential section through follicular cells with no follicular contents visible. v is the resolution of the autoradiograph.

areas of emulsion for analysis which are balanced with respect to the radiation they have received from surrounding structures. If it is not possible to find a large number of areas which are similar in the radiation they have received, one may have to use a number of paired observations in which crossfire is similar in the two areas that are paired, though different from one pair to another[4].

Obviously, any alteration in the conditions of autoradiography that makes the resolution better (the HD smaller) will help to reduce the severity of crossfire effects.

In complex situations where the HD is large by comparison with the size of structures resolved in the specimen, it may be necessary to go to modifications of the various techniques that have been proposed for use at the electron microscope level.

THE ANALYSIS OF ELECTRON MICROSCOPIC
AUTORADIOGRAPHS OF WIDELY SEPARATED SOURCES

The starting point for this type of analysis is the curve of distribution of silver grains around a linear source of radioactivity (Fig. 21). As is explained on p. 66, the *shape* of this curve is the same for autoradiographs produced under widely varying experimental conditions, though the HD may vary in absolute terms[12]. It should be noted, incidentally, that the shape of the distribution curve does change if one switches from one radioisotope to another: the curves for iodine-125, tritium and carbon-14 are significantly different[24]. For any one isotope, however, such a curve calibrated in units of HD is universally applicable, whatever the detailed method of preparing the autoradiograph.

Given this universal curve for the distribution of silver grains around a linear source, it is possible to calculate the expected distribution of grains around a point source, or a disc of known radius, or a radioactive band, or any one of a number of simple geometrical shapes, the distribution always measured in units of the HD of the autoradiographic system in use. Salpeter, Bachmann and Salpeter[12] have presented distribution curves for hollow and solid circular and band sources of various sizes, some of which are shown in Fig. 22.

If one is using a combination of autoradiographic variables for which the HD value has already been found (Fig. 29), one can immediately translate this family of distributions into distances in Å, and a whole series of predicted grain distributions around sources of different sizes and shapes becomes available. If some other combination of variables is in use, it will be necessary to determine experimentally the HD value appropriate to the autoradiographs.

The impact of this data on the analysis of electron microscope autoradiographs is difficult to overestimate. It is possible to compare observed with predicted grain distributions around particular structures to confirm whether or not they are the only source of radioactivity in the specimen. It is possible to attribute to a labelled structure silver grains lying at a considerable distance from it with confidence, if the observed and predicted distributions match. It may be possible to differentiate between a solid structure which is uniformly labelled and a similar structure labelled only at its rim by comparing the grain distribution observed with those predicted for solid or hollow circles. The predicted distributions, in addition, provide a firm basis on which to decide which areas of emulsion to examine if sources which differ in size or shape are to be compared.

Some of the problems involved if the sources to be compared vary in size are illustrated in Figs. 68 and 69. As with light microscopic autoradiographs, variations in size are of little consequence if the source is very large relative to the HD. As the radius of the source approaches the HD, however, it becomes necessary to count silver grains in a wider and wider rim of emulsion outside the source profile to correct for the increasing proportion of grains lying outside smaller sources.

It will be seen from Fig. 68 that counting within 1 HD of the source may be quite adequate if the two sets of sources to be compared do not differ greatly in size. Even sources of radius 4 and 1 HD can be compared in this way with a certain degree of accuracy: the grain density estimated will be 1.2 times higher for the larger source if the initial radioactivity is the same for both sets of sources. As the difference in size increases, so does the error introduced by the higher percentage of grains from the smaller source that fall outside the counting area. The universal distribution curves, however, provide a basis for predicting the precise percentage of the total grains produced that lies within a stated distance of the source, and the autoradiographer must either select areas of emulsion around his sets of sources which contain similar fractions of the total grains, or else apply a correction to his observed counts from the smaller sources.

Sources that vary in shape can also be dealt with on the basis of the universal curves. If one compares a radioactive disc with a solid band source (Fig. 72), the fraction of the total grains produced that lies directly over the source is higher for a band source of a given width than for a disc of the same diameter, and this difference becomes greater as the size of the sources is reduced. Here again, areas of emulsion around each type of source can be chosen so that both are counted at the same efficiency, or a standard area around both types can be scanned, and the counts corrected to allow for the difference in efficiency due to the difference in shape.

In taking the micrographs on which the analysis is to be based, it is

Radius or half thickness of source (in HD units)	% of total grains over source		Source diameter or thickness for 1 HD = 1600 Å
	Disc	Band	
16	90	95	6 μm
4	60	80	1.5 μm
1	25	45	3500 Å
0.25	2.5	15	800 Å

Fig. 72. The percentage of the total grains produced in the emulsion that lies directly over the source varies not only with the size of the source but also with its shape. Here solid disc sources of different radius are compared to solid band sources of different thickness. (Data provided by Dr. M.M. Salpeter)

obviously important to sample the source profiles in a truly random fashion. It is quite unacceptable, for instance, only to photograph those sources with a grain related to them if comparisons of radioactivity are to be made. It is important also to include a wide enough area of tissue around the source to enable grains to be counted up to 5 HD from the edges of the sources. If the micrographs end nearer the source profile, the distribution curves for observed grains will be artificially depressed at further distances from the source, making it much more difficult to detect the presence of other labelled structures which may be in the specimen unsuspected.

The method of constructing distribution histograms of observed grains around labelled structures has been used by Salpeter[25,26], who examined motor endplates labelled with [³H]DFP, and Budd and Salpeter[27], who studied [³H]norepinephrine in the terminals of sympathetic nerves. It depends on the approximation of the labelled sources to some relatively simple geometrical shape, and the generation by a computer of a predicted distribution about this shape, given the universal curve for that isotope. It may not always be easy to describe the sources in such a simple manner, or to find access to suitable computing facilities. In these cases it is relatively easy to construct a predicted distribution from one's own micrographs, using the transparent overlays designed for the analysis of sources producing crossfire effects (p. 246). The origins of the hypothetical β particles on the overlays should be placed on the profiles of the presumed sources in some predetermined, ordered fashion, and the positions of the centres of the resultant hypothetical grains noted.

THE ANALYSIS OF ELECTRON MICROSCOPIC AUTORADIOGRAPHS OF SOURCES WHOSE EFFECTS ON THE EMULSION OVERLAP

The incorporation of a labelled precursor such as an aminoacid may result in the majority of cell organelles becoming radioactive to some extent. At the electron microscope level this can produce an autoradiograph of great complexity, given the close packing and varied shape of cell organelles and their small size relative to the HD obtainable. Simple visual scanning of the micrographs may give only an impression of fairly uniform labelling, while the co-existence in the one specimen of many potential types of source makes it impossible to apply the methods described in the preceding section.

The analysis of this type of autoradiograph is based on the excellent method devised by Blackett and Parry[28,29], which was in turn developed from the earlier work of Williams[30]. Recently, Salpeter, McHenry and Salpeter[24] have published an account of a method which is virtually identical to that of Blackett and Parry. Williams has produced an extensive review of the field in his excellent book[37].

The analysis really depends on the testing of hypotheses. It requires certain assumptions to be made at the start. The tissue must be divided into a number of different items, each one of which is assumed to be uniformly radioactive. The choice of items rests entirely with the experimenter, who will presumably be guided by all the information available from other experiments and other techniques. Items may be relatively large, such as nucleus, extra-cellular space or mitochondrion: they may be junctional regions, such as the boundary zone between mitochondria and cytoplasm: they may be compound items in which very small granules or vesicles are embedded in a larger component, such as the presynaptic vesicles in the cytoplasm of the axon terminal. Such a sub-division of the specimen requires considerable judgement. To use a large number of items complicates the analysis unnecessarily: to lump together in one item structures which differ widely in their radioactivity will produce results which are meaningless and misleading.

Having made this first division of the specimen into items, the first hypothesis to be tested is that the levels of radioactivity in them are identical, in which case the observed distribution of silver grains will be essentially similar to a completely random distribution of hypothetical grains across the same micrographs. First, the observed silver grains are classified. A circle of radius 1 HR is placed around the centre of each developed grain, and the item occurring within the circle is noted. Next, a series of random, hypothetical grains of radius 1 HR are placed over the same micrographs, and the contents of these circles also tabulated. But the crux of the analysis is that the hypothetical origins of these hypothetical grains must also be recorded.

The original description of Blackett and Parry[28] placed these hypothetical sources in a matrix over the micrograph. Then, by means of a computer programme, a distance and a direction were calculated for each hypothetical source to indicate the position of the centre of the hypothetical grain derived from it: the programme ensured that the distribution of distances corresponded to the graph of distribution of grains around a point source, based on the universal curve for that isotope, and that the directions were randomly chosen from the full 360° available. Their more recent paper[29] places the hypothetical grains in a matrix, and indicates the origin of each grain: this change in the form of presentation of the method makes no difference at all to the basic concept.

We have, then, a series of hypothetical sources, each linked to one hypothetical grain. The items beneath each source and each grain are then recorded in a table (Fig. 73), so that each horizontal line represents the position of hypothetical grains from sources within one particular item, and each vertical column represents the hypothetical grains present over each item. The sums of the counts in each vertical column give the total number of hypothetical grains predicted over that item on a purely random distribution. This distribution can then be compared to that seen for the observed grains in the autoradiographs, using the chi^2 test.

In the majority of experiments, the chi^2 test will enable one to reject the hypothesis of random distribution, and this is the stage at which one begins to benefit from the rather laborious method of generating random hypothetical grains. Some columns in the table will contribute more to the total value of chi^2 than others, such as the item Granules in Fig. 73. The high value of chi^2 for this item is due to the low number of hypothetical grains predicted on a random distribution, compared to the number of actual grains found. Without any more ado, the horizontal line in Fig. 73 representing hypothetical sources in granules can be multiplied by a factor greater than one, to attempt to bring the final predicted distribution of grains nearer to the observed one. It is obviously extremely tedious to do this sort of testing and adjustment, and computer routines have been worked out by Blackett and Parry[29] and by Salpeter, McHenry and Salpeter[24] which will adjust and retest until an acceptable level of agreement between predicted and observed grain distributions is found. The factors by which the original lines of the table have had to be multiplied in order to reach this final prediction then give the relative concentrations of radioactivity in each item that would be compatible with the observed grain distribution (Fig. 74).

Suitable masks for tritium are available from Blackett and Parry[29], whose latest paper describes in addition the iterative computer program that they use. The micrographs used in the analysis have to be printed at a suitable magnification, so that the HD of the autoradiographs corresponds to a line marked on the overlay. Salpeter, McHenry and Salpeter[24] present overlays for tritium, iodine-125 and carbon-14 in their paper; these should be photographed and reproduced at a magnification which produces correspondence between the HD of the micrographs and the line on the overlays. They also describe in some detail their computer program.

This analytical method is conceptually simple and surprisingly rapid in application, given access to suitable computing facilities. Several important points have arisen from its use over a number of years. Perhaps the most serious is the difficulty of obtaining any estimate of the error involved in the matching of predicted and observed values. The error of the final estimate of the distribution of radioactivity is not linked to the value of chi^2 itself: it depends on a number of factors which include the numbers of

248

Sources of hypothetical grains	Grains per unit area	Site of hypothetical grains											Total grains from each source
		GP.	Nuc.	Mit.	Gran.	ER.	Golgi.	GP/Mit	GP/Nuc	GP/ER	GP/Gran	GP/Gol	
Groundplasm	1.0	200	5	7	7	0	1	9	7	2	4	2	$\underline{244}$
Nucleus	1.0	4	12	0	0	3	1	0	8	1	0	0	$\underline{29}$
Mitochondria	1.0	3	0	8	1	3	0	6	2	1	2	1	$\underline{27}$
Granules	1.0	16	2	3	45	6	5	3	1	2	25	3	$\underline{111}$
Endoplasmic reticulum	1.0	6	5	3	1	15	2	0	4	30	2	0	$\underline{68}$
Golgi	1.0	4	1	0	1	1	8	0	0	0	1	5	$\underline{21}$
Total hypothetical grains		233	25	21	55	28	17	18	22	36	34	11	500
Observed grains		201	25	15	80	21	25	17	15	30	48	11	488
χ^2		3.1	0	1.5	12.5	2.0	4.1	0.002	1.96	0.7	6.8	0	32.7

Fig. 73. A table of real and hypothetical grain distributions from a series of micrographs analysed by the method of Blackett and Parry. Hypothetical sources have been placed over the prints in a strictly randomised way, and the grains that result have been tabulated according to the structure over which the source lay (first column) and the structure underlying the grain itself. The sum of each horizontal line of the table gives the total number of hypothetical grains produced by one structure: the sum of each vertical column gives the total number of hypothetical grains found over one structure, from whatever source. A comparison of these totals with the observed grain distribution, using the χ^2 test, shows poor agreement. The observed grain distribution is unlikely to have resulted from a uniform distribution of radioactivity in the source. (Data kindly provided by Miss D. Parry)

249

Sources of hypothetical grains	Grains per unit area	Site of hypothetical grains											Total grains from each source
		GP.	Nuc.	Mit.	Gran.	ER.	Golgi	GP/Mit	GP/Nuc	GP/ER	GP/Gran	GP/Gol	
Groundplasm	1.0	200	5	7	7	0	1	9	7	2	4	2	244
Nucleus	1.0	4	12	0	0	3	1	0	8	1	0	0	29
Mitochondria	0	0	0	0	0	0	0	0	0	0	0	0	0
Granules	2.0	32	4	6	90	12	10	6	2	4	50	6	222
Endoplasmic reticulum	1.0	6	5	3	1	15	2	0	4	30	2	0	68
Golgi	1.0	4	1	0	1	1	8	0	0	0	1	5	21
Total hypothetical grains		246	27	16	99	31	22	15	21	37	57	13	584
Observed grains		201	25	15	80	21	25	17	15	30	48	11	488
χ^2		0.06	0.15	0.25	0.09	1.2	2.9	2.4	0.43	0.03	0.28	0	7.8

Fig. 74. The same data as that in Fig. 73 with the relative densities of hypothetical sources over the various structures present adjusted to give a better fit between the distributions of hypothetical and observed real grains. The relative radioactivities given in column 2 are compatible with the observed distribution of grains. (Data kindly provided by Miss D. Parry)

hypothetical and observed grains for each item. To reduce this error as far as possible, Salpeter, McHenry and Salpeter[24] recommend that the numbers of hypothetical grains should be 3–5 times more than the observed, and that none of the totals should be less than 5. This may cause problems if some structure that occurs only very infrequently is believed to be highly radioactive. It is possible to generate hypothetical grains by a stratified method, placing a second matrix with much more closely spaced points over chosen parts of the micrographs, and adjusting the final estimates of radioactivity to take account of this spacing. Blackett and Parry[29] illustrate a situation in which the second matrix consists of points equally spaced along a membrane.

Another problem is posed by the initial assumptions of homogeneity implicit in the choice of items. Given the distribution of radioactivity generated by the analysis, it may be possible to devise a separate test of the homogeneity of the item identified as most radioactive. Alternatively, Salpeter, McHenry and Salpeter[24] suggest the selection of a larger number of items as grain sites than as sources. This subdivision of some items into smaller zones for the purposes of recording the grain distribution builds in some measure of protection: recording the grains in particular zones of some items (i.e. central and peripheral) makes it more difficult to match the observed and predicted distributions with a single level of radioactivity if one or more of those items are not uniformly labelled.

In most other statistical exercises, one is trying to show that two groups of data come from different populations, and a probability value of less than 0.05 is usually accepted as significant. What is an acceptable level in this type of analysis? It is hardly good enough to accept any P value greater than 0.05: Williams suggests 0.5 as the lowest reasonable value. Even this only gives a 1-in-2 chance for the hypothetical and observed distributions to come from the same population.

What if the best fit from the computer optimisation gives a probability very much lower than this, as often happens? The first thing to look at is the item contributing the highest value of chi^2 to the total. If this is a relatively small item, the number of hypothetical grains originating from it may be very small. An "improbable" distribution of a few hypothetical grains from a small but heavily labelled item can distort the analysis significantly. Try the effect of increasing by a factor of $\times 10$ the number of hypothetical disintegrations in this item, plotting the distribution of hypothetical grains between all the items. This distribution, divided by 10, will give a more reliable estimate.

If the probability is still unacceptably low, it is almost certain the original hypothesis was wrong, and that the radioactivity in one or more of the items is not uniform. There is no alternative to dividing the tissue into items once again, guided by the individual chi^2 values in the original analysis. This takes one back to the beginning, requiring the collection of data from the micrographs in terms of the new items. This type of autoradiographic experiment clearly is not quick and simple to analyse.

A new modification has been suggested by Downs and Williams[38,39]. They start with the aim of simplifying the computation, since the need to redefine the items arises so often. They have in fact produced a method which can be carried out on a desk-top computer in the laboratory. From applying hypothetical sources and grains to the various items in their micrographs, they establish the probabilities of crossfire from each item to every other item. Since the number of "real", or observed, grains over each item is known, it is possible to set up a series of linear simultaneous equations which can be solved, giving values for the relative radioactivity of each item. Note that the series of linear equations can always be solved, however inaccurate the original subdivision of the tissue into items. The values for relative radioactivity are then incorporated into a tabulation such as that produced by the methods of Blackett and Parry[28,29] or Salpeter, McHenry and Salpeter[24], in which the same items are listed as sources of radioactivity, but many more items are recognised as possible sites of grains (Fig. 73). The data for this table can be collected at the same time as that used for the simultaneous equations. Chi^2 is calculated, to test the usefulness of the prediction made from solving the equations.

This method avoids the optimising routine needed by the alternative methods, and it is this step in particular that requires considerable computing power. By simplifying the steps involved in computing, it makes easier the task of redefining the items and repeating the analysis. It also can give rather more meaningful error estimates than the other two methods.

In summary, the hypothetical grain method of analysis is highly sophisticated, very flexible and likely to be of immense importance in electron microscopic autoradiography. It is capable of identifying radioactivity at membranes or in very small structures well below the HD of the autoradiographic system, even in the presence of radioactivity elsewhere in the specimen.

THE ANALYSIS OF FLOW RATES BETWEEN COMPARTMENTS

This is a general problem affecting many other experimental techniques apart from autoradiography, and it has generated a considerable literature. Those starting on this type of measurement on the basis of autoradiographic data may be interested in papers by Lajtha, Oliver, Berry and Hell[16] and by Droz[31], which introduce some of the mathematical treatments available.

THE STATISTICAL ANALYSIS OF AUTORADIOGRAPHS
(Written in collaboration with J.M. England)

Statistical techniques are of wide applicability, and are not in general restricted to any one experimental approach in biology. The inclusion of a section on statistics in a book of this nature may seem at first a little difficult to justify. However, in the course of several years' autoradiography, some problems in statistics have recurred frequently enough to make it worth while outlining them and possible methods of tackling them.

Autoradiographers are often dealing with relatively low numbers of observed events — usually silver grains. The levels of radioactivity in biological specimens for autoradiography are often low, the methods of data collection slow and tedious. Experiments may be based on counts of 500 or 1000 grains over each group of sources: with pulse counting techniques and larger biological samples, counts of 10,000 or more per sample are, in contrast, easy enough to obtain. Working with low total counts of silver grains, corrections for radioactive background assume far greater importance, and the level of radioactivity in the source, estimated from the mean grain density, becomes less accurate than with larger numbers of observed events.

THE POISSON DISTRIBUTION

Poisson[32] derived a formula which describes the probability of occurrence of rare events, when a large number of individuals is at risk. This formula has been shown to predict accurately the statistical variations in disintegration rate on repeatedly sampling a population of atoms of a particular radioisotope, for instance. If λ d.p.m. is the mean disintegration rate for the whole population, and λA d.p.m. the mean disintegration rate for repeated samples containing A atoms, the frequency with which a counting rate of i d.p.m. will be observed in samples of size A is given by:

$$\frac{(\lambda A)^i \cdot e^{-\lambda A}}{i!}.$$

The standard deviation (S.D.) of repeated samples of size A will be $\sqrt{\lambda A}$, the square root of the mean value. If the standard deviation is expressed as a percentage of the mean, this is known as the coefficient of variation (C.V.).

It can be seen from Fig. 75 that as the mean number of events counted per sample rises, the C.V. gets smaller: in short, the accuracy with which

Grains/unit area due to source	Sample to background ratio	Coefficient of variation of sample grain density, corrected for background
10	1	55%
20	2	32%
50	5	17%
100	10	11%
200	20	7%
500	50	4.6%

Fig. 75. The variability in grain counts over uniform sources of different levels of radioactivity, assuming the background to be 10 grains per unit area. It is assumed that counts have been made of 1 unit area of background and 1 unit area of emulsion over the source.

the true disintegration rate for the whole population of radioactive atoms is estimated by one sampling increases.

If one takes a large, uniformly labelled source, such as a section of tritiated methylmethacrylate, the number of β particles per hour leaving small areas of equal size on the upper surface of the section will vary according to the Poisson distribution. If a technically satisfactory autoradiograph, free from artefact, is prepared from such a section, the observed grain counts over small areas of section also have a Poisson distribution[14]. The autoradiograph faithfully reflects the radioactive events taking place in the source. If our "source" now becomes a population of cell nuclei, scattered through a tissue, the grain counts in repeated samples will be comparable to the counting areas over the labelled methacrylate. In other words, if the population is really homogeneous and uniform in terms of the number of radioactive atoms per cell nucleus, the observed grain counts will have a Poisson distribution. The accuracy of our estimate of the radioactivity per nucleus will depend, not on the number of nuclei counted, nor on the total area of emulsion scanned, but on the total number of silver grains counted in our sampling of the population.

THE DETERMINATION OF BACKGROUND

The grain counts taken from over the source include a contribution from a variety of factors other than source radioactivity (Chapter 6). Background (B) must be estimated separately, and subtracted from the observed counts of source plus background to give a value for the source alone (S). But repeated counts of background show that this, too, has a Poisson distribution. The observed count, $S_0 + B_0$, therefore, consists of estimates of the "true" values S and B, each of which is likely to be in error, and these errors are additive when $(S_0 + B_0) - B_0$ is used to measure S. The S.D. of estimates of S obtained in this way is given by:

$$\sqrt{[S.D.(S_0 + B_0)^2] + [S.D.(B_0)^2]}$$

There is a limit to the number of B_0 and $S_0 + B_0$ silver grains one is prepared to count, and the optimal allocation of counting effort between these two in order to reach an accurate estimate of S is obviously of considerable interest. Many workers count equal areas of emulsion to estimate B_0 and $S_0 + B_0$: in doing so, their estimates of B are clearly less accurate than of S + B. At very high ratios of S to B, the inaccuracy of estimating B matters little: the lower the ratio of S to B, the more important it becomes to count large numbers of background grains. In general terms, the grain density due to the source should be at least 5 times higher than that due to background if the requirement to scan very large areas of background is to be avoided.

England and Miller[33] have examined the optimal allocation of counting effort between source and background for given levels of accuracy. Their treatment requires that the number of different types of source to be studied in the specimen should be known, and that preliminary counts should give a rough estimate of the ratio in labelling between sources and background. Then, for a chosen level of accuracy, the required number of grains to be accumulated over each type of source and over background areas can be read off from a chart (see Appendix, p. 417). By noting the area of emulsion that needs scanning to reach this total of grains in each case, the appropriate grain density for each type of source plus background can be found, together with the estimated background density.

When photometric measurements of grain density are used instead of grain counts, it has been shown that a well-adjusted photometer does not introduce added variability to the results[14]. The photometer records in arbitrary units, however. One should not substitute the readings in these units for grain numbers in calculations of the accuracy and distribution of observations. It is possible to adjust the photometer to give a very high numerical reading from a field with few grains in it, and quite a spurious idea of the accuracy of the estimation may result. It is the number of events in the emulsion, usually silver grains, that has been observed that governs the accuracy of the sampling. To use the chart on p. 417, then, one should convert the photometric values to the number of silver grains responsible for those observed readings in order to calculate the optimal allocation of counting effort[14].

THE RADIOACTIVITY OF BIOLOGICAL SOURCES

We have seen that a model source can be sufficiently uniform for grain counts over it to have a Poisson distribution[14]. Unfortunately, biological sources are seldom as uniform as this, and there are many reports in the literature of grain counts which fail to conform to the expected Poisson model[15-17]. It may well be that cells or organelles of a particular appearance are not a homogeneous population: perhaps old cells may not incorporate as much radioactive precursor as younger ones of the same tissue, or position in the organ relative to a blood vessel may affect labelling. Any one of a hundred possible reasons may be responsible for the sources, defined in histological terms, having a wider variability of radioactivity than a uniform model source. In this case, the emulsion will accurately mirror events in the popultion of structures making up the source, and the grain counts will have a wider scatter about the mean value than predicted.

In addition to variability in the biological material, technical factors can also introduce added variability into the grain counts. We have seen (p. 139) the variability that inevitably occurs in the thickness of sections cut under comparable conditions. Similarly, variations in emulsion thickness or in the degree of development may increase the variance of grain counts over a homogeneous source.

In short, in most experiments with biological material, the S.D. of the grain counts will be greater than the square root of the mean grain count. That the variance is greater than predicted from a Poisson distribution does not rule out sensible analysis of results, though it may mean that certain procedures and tests are preferable to others. Many of the so-called parametric tests require that the observations from sources that are to be compared have distributions of a particular type, or even that they have S.D.s that are equal. In particular, most parametric tests assume that the grain counts for each source are normally distributed about the mean.

The non-parametric statistical tests, which make few or no assumptions about the distributions of the initial data, are well described in an invaluable book by Siegel[34]. Since they are in general less powerful at discriminating between different levels of labelling than their parametric equivalents, they should not be used if the data meet the rigid criteria of the latter. The use of parametric tests on unsuitable data, however, may well suggest the presence of differences where none exist.

256

THE PAIRED *t*-TEST

If two sets of sources are to be compared, the selection of which individuals in each set are to be counted should not be left to the whim of the microscopist at the time of counting. We have already seen that it is better to work out criteria for selecting the areas of emulsion to be examined before starting counting (p. 207), so that the complete population of sources can be sampled in as thorough a way as possible. There is a lot to be said for applying the same selection criteria to each set of sources, so that each observation on one set is balanced by an observation from the other, made as far as possible under identical conditions. This pairing during grain counting reduces at least some of the variability due to position within the specimen. It also provides data which are suitable for use with the paired *t*-test, which combines considerable power with relative freedom from restrictive assumptions about the distribution of the original grain counts about the mean[35]. Basically, even if the shapes of two distributions differ from each other or from the normal pattern, the differences between paired observations from the two series are often normally distributed.

The calculations are simple. If the grain counts, corrected for background, from two sets of sources are $x_1, x_2, x_3 \ldots x_n$, and $y_1, y_2, y_3 \ldots y_n$, for each pair of counts the difference is found, retaining the sign.

$$x_1 - y_1 = d_1$$
$$x_2 - y_2 = d_2$$
$$x_3 - y_3 = d_3$$
$$\ldots\ldots\ldots\ldots$$
$$x_n - y_n = d_n$$

The number of pairs is n, the degrees of freedom $(n-1)$. The S.D. of the differences (S_d) is given by

$$\sqrt{\frac{n\Sigma d^2 - (\Sigma d)^2}{n(n-1)}}$$

and the statistic t by $\frac{\bar{d}}{S_d}\sqrt{n}$ where $\bar{d}$ is the mean of the differences. The probability associated with the value found for t can be looked up in the appropriate book of statistical tables: it indicates the probability that $d_1, d_2, d_3 \ldots d_n$ do not differ significantly from zero. If a reasonably high number of paired observations is made, such as thirty or more, this test provides a sensitive and powerful way of comparing grain counts, and it is applicable to a great many autoradiographic experiments.

Sometimes, it may not be possible to collect data that are paired in any way that is significant biologically. This does not rule out the use of the paired *t*-test. The results from the two sets of sources can be randomly allocated to pairs, and the test still used. It becomes less powerful when used in this way, however, because random pairing is less effective at reducing the variability of the differences between pairs than pairing based on some genuine similarity between the members of each pair.

ESTIMATIONS OF THE PERCENTAGE OF A POPULATION OF SOURCES THAT ARE RADIOACTIVE

This has already been mentioned in connection with background estimation (p. 126). Many autoradiographic experiments set out to find the percentage of a given population of sources that are radioactive in given circumstances — the proportion of cells synthesizing DNA in a population, for instance. Often a rule-of-thumb is applied, that every cell with 4 or more grains over its nucleus will be considered labelled. This type of rule inevitably introduces errors into the experiment. If the cut-off figure is placed low, the upper end of the Poisson distribution of background grain counts will produce false positives; if the figure is high, the lower end of the Poisson distribution from radioactive nuclei will not be recognized. Often these two distributions overlap, and no arbitrary level can be found which separates background from radioactive successfully. Several treatments of this problem are available in the literature, with different basic assumptions and of varying complexity[18–20,33,36]. The treatment presented here is based on work by England, Rogers and Miller[23].

The only assumptions made are that background, measured over a similar but non-radioactive population of sources to the ones under investigation, has a Poisson distribution, and that grain counts over the radioactive sources also conform to this prediction. To make use of the method, 4 measurements must be made. The proportion of sources with one or more grains over it and the mean grain count per source are determined for both radioactive and non-radioactive populations. In estimating the mean grain counts, sources with no grains at all over them should be included. The proportion of sources that are actually labelled in the radioactive population (P_S) can be found from

$$\frac{(P_{S+B} - P_B)}{(X_{S+B} - X_B)(1 - P_B)} = \frac{1 - \exp\left(\frac{-(X_{S+B} - X_B)}{P_S}\right)P_S}{(X_{S+B} - X_B)}$$

where P_{s+B} is the proportion of sources in the radioactive population with one or more grains, P_B is the corresponding proportion in the non-radioactive population, X_{s+B} is the mean grain count over sources in the radioactive population, and X_B is the corresponding mean from the non-radioactive population. The ratio

$$\frac{(P_{s+B} - P_B)}{(X_{s+B} - X_B)(1 - P_B)}$$

can be calculated from the measurements made on the specimens. Using the graph (Fig. 74), this value is located on the vertical scale, and the corresponding value of $\frac{(X_{s+B} - X_B)}{P_s}$ is read off directly. The value for P_s can be found by dividing $(X_{s+B} - X_B)$, which is known, by the value that has been determined from Fig. 76.

To take an example, let P_{s+B} be 0.25 and P_B be 0.20, and let $(X_{s+B} - X_B)$ be 0.10. The first half of the equation becomes $\frac{0.05}{0.10 \times 0.80} = 0.63$. Locating this on the vertical axis of Fig. 76, the corresponding figure on the horizontal axis becomes 1.0, and P_s is $\frac{0.1}{1.0} = 0.1$. So 10% of the experimental population of sources in fact contain radioactive material.

This method of analysis is quick and simple, and gives accurate estimates even when the levels of grain density over the radioactive sources are so low that some "labelled" cells give rise to no silver grains at all. Earlier treatments of the problem are less accurate at levels of labelling near to background; the formula of Stillström[36], for instance, would only recognise 4% of the population as radioactive in the above example, a substantial underestimate. Where the grain densities over radioactive cells are much higher than background, the results given by the various formulae quoted tend to converge.

STATISTICAL ANALYSIS AND EXPERIMENTAL DESIGN

An experiment should be designed to test a particular hypothesis. Since grains are often in short supply over biological specimens, and grain counting is slow and tedious, the collection of data should always be purposeful and economic. If, instead, many different structures in the specimen are sampled by collecting a few score grains from over each of them, and all are compared to each other in a shotgun series of tests, the chances are high that little useful information will result. To begin with, even if labelling is uniform over every structure, the statistical tests will

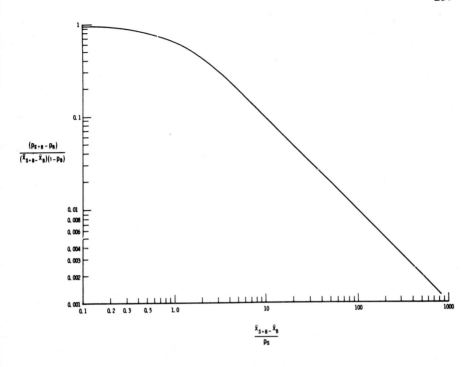

Fig. 76. A graph to assist in the determination of the percentage of sources which are radioactive in a large population. P_{S+B} is the proportion of sources in the population which have one or more developed grains over them; P_B is the corresponding proportion in a control non-radioactive specimen; X_{S+B} is the mean grain count over sources in the population which includes radioactive members; X_B is the mean count from the control, nonradioactive population. From the observed value of $(P_{S+B} - P_B)/(X_{S+B} - X_B)(1 - P_B)$, the corresponding value of $(X_{S+B} - X_B)/P_s$ can be read off on the graph, and the value of P_s (the proportion of sources that are radioactive in the experimental population) can be calculated. For the full treatment of this method, see pp. 257, 258.

indicate a probability of better than 0.05 once in every 20 tests. If genuine differences in radioactivity exist between different structures, the small numbers of grains counted will make these differences difficult to recognise unless they are very big.

Naturally, every effort must be made to exclude technical causes of variability in the observed grain counts — the choice of littermate animals, the selection of sections of a given thickness, the control of emulsion thickness and development, and so on. The question is often asked: "Is it better to count from many sections and a few animals, or from fewer sections and more animals?" The answer depends on the experiment.

260

Whichever stage in preparing the sources for counting introduces the greatest variability into the final counts, that is the step that should receive the greatest investment in terms of the number of observations made. If variation between animals is small, but section thickness poorly controlled, or variation from place to place within an organ or tissue considerable, many sections will be needed from each animal, but relatively few animals. An insatiable curiosity about the reproducibility of one's techniques helps to build up experience, which enables one to allocate effort between animals, slides, sections and counting areas without too much waste. An analysis of variance may assist in indicating the point in the collection of data at which the addition of more material will have the most effect on the precision of the final grain counts. It is precisely at this point where the variance is greatest that the most benefit can be obtained from some system of pairing between the observations.

The more controlled the preparation of the autoradiograph and the more thorough and careful the collection of observations, the more sensitive will be the final statistical analysis in determining whether or not significant differences in radioactivity exist between sources. In general, it is often possible to identify differences in labelling where they are of the order of 20% of the mean grain density: there are occasional examples in the literature where a difference in labelling as small as 10% has been shown to be significant. In view of the many causes of variability, both technical and biological, this represents a considerable achievement.

REFERENCES

1 P. Dörmer, in Y. Neuhoff (Ed.), *Molecular Biology, Biochemistry and Biophysics, Vol. 14, Micromethods in Molecular Biology,* Springer-Verlag, Berlin, 1973.
2 H.G. Boren, E.C. Wright and C.C. Harris, in D.M. Prescott (Ed.), *Methods in Cell Biology, Vol. 8,* Academic Press, New York, 1974.
3 F.W. Flitney, *J. Microscopy,* 90 (1969) 151.
4 D. Darlington and A.W. Rogers, *J. Anat.,* 100 (1966) 813.
5 S.P. Modak, W.E. Lever, A.M. Therwath and V.R.R. Uppulri, *Exptl. Cell Res.,* 76 (1973) 73.
6 R. Perry, M. Errera, A. Hell and H. Durwald, *J. Biophys. Biochem. Cytol.,* 11 (1961) 1.
7 W. Maurer and E. Primbsch, *Exptl. Cell Res.,* 33 (1964) 8.
8 M.M. Salpeter and L. Bachmann, in M.A. Hyatt (Ed.), *Principles and Techniques of Electron Microscopy, Vol. 2,* Van Nostrand Reinhold, New York, 1972.
9 B.L. Gupta, R.B. Moreton and N.C. Cooper, *J. Microscopy,* 99 (1973) 1.
10 M.M. Salpeter, *J. Histochem. Cytochem.,* 21 (1973) 623.
11 H.C. Fertuck and M.M. Salpeter, *J. Histochem. Cytochem.,* 22 (1974) 80.
12 M.M. Salpeter, L. Bachmann and E.E. Salpeter, *J. Cell Biol.,* 41 (1969) 1.
13 P. Dörmer, in U. Luttge (Ed.), *Microautoradiography and Electron Probe Analysis,* Springer-Verlag, Berlin, 1972.

14 J.M. England and A.W. Rogers, *J. Microscopy*, 92 (1970) 159.
15 L.G. Caro and F. Forro, *J. Biophys. Biochem. Cytol.*, 9 (1961) 555.
16 L.G. Lajtha, R. Oliver, R.J. Berry and E. Hell, *Nature*, 187 (1960) 919.
17 B. Chernick and A. Evans, *Exptl. Cell Res.*, 53 (1968) 94.
18 D.J. Moffatt, S.P. Youngberg and W.K. Metcalf, *Cell Tissue Kinet.*, 4 (1971) 293.
19 W. Sawicki, O. Blaton and J. Rowinski, *Histochemie*, 26 (1971) 67.
20 M. Benassi, R. Paoluzi and F. Bresciani, *Cell Tissue Kinet.*, 6 (1973) 81.
21 M. Eisen, *Int. J. Appl. Radiation Isotopes*, 27 (1976) 695.
22 S.E. Shackney, *J. Nat. Cancer Inst.*, 55 (1975) 811.
23 J.M. England, A.W. Rogers and R.G. Miller, *Nature*, 242 (1973) 612.
24 M.M. Salpeter, F.A. McHenry and E.E. Salpeter, *J. Cell Biol.*, 76 (1978) 127.
25 M.M. Salpeter, *J. Cell Biol.*, 32 (1967) 379.
26 M.M. Salpeter, *J. Cell Biol.*, 42 (1969) 122.
27 G.C. Budd and M.M. Salpeter, *J. Cell Biol.*, 41 (1969) 21.
28 N.M. Blackett and D.M. Parry, *J. Cell Biol.*, 57 (1973) 9.
29 N.M. Blackett and D.M. Parry, *J. Histochem. Cytochem.*, 25 (1977) 206.
30 M.A. Williams, *Adv. Opt. Elect. Microscop.*, 3 (1969) 219.
31 B. Droz, *J. Microscop. Biol. Cell.*, 27 (1976) 191.
32 S.D. Poisson, *Recherches sur la Probabilité des Jugements en Matières Criminelles et en Matières Civiles Précédées des Règles Générales du Calcul de Probabilité*, Paris, 1837.
33 J.M. England and E.G. Miller, *J. Microscopy*, 92 (1970) 167.
34 S. Siegel, *Non-Parametric Statistics*, McGraw-Hill, New York, 1956.
35 J.E. Freund, P.E. Livermore and I. Miller, *Manual of Experimental Statistics*, Prentice Hall, Englewood Cliffs, 1960.
36 J. Stillström, *Int. J. Appl. Radiation Isotopes*, 14 (1963) 113.
37 M.A. Williams, in A.M. Glauert (Ed.), *Practical Methods in Electron Microscopy*, Vol. 6, North Holland, Amsterdam, 1977.
38 A. Downs and M.A. Williams, in preparation.
39 M.A. Williams and A. Downs, in preparation.

CHAPTER 12

Absolute Measurements of Radioactivity

THE NEED FOR ABSOLUTE MEASUREMENTS

The basic requirements for relative measurements of radioactivity by means of autoradiographs have already been discussed in Chapter 11, and they make a sobering list. Is it possible to extend these techniques so that the number of radioactive disintegrations taking place in the source during exposure can be found?

In most autoradiographic experiments, such absolute measurements are not necessary. In the usual sort of tracer experiment, where a labelled compound is injected into an animal, and its subsequent distribution between different tissues and cell types studied, the absolute disintegration rate within one cell is not very meaningful, without a great deal of information about the size and turnover rate of precursor pools, which is seldom available. There are situations in which such measurements are of value, however.

One of these is the rapidly growing field of isotope cytochemistry. Labelled reagents can be applied to a tissue, either in vivo or in vitro, in conditions in which they combine with a specific active group. Subsequent measurement of the amount of isotope in a cell or other histological structure can, in controlled circumstances, indicate the number of such reactive groups present. Examples of this approach are provided by studies on motor endplates. Barnard and his co-workers[1,2] have measured the number of acetylcholinesterase sites at endplates using [³H]DFP and [³²P]DFP, and Salpeter's group[3,4] have carried the same approach to the electron microscope level. The acetylcholine receptor has also been labelled, using either curare[5] or bungarotoxin[6] as the radioactive reagent. Although few other experiments of this type have been pushed to such a sophisticated level of quantitation, radioisotope cytochemistry is now widely used, with labelled inhibitors of sodium–potassium ATPase[7], labelled antibodies and even DNA–RNA hybridisation[8] possible. As the possibilities of this approach are more fully exploited and reagents of high specificity and specific activity become available, the ability to determine the absolute number of reactive sites within a structure from its autoradiograph will become increasingly important.

In microbiology, the elegant work of Levinthal and Thomas[9] illustrates a further application for absolute quantitation. They were able to calculate the number of phosphorus atoms in single bacteriophage viruses by means of β-track counts, a technique they called "molecular autoradiography". In many small organisms which can be cultured in controlled conditions, it should be possible to label one particular constituent to a known specific activity by regulating the radioactivity of its precursor in the medium. Subsequent determination of the disintegration rate within the organism would then give a measure of the total number of molecules of that species present.

In radiobiology, the study of the dose rates resulting from internally absorbed radioisotopes and their effects on the tissues depend on a detailed knowledge of the distribution of radioactivity, and of the amounts of isotope present in each target organ. Here, too, the ability to infer the disintegration rate in a structure from its autoradiograph would be extremely valuable.

Absolute measurements of any sort require very high standards of technique. In making relative measurements of radioactivity, comparing the activity of one source with another, it is often enough to assume that the conditions under which the two sources are studied are identical, without investigating them in great detail. Thus the self-absorption need not be known, provided it is similar in the sources being compared. The same is true of many other factors, such as the thickness of the emulsion, its sensitivity, and the possibility of loss of isotope in preparing the sources for autoradiography. In absolute measurements of radioactivity, however, each of the factors that is capable of influencing the efficiency of measurement must be known and corrected for, if the final response of the emulsion is to serve as a basis for calculating the disintegration rate in the source.

It is often said that autoradiographic techniques are fundamentally so unreliable that absolute quantitation is out of the question. This is really a statement of technical failure. One has only to read the use made of nuclear emulsions by particle physicists to see that they have potentialities as a medium for recording the passage of charged particles which biologists have hardly begun to realise. It is possible, with good techniques, to deduce the mass, charge and initial energy of a particle from its track in nuclear emulsion. In fact, Powell, Fowler and Perkins[10] list the fundamental particles which were first identified in nuclear emulsion. As far as β

particles are concerned, Ross and her co-workers[11,12] have used the criteria of track length and grain yield in nuclear emulsion to calculate the energies of the β particles emitted by certain radioactive isotopes. In view of the precision with which emulsions can be used, it seems reasonable to expect that the number of β particles emitted by a source can be accurately measured — a problem far simpler than any of these listed here.

In many instances, autoradiography offers the only means of measuring the disintegration rate in a small source. Nuclear emulsions have a high efficiency for low energy β particles. The record they give is cumulative and, since counting can be limited to extremely small volumes of emulsion in direct contact with the source, the period of counting can be very long indeed, relative to the other methods of detecting charged particles, before the background becomes unacceptably high. For biological sources of cellular dimensions, a disintegration rate well below one per day may be accurately recorded in favourable circumstances.

In this chapter some of the attempts that have been made at absolute measurements of radioactivity will be discussed, and the technical requirements that must be met will be indicated. To do this without repeating too much of the material that has already been dealt with elsewhere, certain assumptions will be made about the conditions of autoradiography. It will be taken for granted that there is no loss or translocation of isotope in preparing the source for exposure, or that if there is any such loss it will be accurately measured and corrected for. The size and shape of the source should be known. The source must be presented to the emulsion in such a way that self-absorption can be ignored, or else calculated and allowed for. It will be assumed that chemography, either positive or negative, has been demonstrably controlled. As far as the emulsion itself is concerned, a suitable exposure time and development routine must be selected to avoid the pitfalls of emulsion saturation or inadequate development. Not until all these conditions have been met can one begin to discuss the possibility of determining the disintegration rate within the source. It is obvious that absolute measurements of radioactivity demand a high level of technical competence, and a number of preliminary control experiments, if they are to be taken seriously.

One further prerequisite to accurate measurement requires consideration in rather more detail — the control of fading of the latent image.

THE CONTROL OF LATENT IMAGE FADING

The creation of a latent image within a crystal of silver halide is a reversible process. The passage of an ionising particle through the crystal results in the deposition of minute amounts of un-ionised silver at preformed sensitivity specks. If these atoms of silver are not too few or too disperse, they can act as a catalyst for the conversion of the entire crystal to metallic silver in the process of chemical development (see p. 17). But the silver deposited at the sensitivity specks may become ionised'to silver bromide once again before the end of exposure, wiping out the latent image (p. 18). High temperatures favour this instability of the latent image, as do the presence of oxidising agents, such as atmospheric oxygen, or of water in the emulsion.

Emulsions differ in their liability to latent image fading. It is usually more severe in emulsions with a smaller grain size, and in emulsions of higher sensitivity.

The duration of exposure clearly influences the severity of fading. Conditions which give quantitative results at 2–3 days may produce very serious fading,in an exposure of 2–3 months. The steps that can be taken to reduce this loss of latent images are mainly concerned with controlling the conditions of exposure. In the first place, the autoradiographs are usually stored in the cold: 4°C is recommended, though the emulsions in general use can be cooled to below $-70°C$ without damage after thorough drying. Adequate drying of the emulsion is important: Ilford recommend drying to a relative humidity (R.H.) of 45–50% and this seems quite adequate in my experience. Messier and Leblond[13] are insistent that the Eastman Kodak NTB emulsions need complete drying to 0% R.H. to ensure freedom from fading. As for the exclusion of atmospheric oxygen, Herz[14] has suggested a very simple method for exposure in carbon dioxide. This seems satisfactory, though there is no direct evidence on the effect of the very low pH that might result from the presence of water in the emulsion. Other inert gases may be used, such as nitrogen, helium, or argon, depending on their availability.

How may latent image fading be recognised in a thin emulsion layer, where the presence of radioactivity is indicated by an increased density of silver grains? A series of slides can be given a uniform exposure either to light or to radiation, and then developed after varying periods of time. If the blackening, measured by grain counting or by microdensitometry, remains constant at all exposure times, there is no loss of the latent images formed by the initial irradiation (Fig. 77). If the grain density falls off with increasing time, fading is present. This is probably the most convenient way of conducting experiments into the optimal conditions of exposure.

As a check with each batch of autoradiographs, to ensure that latent image fading is satisfactorily controlled, grain counts should be taken over similar structures after exposures lasting different lengths of time (Fig. 78). After correcting for radioactive decay, where appropriate, the increase in

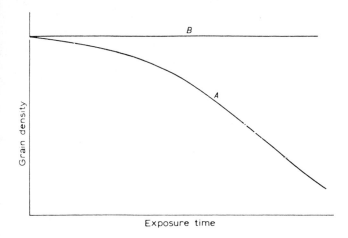

Fig. 77. A graph illustrating the effect of latent image fading on an emulsion layer deliberately fogged before the start of exposure (line A). The observed densities of silver grains will decrease with increasing exposure. Line B shows the unaltered grain density in a similar slide exposed in the absence of latent image fading.

grain count with time should be strictly linear, provided the probability of double hits does not rise to significant levels. Loss of latent images shows itself by the failure of the grain count to maintain its initial rate of increase. In conditions where fading is severe, the graph may flatten off to a plateau, when the rate of latent image formation is balanced by the rate of fading.

In grain density autoradiographs, as mentioned in Chapter 6 (p. 124), latent image fading is often tolerated in an attempt to keep the background as low as possible. With most techniques, the process of applying the emulsion to the specimen and the subsequent drying create a number of background grains. A long exposure with a certain amount of latent image fading results in these grains, present at the start of exposure, being wiped out: in these conditions, the background is likely to consist only of grains due to environmental radiation. Where techniques have evolved with the aim or producing clear autoradiographs with as low a background as possible, it is usually the case that latent image fading is present, even though the autoradiographer may not be aware of the fact. Stringent control of the conditions of exposure may therefore result in higher background levels than are customary, since the emulsion is preserved for viewing with all the grains which were present at the start of exposure, as well as those acquired from environmental radiation during exposure.

In view of the very common occurrence of latent image fading in

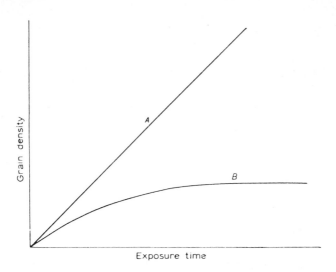

Fig. 78. A graph illustrating the increase in developable silver grains over a labelled source with increasing exposure time. In (A), a linear relation between grain density and exposure time exists, showing that there is no latent image fading. In (B), the rate of increase of grain density falls with increasing exposure times, until a plateau is reached, when the rate of latent image fading has reached an equilibrium with the rate of latent image formation over the source.

autoradiographs that are otherwise excellent and of the difficulty in recognising fading by simple examination of the emulsion, it cannot be emphasised too strongly that no measurement of the disintegration rate within a source can be accepted as valid unless the linear response of the emulsion with increasing exposure times has been clearly shown.

In considering track autoradiographs, the position is slightly different. The units being counted are tracks, not individual silver grains, and even if latent image fading has reduced by one-third the number of grains in a track, the latter may still remain recognisable. With a little experience, the limits of variation in grain size and spacing in tracks of a given length can be recognised under constant conditions of development. Latent image fading will tend to wipe out the smaller grains, and reduce the size of the larger ones. In effect, in a slide affected by fading, a track that has been formed early during exposure will have larger gaps between the grains, and the grains that are present will be smaller, by comparison with a track produced towards the end of the exposure period. So minor degrees of fading may be recognised by examining a single autoradiograph, and yet may not make any appreciable difference to the track count.

Obviously, if fading proceeds further, the spaces between surviving latent images may become so great that it is impossible to follow the course of the track: all that remains of tracks formed early in exposure in this case is a few short runs of small grains, which cannot be satisfactorily linked together. Fig. 37 illustrates this type of appearance. Severe fading like this will vitiate attempts to make absolute measurements, and will be clearly demonstrated by the technique of exposing slides from the same batch for increasing periods, as was the case in grain density autoradiographs.

Track autoradiographs may present the problem of latent image fading in rather a different form. The thick emulsion layers required take a considerable time to dry, during which they have an initially high rate of fading, and it may be difficult to specify accurately the stage at which reasonably permanent track records begin to accumulate. If, as often happens, the total exposure time is about 24 hours, this period of uncertainty may form a significant fraction of the exposure.

Whether tracks or silver grains are the units to be measured, it is more valuable to have counts available for a series of exposure times than to have the same number of observations for one exposure alone, if absolute values are required. In the former case, the absence of latent image fading can be clearly demonstrated; in the latter, it has to be assumed, or inferred from separate control experiments.

ABSOLUTE MEASUREMENTS FROM GRAIN DENSITY AUTORADIOGRAPHS

Let us assume that the many technical problems of preparing an acceptable autoradiograph have all been dealt with, that there is no loss of isotope, no chemography, no latent image fading, and so on. The radioactive source is covered with a thin layer of nuclear emulsion, and the developed grains over it can be clearly counted. Is it possible to relate the observed number of grains to the number of radioactive disintegrations that have taken place in the source during exposure?

Two basically different approaches have been adopted towards this problem. The first is the rigorous, mathematical attempt to predict the grain yield from a given source from basic principles. The second approach is empirical, and involves the preparation of reference sources of known activity, and observation of their grain yield under reproducible conditions. Each of these will be considered in turn.

(a) *Calculations of grain yield*

β Particles have very irregular tracks, so that any prediction based on the distribution in space of their trajectories must be a statistic with very wide limits of variation. The mathematical treatment for the scattering of a collimated beam of monoenergetic electrons passing through a homogeneous medium of known characteristics is already fairly complicated. But, with a radioactive isotope, the β particles have a spectrum of initial energies, and may leave the source in any direction. In addition, the situation represented by Fig. 18c, where the typical grain density autoradiograph is illustrated, involves four different media — the glass of the slide, the source itself, which is usually a tissue section, the nuclear emulsion, and the air or inert gas in which exposure takes place.

Taking all these factors into account, it is, I believe, true to say that accurate prediction of grain densities from sources of known activity is not possible at present from first principles. Not enough is known about the scattering of electrons at interfaces between media of different densities to make the very considerable burden of computation worthwhile. In consequence, any attempt to calculate grain densities has to start from a number of simplifying assumptions.

The most thoroughgoing attempt to date to calculate expected grain densities from known autoradiographic situations is the work of Odeblad[15-17]. It is worth considering it in some detail, if only to illustrate the difficulties of this approach. This is a very interesting attempt to apply matrix theory to the evaluation of grain density autoradiographs. Certain simplifying assumptions are made to begin with. It is assumed that the distribution of radioactive isotope within a tissue section will correspond with the observed pattern of histological structures and that, within each structure — for example, within the nucleus of a cell — the distribution of radioactivity will be homogeneous. It is assumed that the effects of a large number of β particles on the emulsion can be observed, so that statistical statements based on the characteristics of the entire spectrum of energies of the isotope under study are likely to hold good. It is further assumed that precise data are available for the shape and size of the histological structures under study, for the thickness of section, of emulsion, and of intervening material, if any.

Odeblad then introduces a system of coordinates, the X and Y axes in the plane of the specimen, the Z axis a vertical one, at right angles to this plane. Using these coordinates, it is possible to describe the relationship between a given histological structure and a defined volume of emulsion in accurate terms. Finally, it is assumed that the transmission of β particles from specimen to emulsion in the conditions of the experiment can be described by a dimensionless transmission coefficient T(C).

The grains observed in any given volume of emulsion are caused by radiation from the immediately adjacent structure in the specimen, plus a contribution from other nearby structures that reflect both their distance from the observed volume of emulsion and their total content of radioactivity. By examining a large number of volumes of emulsion in this way, it is possible to set up a series of simultaneous equations relating the observed densities of silver grains to the geometrical factors between the chosen volumes of emulsion and each histological structure, to the volume of each structure, and to the concentration of isotope within it.

In order to solve these equations, the transmission function $T(C)$ for the particular isotope must be separately determined. The matrix elements needed for their solution must be either calculated or found by experiment. Finally, a proportionality factor must be determined experimentally for the autoradiographic conditions used, preferably using point or planar sources of known activity, of which the matrix elements are easily calculated.

It is clear that there is a great deal of work involved in the preliminary determination of these factors, and in the many measurements needed for the assembly of the equations. The mathematics involved is also pretty complex. It is nevertheless true that, once these necessary factors have been found for a given autoradiographic situation, it is possible to calculate the amount of radioactivity present in any structure in a biological specimen from observed grain densities.

There is one limitation on the applicability of this mathematical approach which Odeblad himself has pointed out. With isotopes of low maximum energies, the use of the transmission coefficient $T(C)$ becomes more difficult to justify. With tritium, the extreme example, self-absorption in the specimen becomes such an important factor that this generalised coefficient can no longer be used. Odeblad sees his computations as valid and useful only for isotopes of considerably higher maximum energy such as phosphorus-32.

Even with the simplifying assumptions made in these calculations, at least two factors have to be determined empirically before the necessary equations can be set up — the transmission coefficient and the proportionality factor. Odeblad's analysis illustrates very forcibly the difficulties involved in calculating the radioactivity of a source within a complex specimen from the pattern of silver grains in the overlying emulsion.

The parallels between this analytical method and the one proposed by Blackett and Parry[18,19] are interesting. But where Odeblad's method attempts the analysis of the autoradiographs from broad general principles by computation alone, Blackett and Parry use the distribution of silver grains around a point source, determined empirically, together with the distribution of random, hypothetical grains over the specimen, as a basis for computing.

It is not surprising that most attempts at making absolute determinations of radioactivity have been based entirely on the empirical approach, on the preparation of standard reference sources of known activity, and their comparison under controlled conditions with the unknown sources under study.

(b) *Comparisons with standard reference sources*

There are many examples of this method of measuring radioactivity at varying levels of sophistication. Mamul[20] prepared gelatin sources with a known content of sulphur-35 for comparison with his experimental material. Waser and Lüthi[5] used rather similar standard sources containing carbon-14 in their measurements of the uptake of labelled curare by the motor endplates of mouse diaphragm. Probably the most careful attempt to measure radioactivity within the specimen from observed grain densities is the work of Andresen and his co-workers[21,22], who studied the amoeba *Chaos chaos* after feeding it with ^{14}C-labelled material. Possible loss of radioactivity in preparing the autoradiograph, latent image fading, and the reproducibility of grain counts, were all investigated in a most thorough way. The observed grain densities over sectioned amoebae were correlated with the results of Geiger counting from the same amoebae prior to embedding them, and from the sections prior to autoradiography. In addition, standard sources of a solution of labelled glucose were prepared and autoradiographed. More recently, Dörmer[23] has worked out a system for measuring carbon-14 in grain density autoradiographs. This paper should be read by anyone intending to make absolute measurements of radioactivity by means of autoradiography. It sets out very thoroughly the care and the controls that are needed before one can make the leap from observed grain densities to radioactivity in the specimen. In his case, the efficiency of the grain density system was calculated from track autoradiographs of similar specimens. Recently, Thiel and co-workers[29,30] have extended this work to cells labelled with iodine-125. They have used human erythrocytes labelled at the cell surface with ^{125}I-labelled antibodies as standards for comparison, calibrating them by scintillation counting.

Certain basic principles must be observed in any attempt to estimate the amount of a radioisotope in an experimental source from a comparison of its autoradiograph with that of a known reference source — principles that emerge from the discussion on relative measurements of radioactivity in Chapter 11. Ideally, the reference source should resemble the experimental source as closely as possible. Barnard and Marbrook[24] have suggested the acetylation of frog red blood cells with tritiated acetic anhydride under controlled conditions as one way of producing uniformly labelled reference sources of cellular dimensions. Reference has been made already to the labelling of blood cells with iodine-125 antibodies[30]. Cells or bacteria grown in culture in the presence of labelled aminoacids also might provide suitable sources. Caro and Schnös[25] have described labelled phage-infected bacteria which could serve as reference sources for electron microscope autoradiography.

Cells or organisms that can be obtained in bulk and labelled uniformly are probably the best reference sources. Suspensions of such cells can be readily prepared for counting in a haemocytometer, and known numbers of cells taken for estimates of radioactivity by scintillation counting, for example. Smears of these reference sources can then be exposed alongside the experimental sources. Provided the criteria necessary for accurate relative quantitation listed in Chapter 11 are met, the ratio of grain counts over reference and experimental sources can be converted directly into terms of radioactivity.

Reference standards which are uniformly labelled linear sources are the starting point for much work at the electron microscope level (p. 243). From the observed grain distributions around these hot lines, it is possible to compute distributions around sources of many other shapes and sizes. From all that has been said in the previous chapter about sources that vary in size and shape (pp. 233–237), it is clearly not sufficient to use the grain densities over an extended source, such as a section of labelled methacrylate, as a basis for converting grain counts over labelled cells into disintegration rates.

In situations in which radioactive sources are very closely packed, as in many autoradiographs of tissue sections, crossfire effects will complicate the comparison between experimental and reference sources. Grain counts over the experimental sources will include contributions from adjacent parts of the tissue and these must be corrected for before valid comparisons with the reference sources can be made. The methods of Blackett and Parry[18,19] can be very useful in this situation. As outlined on p. 245, their method can give an estimate of the relative contributions of various structures to the specimen under study, and the relative radioactivities of each. If the radioactivity of the whole specimen is known from scintillation counting, or if that of any one component can be found by the techniques described in the following section (p. 274), it then becomes possible to assign an absolute level of radioactivity to each component. Although their method has been developed with electron microscopic autoradiographs in mind, it can be applied very simply to autoradiographs at any magnification.

Absolute measurements of radioactivity by means of grain density autoradiographs are difficult to make with any precision. It is of course always possible to count the silver grains lying directly over a source, and to convert this number into the number of disintegrations taking place within the source during exposure, by using one of the figures for grain yield available in the literature[14,23]. This may be valid if all that is needed is a rough estimate of the radioactivity within the source, and a factor of 3 or 4 in the answer makes little difference. It is clearly not a method of getting a precise measurement, however.

(c) *Absolute measurements by electron microscope autoradiography*

The main problem in analysing electron microscope autoradiographs is the great disparity between the dimensions of many of the structures observed, and the relatively wide scatter of silver grains around them. In Chapter 11, the methods that can be used to define the grain distributions about labelled sources of various sizes and shapes have been discussed already[26]. Provided only that the HD value for the autoradiograph is known, it should be possible to approximate the radioactive structure of interest to a simple geometrical shape of standard size, and to calculate the predicted distribution of silver grains over and around it. If the observed distribution matches the predicted one, the next step is to count all the developed grains within a defined distance from the source, and calculate from the distribution curve the fraction of the total grains produced by the source that they represent. It is an easy step then to calculate the total number of grains attributable to the source, even though many of them will lie over adjacent structures.

To derive from this grain number the disintegration rate in the source, the efficiency of the autoradiograph must be known. Fig. 26 presents a series of efficiency determinations for various combinations of emulsion and developer; in view of the considerable influence of the precise conditions of exposure and development on efficiency, it may be better to determine the efficiency of one's own system directly with a model source of known radioactivity.

These steps in the measurement of radioactivity by electron microscope autoradiography may be clearer with a practical example. Reference has already been made to radioisotope cytochemistry, in which a labelled reagent is titrated against a compound in the tissue, and to the use which has been made of this technique in measuring acetylcholinesterase in motor endplates by reaction with the enzyme inhibitor, DFP[1,2]. Working with sternomastoid endplates of the mouse, Salpeter[3] first plotted the distribution of silver grains relative to the presynaptic membrane, and compared this with the distribution predicted around a radioactive line: as can be seen from Fig. 79, these did not agree. Next, the same grains were analysed for their distribution relative to the post-synaptic membrane (Fig. 80), and a satisfactory fit obtained. The possibility remained that both pre- and post-synaptic membranes were labelled. This has now been tested on

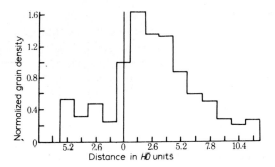

Fig. 79. The distribution of silver grains relative to the presynaptic membrane of motor endplates of mouse sternomastoid muscle labelled with [³H]DFP. The line at zero represents the membrane: values to the left lie over the nerve terminal, to the right over the post-synaptic region. The hypothesis that all the radioactivity is on the presynaptic membrane is clearly untenable. (Data derived from Salpeter, 1967)

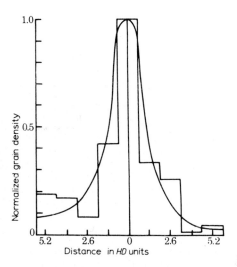

Fig. 80. The distribution of silver grains relative to the post-synaptic membrane of motor endplates of mouse sternomastoid muscle, labelled with [³H]DFP. The smooth curve represents the distribution predicted for a linear source following the course of this membrane: the histogram shows the observed distribution. These data are consistent with labelling confined to the post-synaptic membrane. (Data derived from Salpeter, 1967)

external ocular muscle endplates in which foldings of post-synaptic membranes are much reduced by comparison with the sternomastoid endplates. The membrane profiles on the autoradiographs were divided into two zones: one in which pre- and post-synaptic membranes ran parallel, the other in which folds of post-synaptic membrane were present. A simple comparison of the length of presynaptic, post-synaptic, and total membranes in these two zones with the numbers of silver grains in each showed reasonable support for the hypothesis that both the membranes had equal densities of acetylcholinesterase (Fig. 81)[4].

On the basis that pre- and post-synaptic membranes were equally rich in the enzyme, the total membrane length was measured on a series of autoradiographs of endplates, and the total number of developed grains attributable to them counted. Since the section thickness was known, the total area of membrane responsible for these grains was known. From the efficiency of the autoradiograph, measured with a standard radioactive section, the disintegration rate in the source was obtained and, from this, a statement of the density of molecules of acetylcholinesterase on the pre- and post-synaptic membranes[4].

In these admirably precise experiments, Salpeter not only demonstrated that quantitative measurements are possible by electron microscope autoradiography, she also showed that, with ingenuity and patience, it is possible to identify as labelled two structures which are much smaller than the resolution. The values she obtained for mouse sternomastoid endplates

Region of endplate	Total grains	Grain density divided by:			
		Axonal membrane area	Muscle membrane area	Both membranes area	Cleft volume
Zone I (no folds)	102	1.26	1.14	0.59	3.50
Zone II (folds)	107	2.26	0.69	0.53	2.48

Fig. 81. The analysis of the distribution of silver grains relative to the pre- and post-synaptic membranes of the motor endplates of extraocular muscles of the mouse, after labelling with [^{3}H]DFP. In these endplates, post-synaptic folds are infrequent, and the endplate can be divided into zones in which both membranes run parallel (Zone I), and zones in which there are junctional folds (Zone II). An analysis of the relative areas of presynaptic, post-synaptic and total membrane and of synaptic cleft, in these two zones shows that the observed grain densities in these zones are compatible with uniform labelling of both membranes, but not with uniform labelling of any other structure. (Data provided by Dr. M.M. Salpeter)

agreed well with those of β-track autoradiography and liquid scintillation counting[2,27].

The results obtained from autoradiographing a standard radioactive line source permit one not only to calculate the efficiency of the technique used, but also to measure the HD, and to construct a model source of the same size and shape as the suspected structure in the biological specimen[26]. From this, grain distributions and densities in defined areas of emulsion relative to the source can be predicted, and compared with those over the biological structure. Although the special conditions of electron microscope autoradiography have produced this approach, there is no reason why it may not be applied equally to grain density autoradiographs at the light microscope level, instead of attempting to label reference sources of the same approximate shape and size as the biological structures of interest.

The other possibility at the electron microscopic level is applicable mainly to situations where several or all components of the specimen are radioactive to varying degrees. Here, the crossfire may make it impossible to use the techniques just described. Instead, as mentioned above in connection with light microscopic work, the methods of Blackett and Parry[18,19] can be used. In their analytical method, the specimen is divided up into a number of items, selected by the experimenter, and the assumption is made that each item is uniformly radioactive. Their analysis determines the contributions made to the total volume of the specimen by each item, and their relative radioactivities. From this data, one only needs to know the radioactivity of the specimen as a whole in disintegrations per minute per milligram, from scintillation counting, for instance, to calculate the radioactivity of each item in absolute terms. Alternatively, if one item is much more radioactive than the others, or surrounded by an unlabelled zone so that it is relatively free from crossfire, it may be possible to determine its radioactivity from the autoradiograph alone, using the methods described in the preceding paragraphs. Since the relative radioactivities of all items are known, the absolute values for all the other items can then be calculated.

ABSOLUTE DETERMINATIONS OF RADIOACTIVITY FROM TRACK AUTORADIOGRAPHS

Many of the uncertainties that arise in attempting absolute measurements by grain density techniques stem from the difficulty of relating the observed silver grains in the emulsion to the disintegration taking place in the specimen. In a track autoradiograph, this problem is very much simplified. One track represents one β particle.

As long ago as 1957, Levinthal and Thomas[9] published a description of absolute measurements carried out with track techniques. They labelled bacteriophage virus with phosphorus-32, and prepared a suspension of virus particles in molten Ilford G5 emulsion. This was gelled in a thick layer, and, after suitable exposure and development, was scanned for β tracks. The viruses could not be seen, but their positions in the emulsion could be inferred because each one was the centre of a star of β tracks, radiating out into the emulsion from a common origin. The number of tracks forming each star gave an immediate and direct figure for the number of disintegrations taking place within the virus during exposure. This very elegant experiment is the only one I know in which the absolute radioactivity of the source was calculated directly from its autoradiograph, without any simplifying assumptions or empirical factors, as are needed in the interpretation of the conventional grain density autoradiograph.

In some ways this experiment was a special case. There was no self-absorption in the sources. They could be suspended in emulsion, giving the best and simplest geometry possible between source and emulsion. The isotope, phosphorus-32, gave long tracks, most of them straight at their start, simplifying the problems of track recognition and counting.

The fact remains, however, that the type of measurement made by Levinthal and Thomas could not have been carried out by any other technique at present available, and it is rather surprising that the sensitivity and simplicity offered by β-track autoradiography have not been widely exploited.

Two complicating factors arise in many of the experiments in which one would like to use track autoradiography for absolute measurements — both of them absent in Levinthal's case. The first concerns the proportion of β particles in the energy spectrum of the isotope that would be expected to give rise to a recognisable track. The second arises when the source cannot be suspended in emulsion, but has to be mounted on a slide, and covered only on one side by the recording medium. Each of these introduces uncertainty into the otherwise simple relationship between observed tracks and β particles, and each will be discussed in turn.

(a) *The problem of unrecognised β particles*

A β-track is usually considered as 4 or more silver grains, arranged in a linear fashion. The characteristic way in which they are arranged is described fully elsewhere (pp. 50–53). It is necessary to specify a minimum number of grains in order to distinguish tracks from a random arrangement of single background grains in the emulsion. The energy spectrum of all

β-emitting isotopes is continuous from the maximum energy right down to zero, and it is clear that, below a certain energy level, the particles will have very little chance of producing 4 grains in a row. With phosphorus-32, the proportion of these unrecognisable β particles is likely to be very small, since the maximum energy is high (1.4 MeV), and the curve of energy distribution rather bell-shaped. With carbon-14, however, the maximum energy is only 155 keV, and the shape of the energy spectrum (Fig. 10) indicates that a high proportion of the particles have initial energies less than 40 keV. What percentage of these particles will fail to cause 4 grains in a row?

This problem has been investigated in some detail by Levi, Rogers, Bentzon and Nielsen[28]. They took the isotopes carbon-14, calcium-45 and chlorine-36 and prepared track autoradiographs with each one using Ilford G5 emulsion. After processing, the emulsions were carefully re-swollen to their thickness during exposure. In this way, a large number of tracks was available for study, arranged three dimensionally in the emulsion in the original patterns produced by the β particles. The grain-to-grain track length was measured, and the number of grains counted, for many tracks. The initial energy of each particle was not known, but could be calculated from the track length. In this way, the relationship between initial particle energy and the mean number of grains produced by the particle in G5 emulsion was arrived at (Fig. 15). The variation coefficient about this mean value appeared to be about 20%.

From these relations, it is possible to calculate the percentage of β particles in any known energy spectrum which will give rise to less than 4 grains per track. Levi, Rogers, Bentzon and Nielsen[28] calculated that 14% of all the β particles emitted by carbon-14 would give rise to less than 4 grains. For calcium-45, the percentage of unrecognised β particles was 10%.

Using the convention that 4 or more grains constitute a β-track, it is possible, with any isotope with a maximum energy of 150 keV or higher, to calculate the percentage of particles that might be expected to go unrecognised through failure to produce 4 developed grains. The observed track counts can therefore be corrected to take account of this factor. At maximum energies below 150 keV, the percentage of unrecognised particles becomes rather high, and the certainty with which it can be calculated falls off, so that this correction factor becomes both less accurate and more significant.

(b) *The effect of different geometrical relations between source and emulsion*

In the experiment of Levinthal and Thomas[9] which was described above, a point source of phosphorus-32 was suspended in emulsion. All the β

particles leaving the source were therefore recorded. This very favourable situation cannot always be achieved. How can one relate the number of tracks observed to the number of β particles when the source is mounted on a glass slide, and covered on one side only by emulsion?

Fig. 30 illustrates this situation, which is frequently encountered. Let us assume that the source is very small, and is isolated from other sources. It is reasonable to believe that 50% of the β particles will be initially directed upwards into the emulsion, and the other 50% downwards into the slide. Some of those entering the emulsion will be scattered back into the slide at some point along their trajectory. Similarly, some that initially enter the slide will be reflected back into the emulsion at a variable distance from the source.

The mean distance between adjacent silver grains at the commencement of a track at the high energy end of the spectrum for the isotope concerned can readily be estimated from data given by Levi, Rogers, Bentzon and Nielsen[28]. For carbon-14, the mean figure is about 1.6 μm: for chlorine-36, it is 2.0 μm, and it is not likely to be more than 2.5 μm for any isotope of higher energy. Gaps as big as twice the mean figure may occur, but are unlikely to be exceeded. It follows that the first silver grain in a track has a very high probability of lying within 3.2 μm of the source with carbon-14, 4.0 μm with chlorine-36, and 5.0 μm with particles at minimum ionisation. If one views the source from above, in the usual way, a circle of this radius around the source will contain all but a negligible fraction of the first silver grains in the tracks of the particles that enter the emulsion directly.

When examining such a source, it is reasonable to count any track that can be traced back to within this radius of the source. It is reasonable also to assume that these tracks represent all the particles that left the source to enter the emulsion directly. In other words, the observed track count within this radius of the source, multiplied by two, will give the number of disintegrations taking place in the source. Clearly, since the initial direction of the β particles leaving the source is randomly determined, there will be a statistical uncertainty associated with this factor of two, which will become less significant as more tracks are counted.

There is up to the present very little detailed work on the scattering of β particles between the emulsion and its glass support. Preliminary counts with phosphorus-32 in this laboratory suggest that fewer than 20% of the particles emitted by a source cross the glass–emulsion interface at any point in their trajectory, and that these points of entry or exit are distributed over a very large area. In considering a small source of phosphorus-32 mounted on a glass slide, over 75% of all the tracks produced by it in the emulsion can be traced back to within 5 μm of the source, while the remaining tracks are distributed over an area several hundred microns in radius.

The curve of distribution of the points at which tracks enter the emulsion from a point source labelled with phosphorus-32 has, therefore, the form shown in Fig. 82. It is far from a bell-shaped curve. It has, rather, a high peak over the source, falling very rapidly to low levels at distances greater than 5 μm. It is therefore relatively simple to recognise and count the tracks that enter the emulsion directly from the source. This distribution tends also to simplify problems due to crossfire, since the probability of finding a point of entry of a particle that has been scattered back into the emulsion from the glass slide is uniformly low over wide areas around the source.

(c) *Experimental validation of β-track measurements*

Two experiments will illustrate the application of β-track measurements to biological specimens, and show the very close agreement that can be obtained between these techniques and the results of liquid scintillation counting. In the first experiment, leucine labelled with carbon-14 was injected into a newborn rat and portions of the liver embedded and sectioned. Protein synthesis was proceeding so fast in this tissue that practically every cell was labelled, and the labelling was reasonably uniform from one part of the liver to another. Serial sections were cut at 5 μm; some were coated with Ilford G5 emulsion for track autoradiography, some were mounted and stained for measurement of the area of each

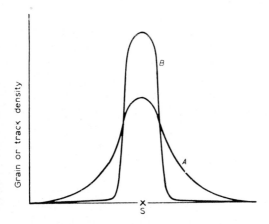

Fig. 82. A graph illustrating the differences between the distribution of silver grains around a source in a grain-density autoradiograph, and the distribution of points of entry around a source in a β-track autoradiograph. The source (S) is presumed to be labelled with phosphorus-32. The grain densities in a thin emulsion layer (line A) fall off relatively slowly with increasing distance from the source. By contrast, the points of entry of tracks into a thick emulsion layer (line B) fall very sharply to a low figure of 4–5 μm from the source.

section, and some were deparaffinised in xylene and suspended in scintillating fluid for counting in a liquid scintillation counter.

The scintillation counting gave a mean figure of 245.9 cpm from samples of 5 sections each, after subtracting background. There was no measurable quenching and the efficiency of counting, estimated with a weighed amount of ^{14}C-labelled toluene of known activity, was 74.5%. This represented 330 disintegrations per minute per 5 sections.

Track counts from autoradiographs with 3 different exposure times showed no latent image fading. The tracks from the areas counted were multiplied up to give a figure for an area equal to 5 sections, and this was converted into tracks per minute per 5 sections. This figure was then corrected for the 14% of carbon-14 β particles which would have failed to give rise to recognisable tracks of 4 or more grains (see p. 279). Finally, the corrected track count was multiplied by 2, on the assumption that 50% of the particles entered the emulsion. This method of estimating the radioactivity of the liver gave an answer of 325 disintegrations per minute per 5 sections.

The agreement between these two answers is very encouraging. It is interesting to compare the simplicity and directness of this method using track autoradiography, with the very complex and uncertain attempts to calculate the number of disintegrations in a source from grain density autoradiographs.

Further confirmation of the validity of estimates of radioactivity based on the tracks entering the emulsion directly from a source is provided by Rogers, Darzynkiewicz, Ostrowski, Barnard and Salpeter[2]. They measured the number of molecules of DFP labelled with phosphorus-32 that were bound to the enzyme acetylcholinesterase in motor endplates of the sternomastoid muscle of the mouse. β-Track autoradiographs were prepared with Ilford G5 emulsion from single, microdissected endplates (Fig. 59). The number of tracks entering the emulsion directly from the endplate was taken to be half the number of disintegrations taking place in the endplate during exposure. In this way, a value was obtained for the number of molecules of acetylcholinesterase present at these motor endplates.

Using a method of labelling that was basically similar, employing tritiated DFP, measurements were obtained by liquid scintillation counting which were in very close agreement with the results of β-track counts[27], and also of quantitative electron microscope autoradiographs of the same material[3].

The possibilities of making absolute measurements of radioactivity with nuclear emulsions may be summarised in the following way. For carbon-14 and all isotopes of higher maximum energy, the rather neglected technique of β-track autoradiography offers the best approach. The correlation between the number of tracks entering the emulsion from the source and the disintegrations taking place within the source is simple and direct.

For tritium, iodine-125, and other low energy isotopes, absolute measurements of radioactivity are best made by preparing sources of known radioactivity, which should resemble the experimental sources as closely as possible. Exposure of both types of source together to a thin emulsion layer and comparison of the grain densities in selected volumes of emulsion over them, permit the concentration or the amount of radioactivity present in the experimental sources to be calculated. Alternatively, grain distributions and densities around model sources which resemble the experimental area in size and shape can be computed from the known distributions around linear sources, and the HD and the efficiency of the system, which may require separate determination.

For the electron microscope, where one is restricted to the use of a thin emulsion layer, the method of comparing grain densities with those of standard reference sources is the only satisfactory one, whatever the isotope used. The standard sources will usually be model ones, whose grain distributions will have been calculated from the observed distributions about a radioactive line source.

Given a reasonable standard of technical competence, and careful control of the many factors that can influence the accuracy of absolute measurements, it is undoubtedly possible to measure the disintegration rate taking place in biological sources of cellular and even subcellular dimensions.

REFERENCES

1 E.A. Barnard, *Int. Rev. Cytol.*, 29 (1970) 213.
2 A.W. Rogers, Z. Darzynkiewicz, K. Ostrowski, E.A. Barnard and M.M. Salpeter, *J. Cell Biol.*, 41 (1969) 665.
3 M.M. Salpeter, *J. Cell Biol.*, 32 (1967) 379.
4 M.M. Salpeter, A.W. Rogers, H. Kasprzak and F.A. McHenry, *J. Cell Biol.*, 78 (1978) 274.
5 P.C. Waser and U. Lüthi, *Helv. Physiol. Pharmacol. Acta*, 20 (1962) 237.
6 H.C. Fertuck and M.M. Salpeter, *Proc. Nat. Acad. Sci. U.S.A.*, 71 (1974) 1376.
7 K.J. Karnaky, L.B. Kinter, W.B. Kinter and C.E. Stirling, *J. Cell Biol.*, 70 (1976) 157.
8 M. Buongiorno-Nardelli and F. Amaldi, *Nature*, 225 (1970) 946.
9 C. Levinthal and C.A. Thomas, *Biochim. Biophys. Acta*, 23 (1957) 453.
10 C.F. Powell, P.H. Fowler and D.H. Perkins, *The Study of Elementary Particles by the Photographic Method*, Pergamon, London, 1959.
11 B. Zajac and M.A.S. Ross, *Nature*, 164 (1949) 311.
12 W. Stanners and M.A.S. Ross, *Proc. Phys. Soc. London*, A 69 (1956) 836.
13 B. Messier and C.P. Leblond, *Proc. Soc. Exptl. Biol. Med.*, 96 (1957) 7.

284

14 R.H. Herz, *Lab. Invest.*, 8 (1959) 71.
15 E. Odeblad, *Acta Radiol.*, 45 (1956) 323.
16 E. Odeblad, *Acta Radiol.*, 48 (1957) 289.
17 E. Odeblad, *Lab. Invest.*, 8 (1959) 113.
18 N.M. Blackett and D.M. Parry, *J. Cell Biol.*, 57 (1973) 9.
19 N.M. Blackett and D.M. Parry, *J. Histochem. Cytochem.*, 25 (1977) 206.
20 Y.V. Mamul, *Int. J. Appl. Radiation Isotopes*, 1 (1956) 178.
21 N. Andresen, C. Chapman-Andresen and H. Holter, *Compt. Rend. Trav. Lab. Carlsberg*, 28 (1953) 189.
22 N. Andresen, C. Chapman-Andresen, H. Holter and C.V. Robinson, *Compt. Rend. Trav. Lab. Carlsberg*, 28 (1953) 499.
23 P. Dörmer, in V. Neuhoff (Ed.), *Molecular Biology, Biochemistry and Biophysics, Vol. 14, Micromethods in Molecular Biology*, Springer-Verlag, Berlin, 1973.
24 E.A. Barnard and J. Marbrook, *Nature*, 189 (1961) 412.
25 L.G. Caro and M. Schnös, *Science*, 149 (1965) 60.
26 M.M. Salpeter, L. Bachmann and E.E. Salpeter, *J. Cell Biol.*, 41 (1969) 1.
27 A.W. Rogers and E.A. Barnard, *J. Cell Biol.*, 41 (1969) 686.
28 H. Levi, A.W. Rogers, M.W. Bentzon and A. Nielsen, *Kgl. Danske Videnskab Selskab. Mat.-Fys. Medd.*, 33 (1963) No. 11.
29 E. Thiel, P. Dörmer, H. Rodt and S. Thierfelder, *J. Immunol. Meth.*, 6 (1975) 317.
30 E. Thiel, P. Dörmer, W. Ruppelt and S. Thierfelder, *J. Immunol. Meth.*, 12 (1976) 237.

PART 3: DESCRIPTIONS OF AUTORADIOGRAPHIC TECHNIQUES

CHAPTER 13

The Planning of Autoradiographic Experiments

A cat may be killed by choking it with cream, but only a few enthusiasts would insist that this is the only available method. Similarly, most autoradiographic experiments can be carried out by several different techniques, all of them producing valid results. While there is often no single correct method for doing a particular experiment, there are certainly many wrong ways of going about it. All too often, unfortunately, a research worker who wishes to use autoradiography for the first time "learns the technique" from some other laboratory, without any clear idea of the range of techniques that are available, or of their relevance to particular types of experiment. In this way, a technique that was perhaps ideal for the work of the first is applied in the second to a problem for which it is quite inappropriate.

It is the aim of this chapter to give an overall survey of the techniques at present available for autoradiography, and to indicate, in general terms, the type of problem to which each is relevant. Inevitably, this will involve a certain amount of repetition of material to be found elsewhere in this book, but there appear to be advantages to this type of arrangement for the biologist who wishes to select a method of carrying out an experiment without necessarily reading the whole book first.

Autoradiographic experiments may be broadly classified on the basis of the way in which the finished autoradiograph will be viewed. Macroscopic observation covers the autoradiography of chromatograms, of large and often irregular objects such as complete skulls, or of whole-body sections through experimental animals. It can give a rapid, roughly quantitative survey of the distribution of radioactivity between areas that are relatively large. Autoradiographs intended for viewing through the light microscope are usually sections or smears of biological material. The need to correlate the distribution of radioactivity with the range of structures visible microscopically makes quite different demands on the techniques to be employed. Electron microscope autoradiography again has a different spectrum of requirements, and the techniques are here heavily biased towards obtaining the highest possible resolution. While the underlying principles governing the detection of ionising radiations by photographic emulsions are the same for the three methods of viewing, the techniques required obviously differ considerably.

There is a clear gradient in complexity as one progresses from the macroscopic to the electron microscopic autoradiograph. The technique of preparing a contact autoradiograph by placing a chromatogram against an X-ray film is very different to the formation of a packed monolayer of very small crystals over a section intended for electron microscopy. Nowhere is this progression from the simple to the difficult seen better than in the autoradiography of diffusible materials. The standard technique of preparing and autoradiographing whole-body sections permits the study of diffusible materials: at the light microscope level special techniques are needed to do this; at the electron microscope level, it is still not possible. The same progression is seen in the analysis of the emulsion response. As the mismatch between the resolution of the viewing technique and that of the autoradiograph itself becomes greater, the problems of relating the emulsion response to the structures in the specimen grow, until the analysis of a series of electron microscopic autoradiographs may represent an investment in time and effort far greater than that spent in preparing the autoradiographs. Wherever there is a choice, the lower magnification should be preferred. To give only one example, there are several painstaking projects in neuroanatomy which have involved cutting serial sections of the brain at 5–10 μm, preparing light microscope autoradiographs of them, grain counting over large areas of section and then integrating the grain counts over relatively large regions, such as neuroanatomical nuclei. If the original aim of an experiment is to determine the relative radioactivities of various structures the size of such nuclei, it is a fabulous waste of time and effort to select a technique which resolves all the variations in isotope uptake between one cell type and another, between cell nucleus and cytoplasm, only to integrate it later. Using a carbon-14 label, thicker sections and exposure to an X-ray film, the autoradiograph can be designed to do this integration directly, with a much simpler and less expensive experiment.

There are no prizes for unnecessary technical complexity. There is everything to be gained from a critical analysis of the sort of information desired at the start of an experiment, and the selection of the simplest technique that will give it.

Within this classification, the information that is sought from an experiment also influences greatly the choice of technique. At the simplest level, autoradiography can be used to demonstrate the distribution of radioactivity. "Is the isotope present in a given structure or not?" Techniques that may be sufficient to answer this type of problem are often rather inadequate for accurate quantitative work, where the radioactivity in one cell or structure is to be compared with that in another, or in similar cells after different experimental treatment — the process of relative quantitation. A further step in technical complexity has to be taken if the radioactivity within a structure has to be measured in absolute terms.

Finally, a miscellaneous group of factors influences the selection of technique: the energy and half-life of the isotope; the characteristics of the labelled compound, which may be soluble or insoluble in any or every one of the solvents used in the usual techniques of preparing an autoradiograph; the physical nature of the specimen, which may be a fine particulate suspension, fibres, solid tissue, and so on; the availability of emulsions, and the physical conditions which will be met in the darkroom and during exposure. All these may have a bearing on the final choice of technique.

THE SELECTION OF AN APPROPRIATE TECHNIQUE

(a) *Autoradiographs for macroscopic viewing*

These techniques are relatively simple, and give a rapid visual guide to the distribution of radioactivity in the specimen. Quantitative data can be obtained quite readily from densitometry of the processed X-ray film, and this can be made the basis for estimating the absolute radioactivity if a standard reference ladder has been exposed with the specimen (Fig. 23). Estimates of specific activity more usually rely on scintillation counting of material eluted from chromatograms, or punched out from whole-body sections. The methods of preparing whole-body sections permit the study of freely diffusible radioactive materials.

The ideal isotopes to use have energies of emission equal to or higher than carbon-14 (155 keV). Tritium and the group of isotopes that emit extra-nuclear electrons of low energies are more difficult to autoradiograph. Chromatograms and other techniques that give spatial separation of highly purified compounds labelled with tritium, for example, are usually autoradiographed in the presence of a scintillation fluor, the conversion from β particles to photons resulting in higher efficiencies of recording[2,3]. With whole-body sections it has been more difficult to get reproducible and uniform efficiencies with scintillators. This may in part be due to variable quenching of the scintillation process by chemicals present in the specimen. The tendency is to use the new ^{3}H-Film produced by CEA Verken for whole-body sections labelled with low energy radioisotopes[4]. If scintillation autoradiography is used, the simultaneous exposure of a reference ladder becomes even more important, as the response of the emulsion to photons is not linear at low intensities[3].

High resolving power is generally not so important as the unmistakable recognition of sites of radioactivity. Heavy blackening in active areas in the shortest possible time, combined with a low and uniform level of background is the ideal. This situation is best met by using X-ray films, which have a high sensitivity, a large grain size, and the advantages of reproducibility and simplicity of handling. Provided the source can be prepared as a flat, regular surface, there is little point in looking at alternative techniques.

Irregular surfaces cause difficulties, and may have to be dealt with by dipping in or spraying with a liquid emulsion. X-ray emulsions are not commercially available in gel form so these irregular specimens may have to be autoradiographed with a nuclear emulsion; the larger the crystal size, the greater the blackening for a given radioactivity.

(b) *Autoradiographs for the light microscope*

Higher resolution and clearer visibility of the underlying specimen are needed when the autoradiograph is to be viewed with the light microscope. The smaller grain sizes of the nuclear emulsions make them obviously preferable to the X-ray emulsions for this type of work.

There are two main groups of techniques available for studies with the light microscope — grain density autoradiographs and track autoradiographs. In the former, the specimen is covered with a relatively thin layer of emulsion, in which the β particles record their passage by the production of individual silver grains. The presence of radioactivity is recognised by an increase in grain density over background levels. In track autoradiographs, a thicker emulsion layer is used, and the passage of a particle through it is recognised by the characteristic arrangement of silver grains to form a particle track. Track autoradiographs are more difficult to prepare, and it is much more tedious to extract information from them. Suitable tracks will only be obtained with isotopes of the energy of carbon-14 ($E_{max} = 155$ keV) or higher.

For studying the distribution of radioactivity within a specimen, grain density methods are clearly preferable to track autoradiographs. Apart from their simplicity, the resolution obtainable is in most cases better than with track autoradiographs, and viewing and photography of the specimen are much easier when it is not covered with a thick layer of emulsion.

If the radioactivity in one site is to be compared with that elsewhere, situations can arise in which track autoradiographs are preferable to grain density methods, though the latter will usually be the technique of choice. In a few cases, particularly when self-absorption with tritium makes it very difficult to relate the grain densities over two types of source, it may be better to go to an isotope, such as phosphorus-32, with which self-absorption becomes negligible, and to record the β particles emitted in a thick emulsion layer.

The real advantages of track autoradiography become evident in measurements of radioactivity in absolute terms. The simple relation of one track to the passage of one β particle through the emulsion makes it comparatively easy to determine the number of disintegrations taking place in the source during exposure. By contrast, absolute determinations of radioactivity with grain density autoradiographs are at best extremely difficult. Track autoradiography, then, is a necessary technique if absolute measurements are to be made with isotopes of the energy of carbon-14 or higher. It is a very precise and elegant method, but it is likely to remain a rather specialised one.

For most autoradiography for the light microscope, then, grain density techniques are the appropriate ones to use. The resolution obtainable ranges from 0.3 μm for tritium to perhaps 10 μm for the highest energies likely to be needed in biological studies. It is possible to observe a great deal of histological detail in the underlying specimen, and the comparison of grain densities over different structures is, in controlled cases, a valid measure of their relative content of radioisotope.

Two main groups of grain density techniques exist: the use of stripping film, and of liquid emulsion. Most experiments can be carried out quite adequately by either method, and the selection will often be made on secondary factors, such as the availability of emulsion, or previous familiarity with one or other technique. Stripping film is a compromise product, giving adequate sensitivity and resolving power, and the benefits of reproducibility that come from a factory-made emulsion layer. It can be used for studying the localisation of radioactivity, and for relative quantitation. It can be applied to any isotope that emits α or β particles, and techniques have been worked out for its use with soluble isotopes. It has a long shelf-life before the build-up of background makes it useless, sometimes proving adequate after 6 months. The one product, Kodak AR-10, and the one basic technique can cover the great majority of autoradiographic experiments quite adequately.

In my experience, however, nearly everything that can be done with stripping film can also be done with liquid emulsions, in many cases somewhat better. Many liquid emulsions are available, in a range of sensitivities and of grain sizes, and simple modifications to the technique of applying the emulsion can give controlled differences in emulsion thickness, from a monolayer of silver halide crystals upwards. With a little bit of experience, then, it is possible to select the sensitivity, grain size, and thickness of the emulsion layer to give the optimal results for the experiment in hand. In short, with the liquid emulsion techniques, it is possible to adapt the autoradiograph to the experimental situation more accurately. But liquid emulsion techniques also have their shortcomings, as will be seen from the following comparison of their performance with that of stripping film in the various types of autoradiographic experiment.

For studies of the localisation of radioactivity within tissues, my own experience is that liquid emulsions are capable of giving better results. By matching the grain size and sensitivity of the emulsion to the needs of the experiment, and by using emulsion layers thinner than that provided by stripping film, it is usually possible to obtain better resolution, without any significant decrease in efficiency. Technically the process of dipping in liquid emulsion is very quick and simple, and adhesion of emulsion to specimen is better. In addition, the absence of the overlying layer of gelatin which is present in stripping film simplifies the staining of the specimen, and makes the use of dark-field illumination for viewing and photomicrography much easier.

When one considers relative quantitation, the user of liquid emulsion faces the hitherto unsolved problem of producing an emulsion layer that is uniform in thickness. Many modifications to the simple process of dipping the slide in liquid emulsion have been tried, but I know of no way to produce a consistent and uniform emulsion thickness on a series of slides, particularly if they carry histological sections, which usually have a very irregular upper profile. In these circumstances, variations in emulsion thickness from one section to another will introduce a source of error into comparisons of grain counts that is quite unacceptable. By contrast, stripping film is far more uniform in thickness, even over the irregularities of a tissue section, and variations in grain density over the specimen will not in general be complicated by fluctuations in the thickness of the emulsion.

Isotopes of very low maximum energy, such as tritium, provide an important exception to the difficulties of relative quantitation with liquid emulsions. It is very unlikely that a particle from tritium would travel more than 2 μm through a nuclear emulsion. It is relatively simple to produce an emulsion the thickness of which during exposure will not be less than 3 μm. Any variations in thickness of such a layer, whether on or between slides, will not affect the efficiency with which radioactivity is recorded, since this is a function of only the lower 2 μm of the emulsion. In these circumstances, it is only the density of background grains that will vary with emulsion thickness and, since this should be a very low figure anyhow, it is perfectly reasonable to use this type of preparation for relative quantitation.

Leblond, Kopriwa and Messier[5] have produced interesting figures illustrating this point, using as sources sections labelled with iodine-131, carbon-14, and tritium. With the two former isotopes, fluctuations in grain density of $\pm 10\%$ were observed from one part of the slide to another, after dipping in liquid emulsion. With tritium, the results were remarkably consistent, in spite of comparable variations in emulsion thickness on these slides.

To sum up the position with relative measurements of grain density, therefore, liquid emulsions can be used to give satisfactory grain density autoradiographs with tritium, but not with isotopes of higher energy. Stripping film, on the other hand, can give good results with any isotope.

Absolute measurements of radioactivity will often require track autoradiography. If suitable standardised reference sources are available, however, the problem simplifies itself to relating the grain densities over experimental and reference sources, and the conclusions of the previous paragraph are again applicable.

When, then, is the final verdict? When should one choose stripping film techniques, and when liquid emulsions?

If the volume and variety of autoradiographs expected in a laboratory are not large, and if, as is usually the case, absolute measurements are not required, stripping film techniques are probably the best. The one product, Kodak AR-10, and the one basic technique can be adapted to practically every experiment that suggests itself. In many laboratories autoradiography is not the principal technique in use, but it is required from time to time to provide supplementary information. In these circumstances it is unreasonable to use a selection of emulsions and techniques, and it is unlikely that sufficient experience will be built up to exploit the full potentialities of liquid emulsions.

If, however, the amount of autoradiography justifies the time spent in working out suitable methods for several emulsions, and the expense of keeping up regular deliveries of each product, better results can be expected with liquid emulsions than with stripping film. Grain density autoradiography can be used for studies of the localisation of radioactivity for isotopes of any energy, and for relative measurements with tritium and iodine-125. Track techniques can give relative and absolute quantitation with isotopes of higher energy. The experience gained with liquid emulsions will provide a firm basis for autoradiography with the electron microscope.

It is, of course, quite possible to use liquid emulsion and stripping film — the former for localisation, and for relative quantitation with tritium and iodine-125, the latter for relative quantitation with isotopes of higher energy. This is not a common choice, however, as most workers prefer to limit the time they spend on acquiring technical proficiency, to leave more time and effort available for their principal investigation.

(c) *Autoradiographs for the electron microscope*

There are two principal reasons for attempting autoradiography at the electron microscope level. The first is the need to correlate the distribution of radioactivity with the range of structures that can only be visualised with the electron microscope. The second, related to it, is the wish for the highest possible resolution.

In practice, the available techniques are based on two emulsions. Ilford L4, with a mean crystal diameter of around 1400 Å, is more frequently used at the electron microscope level than any other emulsion. Its developed grains can readily be seen in the light microscope, which makes it easier to control the preparation of suitable material. The resolving power, however, is limited by the crystal diameter, and is unlikely to be better than about 1500 Å, though better results can probably be obtained in optimal conditions, with extremely thin sources labelled with tritium or iodine-125.

The second emulsion is Eastman Kodak 129-01, the successor to NTE. This has a mean crystal diameter of 500 Å, permitting resolutions of 500–1000 Å, depending on the nature and thickness of the source.

It is important to realise what a technical achievement this emulsion represents. When the crystal diameter is around 500–700 Å, the degree of sensitisation necessary to record the passage of β particles is approaching the theoretical limit, with a correspondingly high probability of spontaneous formation of background grains. At the same time, latent image fading is a far worse problem than with the emulsions used for light microscope work, since the severity of fading is related to crystal size. It is consequently unreasonable to expect the same high sensitivity and reliability from this emulsion that one finds with Ilford L4.

Any improvement in resolution involves two related variables — the thickness of the specimen and the thickness of the emulsion layer. The best resolution results from a very thin specimen, and a monolayer of the smallest crystals of silver halide available. But both these factors operate to reduce the number of developed grains produced by the source in a given exposure. Reducing the specimen thickness decreases the number of radioactive disintegrations per unit area of emulsion, while the chances of a β particle traversing the monolayer of crystals without creating a latent image increase as the crystal diameter is reduced. So attempts to get very high resolution may fail to give an acceptably high yield of silver grains without exposures of many months. These may in turn prove abortive if spontaneous formation of background, or fading of the latent image, occur to any significant extent.

It is important to stress these limitations to high resolution autoradiography in the electron microscope at the outset. It is, generally speaking, not difficult to achieve a resolution of 1500 Å for tritium with Ilford L4, provided sufficient radioactivity can be concentrated within the source to give an adequate trace at the light microscope level. This emulsion has a long shelf-life, is easy to handle, and can be exposed for very long periods before the build-up of background or the fading of latent images limits its effectiveness. It will also record β particles of all energies, so that electron microscope autoradiographs can be obtained, though with poorer resolution, with carbon-14 or even phosphorus-32.

With Eastman Kodak 129-01, it is possible to bring the resolution down below 1000 Å for tritium or iodine-125. Its sensitivity is lower, however, so that isotopes of higher energy will only be recorded at extremely low efficiencies. The sensitivity may vary from one batch of emulsion to the next. The exposures that are possible are limited by the build-up of background, which is much faster than in Ilford L4. In both respects, however, 129-01 appears to be an improvement on its predecessor, NTE[6].

It seems reasonable at the present time to make Ilford L4 emulsion the basis for the majority of autoradiographic experiments at the electron microscope level. Eastman Kodak 129-01 should be reserved for situations in which the highest possible resolution is essential, and previous autoradiographs with Ilford L4 have demonstrated that there is sufficient radioactivity in the source to give a reasonable number of silver grains in the limited exposure time that is possible with this emulsion.

Many techniques have been suggested for applying emulsions to a specimen as a monolayer of silver halide crystals. The techniques that will be described in Chapter 18 are based on two methods: that of Salpeter and Bachmann[7,8], which involves dipping the specimen in liquid emulsion, and that of Caro and Van Tubergen[9], in which an emulsion layer is formed in a wire loop, and then placed in contact with the specimen. Both methods are capable of giving satisfactory monolayers of emulsion.

THE SELECTION OF EMULSION

In some cases, there will be no choice. There is now only the one stripping film available, Kodak AR-10. The relative merits of Ilford L4 and Eastman Kodak 129-01 for electron microscopic autoradiographs have already been discussed.

For macroscopic autoradiographs, the ^{3}H-Film of CEA Verken is the only product suitable for low energies of β particles and extra-nuclear electrons: it is a single-sided film of high sensitivity and large crystal size,

296

without a protective layer over the emulsion[4]. For the autoradiography of chromatograms, either with low energy isotopes in the presence of a scintillator or with direct recording from isotopes of energy equal to or higher than carbon-14, Eastman Kodak recommend either Kodak SB film (Code 5B-5), a single-sided film, or Kodak X-Omat R (Code XR-5), the fastest of their double sided films. Kodirex of Kodak, Ltd. (U.K.) and Ilfex, produced by Ilford's, are comparable.

With track autoradiography, Ilford G5 emulsion is probably the best emulsion to use. Its sensitivity is as high as that of any other emulsion available, and its relatively large grain size makes the viewing of tracks through thick emulsion layers relatively simple. But perhaps the most powerful reason for its selection in this context is the detailed information available on many aspects of the recording of β particle tracks in G5[10]. From this data, one can calculate the appropriate correction factor for β particles of too low an initial energy to give rise to a recognisable track (p. 279), for instance, in any experiment to measure the radioactivity of a source in absolute terms. Long and tedious calibration experiments would be required before this type of calculation could be made for another emulsion.

It is with the use of liquid emulsions for grain density autoradiography that there is a really wide selection of possible emulsions. The widest range of products is provided by Ilford Ltd., of Ilford, Essex, England. Their materials that are of interest to the autoradiographer come in a range of sensitivities from 0 to 5: emulsions with the number code 0 will record α particles, but not β particles; those labelled 2 will record β particles up to about 50 keV in energy, as well as α particles; while those labelled 5 will record β particles at minimum ionisation (i.e. at energies above about 1 MeV) as a continuous track as well, of course, as less energetic β particles. Ilford emulsions are also produced in three grain sizes: the G emulsions, with a mean crystal diameter of 0.27 μm, the K emulsions, at 0.20 μm, and the L emulsions, at 0.15 μm. Undeveloped crystals from each of these emulsion types are illustrated in Fig. 1. Each emulsion, therefore, has a letter and number which identify it. G5 will record β particles at minimum ionisation, and has large silver halide crystals; K2 is sensitive only to β particles of low energy, and to α particles, and has a smaller crystal size; L4 is nearly as sensitive as G5, but has the smallest crystal size that Ilford provide.

Eastman Kodak, of Rochester, N.Y., produce the NTB series, with a mean crystal diameter of about 0.20 μm. NTB emulsion will record α particles, and low energy β particles, such as those from tritium and iodine-125. NTB-2 is more highly sensitised, and can be used for β particles up to about 200 keV, though the grain spacing in tracks at this sort of energy is fairly sparse. NTB-3 will record β particles at minimum ionisation.

Since Agfa-Gevaert ceased to manufacture nuclear emulsions, the only other major suppliers are NiiChimPhoto in Russia (these are the successors of the NIKFI emulsions) and two sources in Japan. Details of the emulsions available from all suppliers, together with information on ordering, are given in the Appendix (p. 420).

There are several important differences between the Ilford and Eastman Kodak emulsions, and techniques evolved for the one will not be suitable for the other without modification. For instance, the Ilford emulsions are less sensitive to visible light, and it is possible to work with lighter safelighting than is needed for Eastman Kodak products. This does not matter much with a simple dipping procedure, but if cryostat sections are to be cut in the darkroom, as in the autoradiography of soluble isotopes, the whole process becomes much quicker and easier with Ilford emulsions, due to this one factor. The physical consistency of the emulsions is notably different: Ilford products requiring dilution to give a layer of 30 μm or so, while the Eastman Kodak emulsions give a thin layer of around 5 μm when used undiluted. The Ilford emulsions should be exposed at a relative humidity of about 45%, and they show little latent image fading under these conditions. Eastman Kodak emulsions should be much more rigorously dried[11] to preserve the latent image, to conditions which would produce a high background from stress artefacts (see p. 16) in the corresponding Ilford products. The developing schedules suitable for the one are not necessarily optimal for the other.

It is probably advisable to use either the NTB emulsions, or the Ilford products. So many points of technique are different in the two cases that it complicates the smooth running of the darkroom unnecessarily to use both types. My own choice is for the Ilford emulsions, with their greater range of characteristics, but this decision must depend a little on geographical position. Nuclear emulsions are best stored at a constant temperature of around 4°C, which is difficult to ensure during shipment overseas. Even when packaged in ice, transportation by air may cause a significant increase in background, perhaps due to the cosmic ray intensity at high altitudes. It may, therefore, be better to use a locally produced emulsion, which can be obtained in good condition, than an overseas product which may have a variable shelf-life and background level on arrival. Certainly, in my own experience, the Ilford emulsions had lower and more reproducible levels of background in Birmingham and Oxford, England, than in New York State or South Australia, though they were still quite usable in the latter places.

How does one select the correct level of sensitivity for a given experiment? It should be realised that a high degree of sensitisation produces an emulsion with a higher rate of production of spontaneous background grains, and a higher sensitivity to all the extraneous factors, such as pressure or environmental radiation, which can produce latent images. Clearly, if a less sensitive emulsion will still record the radiation under study satisfactorily, there will be benefits gained from its use in terms of lower levels of background. NTB and the Ilford 0 emulsions should be used for α particles. NTB2 and the Ilford level 2 of sensitisation give excellent results with tritium and iodine-125, but both may fail to record the β particles from the high energy end of the spectrum with carbon-14 and sulphur-35. Ilford level 4 should cope well with carbon-14 and sulphur-35. NTB3 and Ilford level 5 will record β particles of any energy.

When one considers the selection of an appropriate grain size, the method of viewing that will be employed is of paramount importance. If a stained section is to be viewed at low magnifications by transmitted light, a relatively large developed silver grain is needed, such as that provided by the NTB series, or the Ilford G series. High magnification work with transmitted light and stained sections is probably more satisfactory with the Ilford K series. At the highest magnifications, the Ilford L series can be seen satisfactorily by transmitted light.

If, however, dark-field illumination is used to view the autoradiograph, the fine grained Ilford K and L series can be used much more extensively (see Chapter 9). The advantages of this system are that, even at high magnifications, the silver grains can be held to such a small size that they do not interfere with the microscopy of the underlying tissue. They nevertheless give an unmistakeable light signal, even in the presence of heavily stained biological material.

The choice of grain size, then, depends on the final effect that is required. While it may be practicable to produce a small developed grain from a large silver halide crystal by selecting the correct conditions of development, one should not attempt to produce a grain that is larger than usual by increased development, as this will inevitably give a higher background than is necessary.

SETTING UP A NEW TECHNIQUE

There are a few experiments which should be done before any biological investigation is started on with a new autoradiographic technique. It may in some cases be possible to get the required information from the literature, or from another laboratory but, in the majority of instances, there is no real alternative to doing these preliminary experiments oneself. Two factors require investigation — the possibility of loss or displacement of radioactive material in the process of autoradiography, and the optimal conditions of development.

(a) *Loss or displacement of radioactive material*

The techniques of whole-body autoradiography do not displace or extract radioactive material, so that this source of trouble does not require investigation in that experimental approach. Whenever biological tissue is fixed and embedded for sectioning prior to autoradiography, however, the effects of the histological procedures on the distribution of radioactivity are crucial.

In some experiments, such as the incorporation of labelled thymidine into DNA, there is sufficient evidence available in the literature to assume that there will be no significant loss of label on preparation of tissues for autoradiography. In many other cases, however, this type of documentation will not be available. The techniques of histology and of autoradiography that are necessary for labelled material that is soluble in aqueous or organic solutions are described in Chapter 8. They are very different from the routine methods of paraffin embedding and dipping in molten emulsion, for instance.

Conventional histological and autoradiographic methods give a finished product which can and should be of a very high standard when viewed under the microscope. The methods of dry autoradiography inevitably give a poorer histological preparation. It is preferable to reserve the latter techniques for experiments which really require them.

If the distribution of a labelled compound, such as a drug or hormone, is to be studied in a tissue, it should be shown that the methods of tissue preparation and autoradiography do not involve loss of radioactivity at any stage. The biological material can be sampled by pulse counting techniques before, during and after histological processing, or the processing fluids themselves can be examined for radioactivity which has leached out of the tissue. Alternatively, autoradiographs can be prepared by conventional histological methods, and also by cryostat sectioning and dry autoradiography; a comparison of the final grain distributions should indicate if substantial removal or displacement of radioactive material has taken place in embedding and dipping in liquid emulsion. It is worth emphasising that the production of a reproducible pattern of labelling by conventional methods of histology and autoradiography is no guarantee that loss of radioactivity has not occurred[12].

Experiments involving the incorporation of a precursor molecule into some tissue component are rather more difficult to assess. Here one relies on the histological processing to extract all the unincorporated precursor, so that the presence of radioactivity in the solutions through which the tissue has passed is to be expected. The chemical analysis of the radioactive molecules extracted and those retained in the tissue may be needed before a given routine can be accepted as satisfactory. These problems are discussed in greater detail in Chapter 7.

(b) *Development*

Many of the factors that must be examined in setting up a technique are described in the particular section that deals with that technique later. The one factor that is common to all methods of autoradiography is the development of the exposed emulsion. As was seen in Chapter 2, development is a form of amplification, increasing the deposit of metallic silver present at each latent image speck until it reaches the threshold of visibility. This threshold is a function of the methods that will be used to view the finished autoradiograph. High or low magnification, transmitted bright-field or incident dark-field illumination, the intensity and colour of staining which will provide a background against which the grains must be visualised, all these will influence the size to which grains must be developed in order to be recognised. The amount of silver at the original latent image specks themselves determines the time required for the grains to grow up to this threshold, under constant conditions of development.

It is thus rather naive to imagine that the conditions of development for a given emulsion can be specified in a final and definitive way. The ideal conditions should be determined in the context of the particular experiment. The simplest way in which to do this is to take a series of slides with the type of labelled biological specimen on them that will be used experimentally. Cover them with emulsion and expose them, and then separate the slides into perhaps six different groups, each containing at least three slides. With the developer that will be used subsequently, these slides should then be processed, holding constant the dilution of the developer, its temperature, and the amount of agitation given to the slides, but varying the time of development. If, for instance, the data sheet from the manufacturer recommends a time of 4 minutes, it would be reasonable to develop for 1, 2, 4, 6, 8 and 10 minutes. After fixation, staining, and mounting for microscopy, the autoradiographs should be viewed under the conditions chosen for examining the later experimental material, and the most suitable development time selected. This series of autoradiographs can be kept for future reference, and the best development time for subsequent experiments that may require different conditions of microscopy can be found simply by examining the slides from this time series under the new conditions of viewing. Such a series is illustrated in Fig. 3.

The study of development time is an essential prelude to reliable autoradiography. If the best time appears to be very short, either 1 or 2 minutes in the above example, it may be a good idea either to dilute the developer further, or to work at a slightly lower temperature, so that the inevitable small deviations from accurate timing that are bound to occur occasionally in the darkroom do not have too dramatic an effect on the end product. Similarly, there is no point in using times much in excess of 10 minutes with thin emulsion layers, if the process of development can be conveniently speeded up by using a higher temperature, or a more concentrated developer. It is not easy to achieve absolutely reproducible conditions of development. It would be a mistake, therefore, to choose a development time that is only just short of producing an unacceptably high number of background grains, since slight variations in the amount of agitation given to the slides in developer may be sufficient to produce this high background in occasional batches of slides. One should select a time that appears to give a little latitude for slight errors in technique without damaging the experiment.

Each emulsion will require its own time study. In my experience, Ilford G5 and K2 emulsions, for instance, need quite different conditions of development.

It is possible, though not necessary, to place the results of such a study of development time on a quantitative basis (Fig. 4), by plotting the observed density of grains in labelled areas against time, and also the increase in background against time. In most instances, the density of grains in labelled areas will rise to a certain figure, and then level off. The density of background grains will remain at a fairly low level at first, rising rapidly at longer development times. With such a plot available, the choice of the best development time becomes more precise, as it is relatively easy to select a point on the plateau of grain density well short of the rapid climb in density of background grains.

If you are attempting autoradiography for the first time, it is a good idea to look first at the distribution of some compound whose fate in the body is well documented. If your very first efforts are carried out on some material that has never been autoradiographed before it may be difficult to know whether or not the distributions you find are the real thing. Radioactive iodide is a useful test substance for whole-body autoradiography, with a number of extra-thyroidal sites of concentration to supplement the massive uptake by the thyroid. At the light and electron microscope levels, [^{3}H]thymidine incorporation by cells of the epithelium of the small intestine provides a suitable pattern. For diffusible materials at the light microscope level, the uptake of [^{125}I]iodide by the submaxillary salivary glands of male mice[13] is a good model system, which I have often used on training courses.

CONTROL PROCEDURES NECESSARY
FOR EACH EXPERIMENT

It is easy to assume that the vial containing the radioactive material has in it exactly what is written on the label. In the majority of cases, the label is a reliable record of what was in the vial when it left the manufacturer, but radiation decomposition is a very real hazard, particularly if the conditions of storage have not been ideal[14]. In any experiment with radioisotopes, some check on the purity of the labelled compound is advisable. This is particularly important in autoradiography, since preparing the specimen for exposure does not include any controlled purification procedure, and the distributions of impurities in the body are often significantly different from that of the original labelled material. I would strongly recommend checking the purity of the radiochemical by chromatography before investing time and effort in studying its distribution by autoradiography. One particular hazard is the multi-dose vial of radioactive material. If bacteria are introduced by withdrawing one dose with a non-sterile needle, all the remaining labelled material may be rapidly converted into totally different compounds.

But in a sense these are hazards common to all experiments with radioisotopes. What control procedures are needed, that are particular to autoradiography?

It cannot be emphasised too strongly that autoradiography is not a simple staining technique, but a method for carrying out experiments with radioactive isotopes[15]. There is no valid deduction that can be made from looking at a single autoradiograph. The situation is similar to liquid scintillation counting where the counts observed from a single sample are quite impossible to interpret. In both cases, specific controls are needed in each experiment to exclude the presence of spurious counts, and the possible loss of counts through some process reducing the efficiency of the recording medium.

Silver grains can be produced in the emulsion by chemography (the chemical interaction of specimen and emulsion), by heat, by light, or by pressure. The most reasonable control against these false positives is to expose, with every batch of autoradiographs, identical specimens which are not radioactive. It is not sufficient to do this control experiment only once, as the conditions of working may change slowly over a period of weeks without the autoradiographer being aware of the relevance of the change. It takes very little planning and work to have one slide which is not radioactive available for inclusion in every batch of autoradiographs.

The occurrence of silver grains from causes other than radioactivity in the specimen is usually fairly easy to detect. By contrast, the loss of silver grains from areas in which they should be found is very easy to miss. The comparable phenomenon in liquid scintillation counting is quenching, the process whereby the energy lost by the β particles is transformed, not into a pulse of light, but into other forms of energy, or into light of inappropriate wavelengths. No results with liquid scintillation counting are acceptable unless some control step to indicate the presence and extent of quenching has been carried out. In the same way, no attempt to interpret an autoradiograph should be accepted unless it can be demonstrated that the emulsion was truly recording the presence of radioactivity as developed silver grains. The common causes of false negatives of this sort are chemography, and fading of the latent image during exposure: the latter is fully discussed on pp. 265–269.

The best control against such false negatives is to take one experimental slide from each batch of autoradiographs and to expose it to light, or to a standard dose of external radiation. This control slide is then returned to the container holding the rest of the slides, so that exposure and development can proceed under identical conditions. The result of this simple control procedure is often surprising. Fig. 39 illustrates an instance where gross loss of emulsion response took place with a block of tissue, where the rest of the biological material being autoradiographed in the same experiment did not show any fading effect on the emulsion. If these blocks had represented the results of some experimental procedure carried out in vivo, this unexpected artefact might have been misinterpreted in terms of the abolition of the uptake of radioactivity, if adequate controls had not been available.

These two control steps — the specimen that is known to be non-radioactive, and the experimental specimen that has been exposed to light — should form just as vital a part of any autoradiographic experiment as the measurement of background and of quenching in liquid scintillation counting. If you go to the trouble of taking these steps, it is only reasonable to say so in any published account of your work.

If chemography has been found to occur occasionally in a series of specimens, it is possible to cover one section from each block with an impermeable film (these are discussed further in Chapter 7). If there is no significant difference in the distribution of silver grains between this control section and the sections without an impermeable film, it is reasonable to assume that there has been no chemography, and to base one's findings on the material without the film.

The dangers of strange, chemographic artefacts will be much reduced if care is taken to work with high purity chemicals during the stages of preparing specimens for autoradiography, and to maintain a high standard of cleanliness (pp. 134, 135).

At the end of each experiment, you must be in a position to answer two questions before even starting on the process of explaining the distributions of silver grains in terms of biological processes. These are:

"Are the silver grains on your autoradiographs due to radioactivity in the specimen, or to other causes?"

"Has the emulsion recorded the particles reaching it from all parts of the specimen with equal efficiency?"

The controls run with each experiment give you the answers, and permit you to go on to analyse the grain distributions, confident that they reflect the distribution of radioactivity in the specimen.

THE DESIGN OF AUTORADIOGRAPHIC EXPERIMENTS

In planning an autoradiographic experiment, one should always remember that the major commitment in time and effort comes in collecting and analysing the results. In general, the techniques of preparing autoradiographs are not difficult. It makes no sense at all to economise on specimens or on darkroom time if by doing so the problems of collection and analysis are made worse.

Designing an experiment involves making decisions on a number of related problems. How much radioactivity should be given to the animal, for instance? This question is unanswerable without more information, which is often not available in the literature. It all depends on the percentage of an injected dose that is retained in the tissue in question. After giving a test dose, it is not difficult to measure the radioactivity in the tissue, after histological processing if that is likely to affect the result considerably, by scintillation counting. A useful fact is that, at 1 μCi/g, 1000 cubic microns will produce 3.2 disintegrations per day. Having found an approximate figure for the efficiency of the system you intend to use from Chapter 5, calculate a dose which will give you a minimum of 3 grains per 100 square microns in 3 weeks exposure. This will indicate the scale of the dose that will be needed.

The size of the dose that is optimal depends on the length of exposure that is acceptable and, even more critically, on the distribution of radioactivity within the specimen. If the 3 grains per 100 square microns, averaged over the whole specimen, are concentrated in a very small fraction of it, as with labelling of newly synthesised DNA by tritiated thymidine, the dose required may be much less than the original calculation indicates.

Given all the uncertainties, one often saves time by doing a small pilot experiment first, with a test dose calculated on the basis of the scintillation counts, and tissue collected for autoradiographs that are exposed for various times. Such a pilot experiment helps you to find the combination of dose and exposure time which will give grain densities suitable for counting, while if you jump straight in with the main experiment, you may find yourself committed to autoradiographs that are very difficult to count: either the least radioactive sources of interest have such a low grain density that several square metres of emulsion have to be scanned to collect a significant number of grains, or the most radioactive have produced such a high grain density that counting is almost impossible and corrections for double hits become very significant (p. 97).

How many specimens (experimental animals in most cases) should there be in each experimental group? In most well-planned autoradiographic experiments, the major source of variability is in the biological system, not the technique of measurement. That being so, it is not really intelligent to base each group on one or two animals only, or to lump together grain counts from different animals in the same group without checking the contribution of inter-animal variation to the total error. If two groups of animals are to be compared, the simplest non-parametric test calls for at least four in each group to give any reasonable chance of detecting significant differences. Increasing the number of animals in each group need not add much to the task of preparing the autoradiographs. It can make all the difference to the analysis.

The statistical analysis of autoradiographic data is dealt with in Chapter 11, where methods are presented for finding out how much counting is needed for given levels of significance. In order to use these methods, some estimates of approximate grain density are needed. Once again, before starting on the detailed collection of data from the autoradiographs, a small pilot experiment, counting from the experimental series, will allow you to design this part of the procedure in the most economical way possible.

In summary, always plan your experiments around the collection of data, since this is the rate-limiting step in most autoradiography.

EXPERIMENTS WITH TWO ISOTOPES

It is sometimes desirable to autoradiograph material that is labelled simultaneously with two different radioactive isotopes. In this way, the relationship between patterns of synthesis of DNA and RNA, or of RNA and protein can be studied in the same cell population, using precursors labelled with tritium and with carbon-14, for example. It would be extremely useful if the microscopist could determine very simply which cells are labelled with one or the other isotope, or with both at once.

Unfortunately, the basic physics of the situation makes this simple differentiation between two different isotopes difficult. In the case of tritium and carbon-14, for instance, the β particles from tritium have initial energies from 18 keV down to zero; those from carbon-14 are from 155 keV down to zero. In the range from 18 keV to zero, then, β particles will be given off by both isotopes, and no technique exists to differentiate between the two at these energies. This energy range includes about 15% of the β particles emitted by carbon-14. Any attempt to differentiate between these two isotopes must take account of this overlap in energies. The presence of carbon-14 can readily be inferred by a number of techniques which indicate the presence of particles of higher energy than 18 keV. The presence of tritium can only be established by showing that the number of particles of less than 18 keV coming from a source is significantly greater than the figure predicted for labelling with carbon-14 alone.

It is quite possible to autoradiograph the specimen with a single layer of emulsion about 30 μm thick[16]. In this situation, tritium will produce many single developed grains over the source, with decreasing probabilities of finding two, three, or more grains per β particle, up to the maximum of about 7 grains, which would be a highly improbable finding[17]. All these grains would lie within 3 μm of the source, the majority of them within 1 μm. Carbon-14 will give rise to tracks of up to 80 or so grains, at the improbable upper limit, and these grains may occur up to 40 μm from the source, though most of them will lie within 10 μm of the source[10]. It is possible, then, to scan the emulsion for tracks of more than 10 grains, extending more than 5 μm through the emulsion. These can only come from carbon-14. All the cells are then examined for grains lying within 3 μm. Those with grains within 3 μm that are not labelled with carbon-14 tracks are sites of localisation of tritium. All those identified as sites of carbon-14 activity will also contain silver grains within the smaller emulsion volume. Tritium will only be confidently recognised in these cells if the proportion of silver grains in the smaller volume is obviously greater than would be expected from the known yield of carbon-14 tracks. This can simply be determined by comparison with reference cells labelled with carbon-14 alone.

This sounds a difficult procedure, but in fact it can be greatly simplified by controlling the ratio of tritium to carbon-14 in the specimen. If this ratio is deliberately kept high, any cell significantly labelled with tritium will have a higher grain density in the small volume of emulsion over it than would be expected from the heaviest labelling with carbon-14.

Several attempts have been made to devise methods of autoradiography that would simplify still further this differentiation between tritium and isotopes of higher energy. Baserga[18,19] suggested a double exposure method. He dipped his slides in Eastman Kodak emulsion (NTB-2 or NTB-3), giving a presumed thickness of about 5 μm during exposure, and developed this layer after exposure in the usual way. The section was then stained, and a layer of nitrocellulose applied over the autoradiograph. The slide was then dipped a second time in emulsion and exposed and developed once more: this time, only the second emulsion layer could record the passage of β particles. It has sometimes been assumed that silver grains in the first emulsion layer are caused by tritium, those in the second layer being due to carbon-14, an assumption unfortunately strengthened by the diagram that has been used to illustrate the technique[19]. Obviously, tritium cannot affect the second emulsion layer; it should be equally obvious that carbon-14 can affect the first, as well as tritium.

The principle of two exposures is useful, however, as it enables one to exploit to the full the effect of a high ratio of tritium to carbon-14 in the source. The first exposure can be made relatively short: the very high tritium activities in the source can cause sufficient blackening in this time, while the much lower activity of carbon-14 will hardly produce a significant number of silver grains in the first emulsion layer. The second exposure can be made much longer, to collect a reasonable grain density from the low activity carbon-14.

It is possible to use two layers of stripping film for the two exposures, the gelatin support for the first layer providing an effective separation in focal level between the two emulsions. This has the advantage that the various layers of this sandwich are of defined and reproducible thickness. The final thickness of two complete layers of stripping film makes microscopy rather difficult, however, and an unsupported emulsion layer is sometimes used for the second exposure (p. 336). Gelatin is often used as the inert layer when the autoradiographs are made with liquid emulsions: it has to be applied by dipping after the photographic processing of the first layer is complete[20,21]. Many variations are possible to try to improve the ease with which silver grains in the two layers may be differentiated from each other during observation of the completed autoradiograph. Kesse, Harriss and Gyftaki[22] have used Gevaert NUC 715 for the first layer, and Eastman Kodak NTB-2 for the second, producing differences in the size of developed grains as a distinguishing feature of the two layers. Field, Dawson and Gibbs[23] have suggested the development of coloured grains in the first layer as an aid to discrimination.

The sensitivity of the two emulsion layers might also be varied to improve the discrimination between tritium and carbon-14. An emulsion such as Ilford K2 records tritium at relatively high efficiencies, while its sensitivity is not enough to record the initial part of the track of particles from the high energy end of the spectrum from carbon-14, except as sporadic grains. This emulsion could be used for a first exposure, and the size of developed silver grain held down by the method of development. The second emulsion layer could then be Ilford G5, which has a larger crystal diameter and a higher sensitivity. Full development of this second layer would give a clear difference in grain size between the two layers.

However simple the differentiation of the silver grains in the two emulsion layers becomes, the problem of sorting out the relative contributions of tritium and carbon-14 to the first layer remains. This is essentially a statistical problem. The determination of the percentage of cells labelled in a population may require a certain amount of care, even if one isotope only is used: on pp. 257, 258 a technique for doing this is presented.

With two isotopes, the same problem exists, but in more complex form. Cells containing carbon-14 may give rise to grains in the lower emulsion only, forming the lower end of the Poisson distribution of grain numbers per cell when one considers the upper emulsion alone. Schultze, Maurer and Hagenbusch[21] have carried out an analysis of this problem, and succeeded in adjusting their ratios of tritium to carbon-14, their exposure times and the criteria for recognising cells as labelled in the first and second emulsion layers, to achieve a very clear discrimination between cells labelled with tritium alone, with carbon-14 alone, with both isotopes or with neither. They have shown the control experiments that are necessary in setting up this discrimination. These involve the preparation of autoradiographs with each isotope alone, and the collection of counts over cells in the lower emulsion. These are expressed as histograms of grains per cell. From these histograms it is possible to adjust the doses of radioactivity or the exposure times of each layer to achieve the required degree of discrimination.

Van Rooijen[24-26] has suggested a rather different approach to the simultaneous use of tritium and either iodine-131 or iodine-125. This relies on the short half-life of the iodine isotope. The cells are first exposed to AR-10 stripping film[24], applied with the inert gelatin layer next to the specimen. In this situation, the tritium will not be recorded. It may come as a surprise to find that iodine-125 is capable of affecting the emulsion in this geometrical situation, but a glance at the Appendix (p. 419) will show that the very complex decay scheme of this isotope includes some extra-nuclear electrons of over 30 keV. So the iodine is recorded in a first exposure, though at rather poor resolution, due to the separation of source from emulsion by the gelatin layer. The autoradiograph is photographed, and the stripping film gently peeled off. The specimen is then put away for the radioactivity due to the iodine isotope to decay away to a negligible level. It is then autoradiographed a second time with AR-10 applied with emulsion next to the specimen to record the tritium distribution. An alternative technique[25] for smears involves placing AR-10 on a blank slide, emulsion layer against the glass, and putting the cell suspension on the surface of the gelatin layer for the first exposure, which will record the iodine isotope only. After a suitable period for decay, the slide is dipped in liquid emulsion to record the tritium. This method avoids photographing the first layer of AR-10 and having to strip it off before the second exposure.

With macroscopic specimens, it is also possible to try to discriminate between tritium and carbon-14. Gruenstein and Smith[27] exposed thin layer chromatograms to X-ray film to record carbon-14, then a second time in the presence of scintillators to detect tritium. Again, the problem of carbon-14 β particles affecting the film during the tritium exposure has to be considered.

Clearly, tritium can be used with almost any other isotope, provided only that the second one has a higher energy. It is theoretically possible to distinguish between, say, carbon-14 and phosphorus-32 by an extension of the methods described for tritium and carbon-14, but it would be technically difficult.

THE DESIGN AND EQUIPMENT OF THE DARKROOM

Since the darkroom is an essential part of any autoradiographic experiment, a few words about its design are probably in place here. It is the laboratory in which many of the important steps of the experiment are carried out, and a bit of care and thought in its design and equipment will pay dividends.

Ideally, it should be kept for autoradiography. The requirements of a photographic unit are very different. In particular, temperature and humidity can affect the preparation of autoradiographs critically, while normal photographic procedures can tolerate a wide range of conditions. The standards of cleanliness are necessarily far higher in a room where nuclear emulsions for subsequent viewing under the microscope are handled. It is seldom satisfactory to prepare autoradiographs in a darkroom that is routinely used by a photographic unit.

Several workers I know have darkrooms that resemble converted broom cupboards. While this may be necessary if more space is not available, it is not satisfactory. Many of the techniques of autoradiography are more simply and conveniently carried out by two people working together, and some techniques require relatively bulky pieces of apparatus, such as a cryostat, to function in the darkroom. A light trap, enabling people to enter and leave the darkroom without admitting light, is essential, particularly if track autoradiographs are to be prepared, as these may take many hours to process.

Ideally, the temperature and the humidity in the darkroom should be controlled. For Kodak AR-10 stripping film, the ideal conditions are 15–18°C, and 60–65% relative humidity, and considerable deviations from these conditions can make the technique almost impossible (see Chapter 15). For the Ilford emulsions, the temperature should not be much over 20°C, and the relative humidity 45–50%. Stable specified conditions like these are easier to obtain if the darkroom has no wall common with the outside of the building, and hence is little influenced by changes in the weather. By the time a darkroom is lightproof, it is usually also poorly ventilated so that an exhaust fan is a good idea. The air intake should be easily accessible: this makes it a simple matter to alter the humidity inside the darkroom by passing the entering air through water, or through a drying agent. Obviously, a thermometer and humidity gauge are needed.

In general, builders have no idea how dark a darkroom ought to be and they will frequently hand over a room into which the sun sends shafts of light. The user must be prepared to complete its light proofing himself, after a full 20 minutes in darkness to give full dark adaptation. It is easy to miss thin, horizontal beams of light at other than head height. If they occur at or about the height of the working surface, they may elude detection, unless specifically looked for, and cause a high level of background.

Safelight and ceiling light should be on separate switches, so that absolute darkness is possible, and preferably the switches should be well apart, so there is no danger of confusing them. It is usually a good idea to have a small bench safelight, in addition to the central one, for special techniques such as cryostat sectioning, and for working with the light reflected from surfaces, as with stripping film. The bulbs in the safelights should be 15 W or less.

Adequate working surfaces are essential. It is a good idea, if possible, to designate one area for preparing autoradiographs, and another for developing and fixing them. In Dr. Leblond's department at McGill University, separate darkrooms are used for these purposes[28]. It is useful to have a plentiful supply of electric points, so that driers, stirrers, waterbaths and so on can be used as required without elaborate tangles of wires.

A refrigerator is essential, both to store the emulsion prior to use, and to expose autoradiographs. Once again, this should be kept for autoradiography. It is a good idea to place some lead or steel shielding, either inside the refrigerator or around it, to reduce the level of extrinsic radiation. Naturally the darkroom should be situated well away from any known source of radiation. As far as cosmic rays are concerned, a large building will provide a certain amount of screening: given the choice, the basement of a multistorey building is a better place for a darkroom than the roof, or a single-storey structure.

A sink with water supply is necessary in the darkroom, preferably in the area chosen for developing and fixing.

Cleanliness is extremely important. Wet nuclear emulsion will collect dust very fast, so that any step taken to make the darkroom easy to clean and dust is a good idea. Any shelves or cupboards should have doors, and apparatus not in use should be put away.

As far as equipment is concerned, the items needed for each technique will be indicated in the appropriate chapter. One general point that requires emphasis is that all electrical apparatus, including the switches on the light circuits, should not emit flashes of light. Many items of equipment that are perfectly satisfactory in normal laboratory use can be seen to emit showers of sparks when used in absolute darkness.

RECORD OF EXPERIMENTS

I am amazed at how many of my colleagues do not keep adequate notes of their autoradiographs. An autoradiograph is a permanent record of the distribution of radioactivity in a specimen, and is just as valuable in 10 years time as it is today.

It is often possible to return to an old series of autoradiographs to test a new hypothesis. That being so, it is sensible to record clearly how the specimen and autoradiographs were prepared and to preserve these notes, together with the completed autoradiographs. Without doing this, it will be quite impossible to remember in several years time exactly how long the exposure was, or how much radioactivity was injected.

Records have another importance, in diagnosing trouble. I note the date, the method and emulsion used, the drying sequence, the temperature and humidity in the darkroom, the length and conditions of exposure, the development conditions and the batch number of the emulsion for every set of autoradiographs I produce. This book of autoradiographs has been very useful over the years. If some batches of slides have not been satisfactory, it is an easy matter to check back and see if one delivery of emulsion was involved or several, if there is any correlation with darkroom conditions or development. These comments may seem trivial, but if several experiments with different exposure times are running simultaneously, it may be quite impossible to remember which autoradiographs were made with which batch of emulsion, and so on. The diagnosis of faults is made much easier by the existence of such a record.

312

REFERENCES

1 S.A.M. Cross, A.D. Groves and T. Hesselbo, *Int. J. Appl. Radiation Isotopes*, 25 (1974) 381.
2 U. Lüthi and P.G. Waser, *Advances in Tracer Method*, 3 (1966) 149.
3 R.A. Laskey and A.D. Mill, *Eur. J. Biochem.*, 56 (1975) 335.
4 B. Larsson and S. Ullberg, *Science Tools*, Special Issue, 1977.
5 C.P. Leblond, B.M. Kopriwa and B. Messier, in R. Wegman (Ed.), *Histochemistry and Cytochemistry*, Pergamon, London, 1963.
6 M.M. Salpeter and M. Szabo, *J. Histochem. Cytochem.*, 24 (1976) 1204.
7 M.M. Salpeter and L. Bachmann, *J. Cell Biol.*, 22 (1964) 469.
8 M.M. Salpeter and L. Bachmann, in M.A. Hayat (Ed.), *Principles and Techniques of Electron Microscopy*, Vol. 2, Van Nostrand–Reinhold, New York, 1972.
9 L.G. Caro and R.P. van Tubergen, *J. Cell Biol.*, 15 (1962) 173.
10 H. Levi, A.W. Rogers, M.W. Bentzon and A. Nielsen, *Kgl. Danske Videnskab. Selskab. Mat.-Fys. Medd.*, 33 (1963) No. 11.
11 B. Messier and C.P. Leblond, *Proc. Soc. Exptl. Biol. Med.*, 96 (1957) 7.
12 W.E. Stumpf and L.J. Roth, *J. Histochem. Cytochem.*, 14 (1966) 274.
13 A.W. Rogers and K. Brown-Grant, *J. Anat.*, 109 (1971) 51.
14 E.A. Evans, *J. Microscopy*, 96 (1972) 165.
15 A.W. Rogers, *Ann. Chir. Gynaecol. Fenniae*, 58 (1969) 269.
16 S.W. Perdue, R.F. Kimball and A.W. Hsie, *Exptl. Cell Res.*, 107 (1977) 47.
17 H. Levi, *Scand. J. Haematol.*, 1 (1964) 138.
18 R. Baserga, *J. Histochem. Cytochem.*, 9 (1961) 586.
19 R. Baserga, *J. Cell Biol.*, 12 (1962) 633.
20 J.A. Shand and M. de Sousa, *J. Immunol. Meth.*, 6 (1974) 141.
21 B. Schultze, W. Maurer and H. Hagenbusch, *Cell Tissue Kinet.*, 9 (1976) 245.
22 M. Kesse, E.B. Harriss and E. Gyftaki, in *Radio-Isotope Sample Measurement Techniques in Medicine and Biology*, IAEA, Vienna, 1965.
23 E.O. Field, K.B. Dawson and J.E. Gibbs, *Stain Technol.*, 40 (1965) 295.
24 N. van Rooijen, *J. Immunol. Meth.*, 2 (1972) 197.
25 N. van Rooijen, *J. Immunol. Meth.*, 3 (1973) 71.
26 N. van Rooijen, *Int. J. Appl. Radiation Isotopes*, 27 (1976) 547.
27 E. Gruenstein and T.W. Smith, *Analyt. Biochem.*, 61 (1974) 429.
28 B.M. Kopriwa, *J. Histochem. Cytochem.*, 11 (1963) 553.

CHAPTER 14

The Autoradiography of Macroscopic Specimens

Autoradiography is not limited to microscopic specimens. Whole animals that have been sectioned, whole bones, or large and complex structures such as the brain may require scanning to find the distribution within them of radioactive material. In addition, many methods of chemical analysis produce specimens in which compounds which may be radioactive are separated in space from other related compounds. In fact, the autoradiography of chromatograms is so widely practised that it probably produces a greater volume of autoradiographs annually than does the study of sectioned biological material. Not only chromatograms may be autoradiographed: electrophoresis gels, precipitin reactions, in fact any method of analysis that gives spatial separation of the compounds studied may be combined with exposure to photographic emulsion.

There is no fundamental difference between an autoradiograph produced for viewing naked-eye and one for examination under the microscope. The geometrical factors influencing efficiency and resolution are common to both, as are the causes of background and the sources of artefact. The differences that arise in the autoradiography of large specimens are practical ones. The resolution that is required is usually measured in millimetres rather than microns. The presence of radioactivity in the source is now recognised by blackening of the film, not by a higher density of individual silver grains. Quantitative measurements of radioactivity are technically much simpler by electronic pulse-counting methods than is the case with sources of cellular dimensions: it may be possible, for instance, to use a Geiger counter with a small aperture directly over the surface of the specimen, to elute the radioactive area from a chromatogram for scintillation counting, or to dissect out the labelled organ for counting from the animal that has been examined by whole-body autoradiography.

Macroscopic autoradiographs, then, are usually required to give a quick, simple indication of the distribution of radioactivity between relatively large areas. In this situation, the emulsions produced commercially for

X-ray films have several important advantages over the nuclear research emulsions used for autoradiography for the light and electron microscopes. These nuclear emulsions have a small crystal size, and hence a small developed silver grain. Although their efficiency is high in terms of the number of developed grains per incident particle, a very high density of developed grains is needed to produce blackening of the emulsion visible to the unaided eye. X-ray emulsions have mean crystal diameters an order of magnitude larger than the nuclear research emulsions. Fewer developed grains per unit area are therefore needed to give recognisable blackening. If one defines the efficiency of a macroscopic autoradiograph as the blackening, or optical density of a layer of emulsion of constant thickness produced by a given exposure to radiation, X-ray emulsions are many times more efficient than nuclear emulsions.

The choice of X-ray film for this type of autoradiograph has many advantages. The films available commercially are very uniform in thickness and in emulsion response, they have a reasonably long shelf-life before the build-up of background reduces their usefulness appreciably, and the optimal conditions for development are usually stated by the manufacturer.

The preparation of large samples for autoradiography should therefore be designed with the use of X-ray film in mind. In other words, wherever possible, the specimen should present a flat surface to the film. In the case of chromatograms this is easily managed. Solid specimens can often be sectioned, ground, or otherwise flattened to achieve the same end. Irregular surfaces that cannot be reduced to a convenient shape may require coating with molten emulsion.

Specimens labelled with tritium are a special case. The specimen itself is almost always much thicker than the maximum range of the β particles from tritium, there may be a small air gap between specimen and film, and all but one of the types of X-ray film at present available carry a surface layer of gelatin 0.5–1.0 μm thick to protect the emulsion from abrasion during handling. These three factors combine to produce an extremely low efficiency for tritium. The several methods that have been proposed to get around this problem will be discussed later in the chapter.

THE AUTORADIOGRAPHY OF CHROMATOGRAMS

Chromatograms present few geometrical problems, as they are planar already. A few points arise in connection with their preparation, however. In most cases, 10 mg of compound is needed to enable a spot to be identified by staining or by spectrophotometry: at high specific activities, much less than this can give significant blackening on an autoradiograph. Tsuk, Castro, Laufer and Schwartz[1] have discussed the implications of this for the autoradiography of chromatograms. When microgram samples are chromatographed, artefacts may arise from such causes as the complexing of compounds with components on the paper, or even with trace metals. The precise method of drying the spot may influence the chromatogram. It is obviously a good idea to select the conditions of chromatography that will give the most concentrated spot: for the same total radioactivity, the blackening over it will be more intense.

Thin film chromatograms, with either silica or alumina as the stationary phase, are very powdery, and it is easy to damage them in the darkroom manipulations of preparing an autoradiograph. A light spraying with one of the commercially available materials such as Quelspray (polyvinyl chloride in an aerosol) will make the chromatogram much more robust without increasing self-absorption significantly. Care should be taken to allow the solvent to evaporate completely before autoradiography, as it is possible to desensitise the emulsion if this step is omitted. Polyacrylamide gels may distort or crack on drying, producing a surface which is rather uneven and far from ideal for exposure to an X-ray film. If the gel is bonded to a rigid base before drying, some of these difficulties can be avoided[2].

With all isotopes that emit β particles with a maximum energy above about 100 keV, such as carbon-14, sulphur-35, phosphorus-32, apposition to an X-ray film is the method of choice. X-ray films consist usually of a supporting layer, coated on both sides with emulsion. With carbon-14 and sulphur-35, the β particles can barely reach the layer of emulsion away from the specimen, and the autoradiograph is restricted to the nearer emulsion. In this case, it makes sense to use a film coated on one side only, such as Kodak Single-Coated Medical X-ray Film — Blue Sensitive (Code SB 54), which has been found to have the highest sensitivity for low energy β particles in a recent study at the Eastman Kodak Laboratories[3]. Alternatively, Eastman Kodak X-Omat R or Kodak (U.K.) Kodirex film can be used, and the emulsion removed from the side of the film furthest from the specimen when development and fixation are over (p. 332). With isotopes of higher maximum energy, such as chlorine-36 or phosphorus-32, the double-sided films produce more blackening for a given amount of radioactivity. Any separation of specimen from film will reduce both overall efficiency and the resolving power, so that it is best to place the film in direct contact with the specimen, if chemography is demonstrably absent. Efficiency can often be slightly increased by placing a layer of material of high density, and hence good back-scattering characteristics, on the side of the film away from the specimen. Direct pressure on the film, in order to keep it in close contact with the specimen, will not damage the film, unless it is excessive or combined with an uneven specimen surface, giving an uneven distribution of pressure. Sliding pressures should be avoided, as these cause scratches on the emulsion. Richardson, Weliky, Batchelder, Griffith and Engel[4] have described a very simple exposure box for maintaining a gentle, evenly distributed pressure during exposure.

The X-ray film will have to be separated from the specimen for development and fixation. Usually, realignment of film and specimen to identify the position of the labelled areas present few problems. If this proves difficult, marking the specimen with radioactive ink prior to exposure may help. A suitable ink can readily be made by dissolving any water-soluble compound labelled with carbon-14 in Indian ink. Development and fixation should follow the recommendations of the manufacturer. These steps are often handled more easily if the films are placed in the special holders used in all hospital radiography departments.

Tsuk, Castro, Laufer and Schwartz[1] have published a guide to exposure times for carbon-14 on paper chromatograms, which is reproduced as Fig. 36. The size of the spot will clearly have some modifying effect on the minimum exposure time needed to give a positive autoradiograph. It is sometimes possible to recognise blackening that would otherwise be too weak to distinguish from background by tilting the film, viewing it at an oblique angle. This in effect superimposes the blackening in adjacent areas, and may reinforce slight degrees of signal above noise to the level at which it can be confidently recognised.

It is possible to hazard a guess about the relative activities of two sources from just looking at the films under standard conditions and, with experience, this visual quantitation can become surprisingly accurate[1]. Greater accuracy can be obtained by measuring the blackening of the film with a microdensitometer. A very convenient instrument has been described by Cross, Groves and Hesselbo[5]. Working with electrophoresis gels, Lim, Huang and Davis[6] concluded some time ago that, "within a limited range, a linear relation is obtained between radioactivity and optical density in the autoradiogram, as well as between protein concentration and optical density in the gel". For those for whom densitometry is not sensitive enough, a method has been described for toning the silver in the developed X-ray film with mercury-203, followed by pulse counting from the surface of the film[7]. Most people would prefer to elute the sample from the chromatogram for scintillation counting.

Artefacts due to chemography (see p. 116) can occur with large specimens in contact with X-ray film just as in histological material covered with nuclear emulsion. Fig. 40 illustrates a case of positive and negative chemography produced by a section of human femur. Even with relatively pure materials on chromatograms, Richardson, Weliky, Batchelder,

Griffith and Engel[4] and Chamberlain, Hughes, Rogers and Thomas[8] have drawn attention to this type of artefact. Chemography should always be looked for and if possible excluded, by exposing unlabelled but otherwise similar specimens to normal and to fogged emulsion (p. 302). If chemography is found, it is usually sufficient to insert a thin layer of inert material, such as Saran, between specimen and film. Alternatively, the specimen can be coated with a thin layer of polyvinyl chloride (p. 146) or nitrocellulose before exposure. Chemography may be reduced in severity by exposure at low temperatures.

Biochemists are always in a hurry, and a number of methods have been proposed for shortening the exposure time. Polaroid film has been used in several laboratories[9-12]: Type 57 film, with an ASA rating of 3200, is the most sensitive. The chief advantage of Polaroid film is convenience rather than speed: darkroom facilities are hardly needed. Specimens labelled with carbon-14 or sulphur-35 should be introduced into the pack, in contact with the film. Phosphorus-32 or iodine-131 can give satisfactory autoradiographs from outside the pack.

Brisgunov, Rebentish, Gordon and Debabov[13] have described the use of a zinc sulphide screen and image intensifier to produce rapid images from polyacrylamide gels which are still wet. The dried and sectioned gels give clearer and more intense autoradiographs, but only at the expense of a longer delay. Finally, Liss, Lindsey and Catchpole[14] have used a microchannel plate and phosphor screen to generate a rapid image of the distribution of radioactivity in whole-body autoradiographs, and the system is equally applicable to chromatograms. An electron amplification of about $\times 10^4$ is obtainable, with a resolution of about 30 μm.

THE AUTORADIOGRAPHY OF TRITIUM ON CHROMATOGRAMS

Tritium has become such an important radioisotope to biologists that the purification of labelled compounds by the manufacturer, and their separation and identification by the research worker, have produced a considerable literature on the detection of tritium on chromatograms. A bibliography covering the period up to 1968 is given in a supplement to the Journal of Chromatography[15]; more recent papers are listed in the bibliographies published in current numbers of the same journal.

It is possible to autoradiograph tritium on a paper or thin layer chromatogram by placing it in contact with an X-ray film. The efficiency is very low, however. Chamberlain, Hughes, Rogers and Thomas[8] have calculated that 0.3 μCi/cm^2 for 24 hours is needed to give blackening at the very limit of detection above background; a study from Eastman Kodak Company found an optical density in the developed film of 0.10 after 8.0 μCi/cm^2 in 24 hour exposure[3]. The reasons for this low efficiency are the high self-absorption within the chromatogram, and the anti-abrasion coating of gelatin over the emulsion, which provides a very effective barrier to those β's that do leave the surface of the specimen.

Several approaches have been tried to improve efficiency. If the chromatogram could be actually impregnated with emulsion, the improved geometry would result in higher efficiencies. Rogers[16] tried dipping paper chromatograms in molten Ilford K2 emulsion, but the small grain size relative to X-ray film removed most of the advantage gained by better geometry. Markman[17] modified the method by spraying molten emulsion on to the chromatogram, which carries less risk of eluting compounds from the paper or the thin layer. Chamberlain, Hughes, Rogers and Thomas[8] obtained an X-ray emulsion in gel form, Ilford XK, and impregnated the chromatogram with that. They found an improvement in efficiency of about 10 times by comparison with contact exposure to an X-ray film.

This impregnation technique has never been satisfactory, and XK emulsion is no longer commercially available. After impregnation with emulsion, the chromatogram cannot be recovered for a second exposure, nor is elution for scintillation counting possible. Spreading of the spots on the chromatogram is almost inevitable, even with spray application of emulsion, and chemographic artefacts are common with the very close contact of emulsion and specimen. It is also difficult to coat a chromatogram evenly, and high and variable backgrounds combine with efficiencies that often vary within one experiment by a factor of 5.

A much more satisfactory procedure has developed from a method proposed by Wilson[18]. He suggested immersing the chromatogram in a liquid scintillator, converting the β particles into light emission, and detecting this on an X-ray film attached to the chromatogram. As originally proposed, this method also had several major disadvantages. Compounds soluble in the organic solvents in which scintillators are dissolved were eluted from the chromatogram, while the efficiency that was claimed was not dramatically better than that achieved by apposition to X-ray film alone. Parups, Hoffman and Jackson[19] suggested dipping the chromatogram in a saturated solution of anthracene in benzene, and evaporating off the benzene, thus impregnating the paper with anthracene microcrystals, before exposing against X-ray film. Again, the improvement in efficiency was not remarkable.

The advance which permitted this approach to be useful was the discovery that efficiencies could be greatly improved by exposing the chromatogram, scintillator and film at low temperatures. Lüthi and

Waser[20], for instance, made thin layer chromatograms from a mixture of equal parts of silica gel G and finely divided anthracene. For tritium exposed by contact to X-ray film, reducing the temperature from $+4$ to $-30°C$ doubled the efficiency: reducing the temperature to $-70°C$ increased the efficiency 30 times. Randerath[21] worked with cellulose or PE1-cellulose sheets. These were dipped in a 7% solution of the scintillator PPO in diethyl ether, which was then evaporated off. On exposure to X-ray film at $-79°C$, $0.006-0.008$ $\mu Ci/cm^2$ could be detected in 24 hours, while 15–20 times as much was needed at room temperature to give a comparable autoradiograph.

The whole technique has been thoroughly investigated in a series of papers by Prydz and his co-workers in Oslo[22–25]. They have suggested the designation "β-radioluminescence" for the process by which β particles are converted to light photons for the purposes of detection. Several important observations emerge from their work.

In the first place, it was possible to incorporate up to 50% anthracene in the form of small crystals in thin layer chromatograms made of kieselguhr, silica-gel or cellulose without appreciably affecting the R_F values of a range of compounds tested[24]. If the light output of the chromatogram was measured with a photomultiplier tube instead of an X-ray film, a number of scintillators showed an increase in light output down to $-190°C$. While anthracene showed the best light output for a given radioactivity at $-80°C$, benzene was better at $-190°C$. Many scintillators gave different curves of light output against temperature on rewarming than were found on cooling: thermoluminescence peaks were often seen on rewarming at between -60 and $-80°C$ (ref. 22). However, the improvements in light output on cooling were nowhere near as dramatic with anthracene as scintillator, for instance, as were expected from the previous reports in which X-ray film was used as a detector[20,21]. Cooling produced an improvement in light output from a scintillator incorporated in a tritium radio-chromatogram by a factor of 2–5 times, depending on the precise conditions.

The possibility remained that the much greater improvements in sensitivity found with X-ray film were due to some effect of cooling on the photographic process. This was then investigated, using a controlled light source rather than a scintillator–radioisotope mixture[23]. Low temperatures produced very significant improvements in blackening on X-ray film, even with this standard source of photons. It appears that trapping of single electrons in the silver halide crystal in response to photons often proves transient at room temperature, but can be stabilised at very low temperatures, with a greatly increased probability of forming a developable latent image.

There are thus two components contributing to the high efficiency of detection of tritium on chromatograms, when β-radioluminescence and X-ray films are used at low temperatures. The first is a small but significant improvement in light output by the scintillator; the second is a dramatic increase in the sensitivity of the X-ray film for the photons emitted. These combine to lower the threshold of detection for tritium by factors of 15–30 times, depending on the conditions of chromatography and exposure (Fig. 7).

It would seem natural to try this detection system with other β-emitting isotopes, but a number of experiments with carbon-14 on chromatograms have shown that the efficiency of detection is not greatly improved, either by the presence of the scintillator or the additional step of exposure at low temperature[20-22]. One can conclude from this, that the detection of β particles by X-ray film is more efficient than the detection of the photons of β-radioluminescence, and that the latent images produced by single β particles are so large that the stabilisation of trapping of single electrons by lowering the temperature contributes nothing to the overall efficiency of detection.

In summary, then, β-radioluminescence has little to offer to the autoradiography of chromatograms labelled with carbon-14 or other isotopes of similar or higher maximum energies. For tritium and other isotopes of similar energies, where the majority of the β particles never reach the emulsion layer, the conversion of the energy of the particles into light photons by a scintillator enables their presence to be detected. The efficiency of detection can be greatly improved by exposure at temperatures of $-70°C$ or below. Half a loaf is better than no bread.

The substitution of photons for β particles has one interesting consequence. While the deposits of silver at sensitivity specks of silver bromide crystals that have been hit by one β particle are usually large enough to be developed, a single photon may only produce a single atom of silver, and such a single-atom nucleus of silver is very unstable. Multiple hits by photons are needed to build up silver deposits with a high probability of development. In consequence, such scintillation autoradiographs will not have a linear relationship between radioactivity and emulsion response, but a much more complex relationship, with the low intensity reciprocity failure that is characteristic of exposures to light. Laskey and Mills[26] have drawn attention to this problem, and also suggested an ingenious solution. If the X-ray film is given a carefully controlled exposure to light before autoradiography, silver deposits will be formed in crystals throughout the emulsion. At low enough light intensities, these deposits will only increase the probability of development of background crystals by a small amount. Each crystal will become a more efficient detector for further photons, however, since the small silver deposit produced by the pre-exposure acts as a very efficient trap for further photolytic electrons in the crystal, and relatively small increases in the numbers of silver atoms at sensitivity specks can raise the crystal to developability. Fig. 7 shows that this controlled pre-exposure to light can increase the sensitivity of detection of tritium by a factor of 10. It can also result in a linearity of emulsion response which is not possible with scintillation autoradiography alone. Laskey and Mills give a detailed description of their technique in an appendix to their paper[26].

Users of liquid scintillation counting will be all too familiar with the problems of quenching. Energy transfer through the scintillation fluid to the primary scintillator molecules can be prevented by a number of chemical compounds, while the photons produced by the scintillator molecules can be absorbed before reaching the detector by another range of compounds — the processes of chemical and colour quenching, respectively. Quenching can be produced by a wide range of compounds, which need only be present in trace amounts. No data is acceptable in liquid scintillation counting unless quenching has been measured and corrected for, and quenching is so variable from specimen to specimen that corrections have to be found for each sample separately. It is interesting that the literature on the use of scintillators in the autoradiography of chromatograms labelled with tritium contains hardly any references to quenching. It may be that sizeable fluctuations in autoradiographic efficiency are not too important when the measurement of radioactivity depends ultimately on the elution and scintillation counting of the hot spots on the chromatogram. It may also be true that relatively pure compounds in a fairly inert matrix are unlikely to cause very severe or variable quenching. It is interesting to note that scintillation autoradiography has found little favour for whole-body sections, in which a wide variety of reactive groups is present in variable and unpredictable amounts.

A Swedish manufacturer, CEA-Verken, of Strängnäs, has recently produced a special film for work with whole-body sections labelled with tritium[27]. This ^{3}H-Film has a single emulsion layer with a high silver-to-gelatin ratio and large crystal size; it also lacks the customary anti-abrasion coating. In trials with whole-body sections, it has proved to be about 12 times more sensitive than Kodirex or X-Omat R (Fig. 83). It has not so far been evaluated for work with chromatograms. It requires very careful handling if scratches on the surface of the emulsion are to be avoided.

The main emphasis in attempts to autoradiograph chromatograms and similar preparations labelled with tritium has been to increase the efficiency of detection by the X-ray film, without modifying the specimen in any way. Recently, however, Amaldi[28] has proposed the use of poly-

Film type	Relative sensitivity
^{3}H-Film (CEA-Verken)	1
Kodirex (Kodak)	1/12
X-Omat R (Eastman Kodak)	1/12
Singul X-RP (CEA-Verken)	1/16
G5 (Ilford)	1/16
Industrex C (Kodak)	1/64
Structurix D7 (Agfa-Gevaert)	1/64
Kodalith Royal-Ortho (2569) (Kodak)	1/256

Fig. 83. The relative sensitivities of different X-ray films to whole-body sections containing tritium. (From Larsson and Ullberg, 1977)

acrylamide gels which are very thin. Since the major loss of efficiency comes from self-absorption within the specimen, this approach offers considerable scope for improving efficiencies, particularly in association with ^{3}H-Film.

Scintillation autoradiography makes possible the separate detection of carbon-14 and tritium on the same specimen. Gruenstein and Smith[29] have detected carbon-14 on thin layer plates by routine exposure to X-ray film, followed by detection of both isotopes in a second exposure in the presence of scintillation fluid.

There is, of course, no necessity to use X-ray film as the final detector for β-radioluminescence. Scanning devices based on photomultiplier tubes have been described[30,31]. They give a quantitative, two-dimensional picture of the distribution of tritium over a chromatogram. They are at present expensive and complex, however, and there will for many years be a place for the much simpler method of exposure to an X-ray film.

WHOLE-BODY AUTORADIOGRAPHY

This technique was first developed by Sven Ullberg[32] to study the distribution of ^{35}S-labelled benzylpenicillin in mice. From the first publication in 1954, it has continued to grow in importance, and is now widely used in the evaluation of new drugs and in studies on the distribution and effects of environmental pollutants. It might seem strange that the quantitatively accurate method of sampling tissues for scintillation counting should have been successfully challenged by this technique. The sampling implies a selection of those tissues in which radioactivity is expected, while whole-body autoradiography makes no prior assumptions, but surveys all the tissues and organs impartially. A further point in favour of autoradiography is the ease with which the eye and brain can see the patterns of distribution and their variation with time in a series of pictures, whereas the same information presented numerically as a succession of tables and graphs takes longer to assimilate.

Though the technique is called whole-body autoradiography, it is equally applicable to restricted regions of the body. The brain, in particular, is large, complex in structure, and described in terms of neuroanatomical nuclei and fibre systems which are large relative to the resolution of autoradiographs at the light microscope level. If the distribution of radioactivity between such structures is of interest, the exposure of several sections cut at 20 μm to X-ray film, and the use of isotopes such as carbon-14 and sulphur-35 produce autoradiographs of surprisingly good quality (Fig. 84).

Working at the level where the autoradiograph is viewed by the unaided eye has considerable advantages. The techniques of autoradiography are

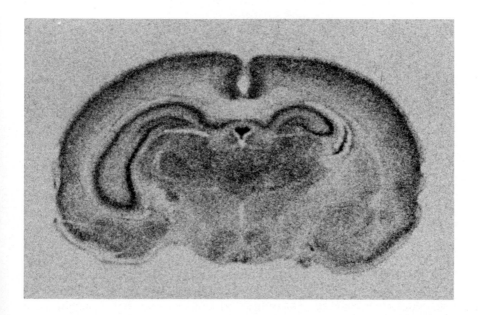

Fig. 84. An autoradiograph to show the relative rates of synthesis of protein in different parts of the brain of the 4-day-old rat. After the injection of a ^{14}C-labelled aminoacid, the brain was fixed, sectioned at 20 μm, and autoradiographed by contact against X-ray film. ($\times$ 10). (Material prepared by Dr. H. Darrah)

324

simple and the emulsion is prepared as X-ray film by manufacturers who ensure reproducible thickness and sensitivity. The standard technique of section preparation preserves diffusible materials in situ. Finally, the analysis of these autoradiographs is relatively uncomplicated, as the resolution in autoradiographic terms is good relative to the size of structures that can be seen in the specimen.

The specimen can be as large as a small monkey. Animals that are pregnant can be studied to see whether the labelled compound enters the foetus[33]. With only slightly modified techniques, the distributions of volatile substances such as chloroform can be studied[34]. It is possible to prepare sections first and label them in vitro before autoradiography, making possible a number of experiments on human tissue.

The technique is designed for isotopes with a maximum energy of β particle equal to or higher than carbon-14 (155 keV). The modifications needed to enable tritium to be recorded are discussed in a separate section later.

It is inevitable that the description of techniques that follows should rely heavily on the work of Ullberg. He has published a recent account of the method in some detail in Science Tools, the journal of the LKB Instrument Company, with a useful bibliography[35].

(a) *Preparing the sections*

Basically, the method relies on freezing the animal and cutting frozen sections. The ideal machine for sectioning is the LKB Cryomicrotome PMV 2250. This has been designed with the needs of the technique in mind, in collaboration with Ullberg. Its only disadvantage is the price. A less elaborate but quite adequate machine is produced by Bright and Co. (Huntingdon, U.K.). For those who find this alternative still too expensive, it is always possible to work with a heavy sledge microtome in a cold room at $-20°C$ or below.

The specimen, usually a mouse, is first frozen on to the stage of the microtome. Ideally, the stage should have a number of projections, like studs, from its upper surface, to give a firm anchoring to the block. The stage should be first covered with a slurry of carboxymethyl cellulose in water (CMC gel), which is frozen into a firm base by immersing the stage in hexane cooled with dry ice, at a temperature of about $-75°C$. The mouse is then positioned on the stage and packed round with more CMC gel until it is completely covered. The stage and specimen are frozen again for a further 20 minutes by immersion in hexane in dry ice.

This procedure is satisfactory for most specimens. If more rapid freezing is required, isopentane cooled with liquid nitrogen may be used, or the mouse may be frozen separately before placing it on the stage in CMC gel.

Cracking of the specimen on freezing may be troublesome: it can be minimised by slow, stepwise immersion in the freezing mixture.

Once frozen, the specimen block is attached to the microtome and allowed to reach the temperature of the microtome cabinet, usually about $-20°C$. The block is then cut down in 100 μm sections until a level is reached which is of interest. Large, frozen sections are very fragile, and the key step in obtaining complete sections is to attach a sheet of transparent adhesive tape to the surface of the block before cutting: the section is then removed on the tape. The most widely used tapes are produced by the Minnesota Mining and Manufacturing Co., St. Paul, Minnesota, U.S.A. The 3M Type 800 and Type 810 have a synthetic adhesive and a cellulose acetate backing: Type 800 is clear, while Type 810 is opaque. Both adhere well to the surface of the frozen block and remain flexible at low temperatures. To cut a whole-body section, attach the tape to the surface of the block, making sure that it is uniformly fixed without intervening air bubbles. Cut the section at 20 μm with a slow but steady cutting speed, pressing the tape gently against the block face just ahead of the knife with a plastic ruler and lifting the tape and section clear of the block behind the knife.

It may help to expose a few thicker sections (say at 100 μm) to get a more rapid picture, though with poorer resolution. It is probably better to take a section adjacent to the one used for autoradiography for staining as a permanent record, rather than using the autoradiographic specimen itself. Sections have been successfully cut as thin as 5 μm; while the resolution will improve slightly, this will be at the expense of longer exposures. Twenty microns is a reasonable compromise. The quality of the knife is critical, and it may be advisable to replace the knife that was used for trimming the block with a fresh one before sectioning starts.

The sections should be kept frozen until they have freeze-dried, before placing them in contact with the X-ray film. It may be convenient to attach them to light frames of plastic at this stage. They can be freeze-dried in the cryostat cabinet, or transferred to a cold room at $-20°C$. The thinner sections (20 μm or less) will freeze-dry overnight, but sections of 100 μm may take 2–3 days. After freeze-drying, the sections should be placed in a desiccator to reach room temperature without condensation producing moisture on them.

Whole-body sections have proved remarkably free from chemography. This is due to three factors, one of which is the preliminary freeze-drying. The low temperature of exposure and the presence of the anti-abrasion coating between specimen and emulsion are the others.

Incubations in radioactive media or the extraction of particular classes of labelled compound can be carried out on the sections while they remain attached to the tape. The two tapes mentioned above will withstand 1–2 hours in aqueous media before softening.

(b) *Autoradiographing the sections*

The choice of X-ray film for isotopes such as carbon-14 has already been discussed in connection with the autoradiography of chromatograms (p. 316).

The dried section can have identifying data written on the tape with radioactive ink before exposure. It is very useful to fix a radioactive scale or ladder to the tape. Several methods of preparing such a ladder have been described[5,36,37].

In the darkroom, the section is placed on the X-ray film, taking care to avoid sliding, scratching movements. Absorbent, soft material such as blotting paper is placed on each side of the sandwich formed by tape, section and film. A block of aluminium is then placed on the outer side of each layer of blotting paper, and spring clips are used to clamp the whole package firmly together. With carbon-14 and sulphur-35, it is possible to clip together several sections and films in the same stack, separated from their neighbours by a single thickness of cardboard. With phosphorus-32 or any of the isotopes that emit γ rays, sections should be singly packaged to avoid crossfire.

Exposure should take place in a lightproof box at -10 to $-20°C$.

After exposure, the film is separated from the section, and development and fixation should follow the recommendations of the manufacturer of the film. The resulting autoradiograph can be used as a negative in a standard photographic enlarger.

(c) *Staining the section*

As mentioned earlier, it may give better results if a section adjacent to the autoradiographic specimen is taken for staining. The section, still on the tape, can be fixed histologically in buffered formalin (4% formalin at pH 7.4 in 0.02 M phosphate buffer), and then stained by dipping for 5 seconds in Mayer's haematoxylin, soaked in water for 15 minutes, and counterstained in eosin (0.4% in 70% ethanol) for 1 second, before dehydration in increasing concentrations of ethanol. The stained and dehydrated section can be mounted on a glass slide under a coverslip, using Euparal (Flatters and Garnett, Manchester, U.K.) as a mounting medium. The presence of the tape does not affect the final product, providing Euparal is present on both sides of the tape.

Alternatively, the stained and dehydrated tape and section can be sprayed with Trycolac (Aerosol Marketing Co., London)[38]. The finished preparation is much simpler to prepare, and is much lighter and less bulky for storage; it is also less expensive.

Many stains may be used on whole-body sections, including histochemical reactions.

(d) *Whole-body autoradiography with tritium*

The technical difficulties of working with tritium are no different to those already described in connection with the autoradiography of chromatograms (p. 317). Self-absorption in the specimen and external absorption by the anti-abrasion coating of the X-ray film combine to give very low efficiencies. Ullberg used to use glass plates coated with Ilford G5 emulsion as an alternative to X-ray film[35]. The new ³H-Film from CEA-Verken (Strängnäs, Sweden) offers efficiencies 10–12 times higher than the best previously obtainable[27]. This ³H-Film has no anti-abrasion coat, and is a single-sided film made from an X-ray emulsion with large crystals and a high silver-to-gelatin ratio. It requires careful handling if scratches are to be avoided, but is clearly the technique of choice at present for tritium-labelled specimens (Fig. 83). Fig. 85 enables one to calculate approximate exposures for whole-body sections containing tritium, autoradiographed with Kodak RP X-Omat film. Exposures for other emulsions including ³H-Film can be calculated with correction factors from Fig. 83.

Farebrother and Creasy[39] have experimented with sections sprayed with 7% PPO in ether, followed by exposure to Kodak Blue Brand X-ray film (one of the screen, or light-sensitive, X-ray films). By exposure at $-70°C$, they have demonstrated that it is possible to reduce exposure times by a factor of between 30 and 60. This application of scintillation autoradiography has its drawbacks, however. The increased efficiency occurs at the expense of resolution, and the distribution of blackening on the developed film is different to that found with ³H-Film. The non-linearity in response to photons, commented upon earlier (p. 320), means that the blackening over highly radioactive areas appears fast, but lightly labelled organs may fail to be detected. Possible effects of quenching have not been commented upon yet in this system: they are likely to be relatively minor since this method effectively covers the specimen with fine crystals of PPO, without any solvent present, so that chemical quenching is unlikely. I feel this approach needs more study before it is accepted as a standard technique in whole-body-autoradiography.

(e) *The interpretation of whole-body autoradiographs*

This presents few problems, by comparison with analysis at the light and electron microscope levels. Self-absorption in the specimen is relatively uniform from tissue to tissue with carbon-14 and sulphur-35, with the one exception of bone. The density seen over compact bone in a section 100 μm thick is about half that over soft tissues with a similar radioactive content[32].

328

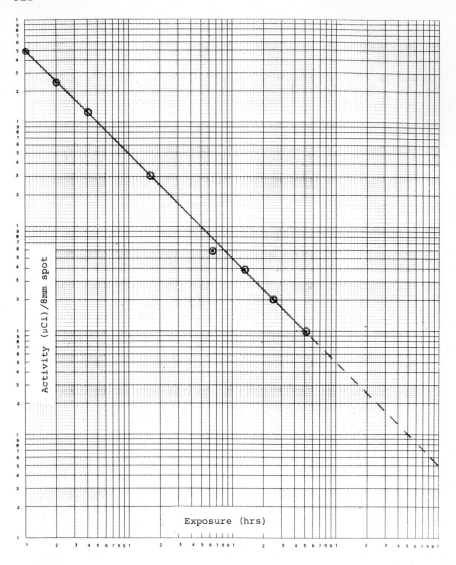

Fig. 85. A graph relating the radioactivity of tritium in a whole-body section to the exposure time in hours needed to give a recognisable blackening in Kodak RP X-Omat film. (Data provided by Mr. R. McCullogh)

It is usual to relate the observed densities on the processed film to that seen over blood in the heart and great vessels. It is relatively simple to withdraw a blood sample for scintillation counting immediately before freezing the animal. The simultaneous exposure of a radioactive ladder to the film permits the densities observed over tissues to be converted into units of radioactivity (Fig. 23). The pencil densitometer of Cross, Groves and Hesselbo[5] is a convenient instrument for measuring these densities rapidly. It is possible, of course, to punch out or cut out from the section small areas for scintillation counting.

One should always remember that the emulsion detects radioactivity, irrespective of the chemical compound in which it is present. Observed distributions following injection of a drug or hormone frequently demonstrate the localisation of metabolites rather than the compound itself. It may be possible to extract one or other source of radioactivity from the section before exposure, or to punch areas out of a 100 μm section for analysis.

(f) *Difficult compounds and tissues*

Compounds which are fat soluble may give poorly localised autoradiographs by the standard methods. This diffusion can be prevented by keeping the section at $-20°C$ from cutting right through to the end of exposure. This means placing it against a film in a cold room or refrigerated cabinet under darkroom conditions[35].

Cohen and Hood have been able to autoradiograph the distribution of volatile anaesthetics[34,40,41]. The animal is frozen in liquid nitrogen, embedded in CMC gel and refrozen in liquid nitrogen. The block is then cut in half with a special saw and the cut face of the block exposed to Agfa-Gevaert Structurix X-ray film at the temperature of either liquid nitrogen or dry ice, depending on the boiling point of the compound under study. The autoradiograph is then compared to a stained section, cut subsequently from the face of the same block. Exposure of such a section, after freeze-drying, will demonstrate the distribution of non-volatile metabolites.

Very hard specimens, such as bones or teeth, may have to be sawn or ground to provide a flat surface for autoradiography. With carbon-14 or sulphur-35 the resolution obtainable from a surface of a solid object is acceptable. With β-emitting isotopes of much higher energy, the advantages of a section become greater.

THE AUTORADIOGRAPHY OF IRREGULAR SPECIMENS

It is sometimes necessary to autoradiograph specimens which cannot be reduced to a flat surface, such as whole skulls in the study of bone deposition during growth. In these circumstances, the choice lies between dipping the specimen in molten emulsion, or spraying emulsion on to it. Personally, I am very reluctant to spray: the fine aerosol of emulsion produced in the confined space of a darkroom is most unpleasant, and probably the silver hairs on the autoradiographer's head that result from it are the least damaging of its effects.

Once again, there is a gain in overall efficiency from the use of a large-grained emulsion. Now that Ilford XK in gel form is no longer available, Ilford G5 or one of the Kodak NTB series would seem to be the emulsion of choice. In addition to the possibility of chemography, an irregular specimen may produce stress artefacts, particularly at abrupt changes in surface contour, such as the sutures of the skull. These can only be guarded against by the autoradiography of non-radioactive controls. Techniques for applying the emulsion, both by spraying and dipping, are described at the end of this chapter.

FUTURE DEVELOPMENTS IN THE AUTORADIOGRAPHY OF CHEMICAL SPECIMENS

The biologist is very apt to associate autoradiography with the localisation of radioactivity in tissue sections. To the biochemist, on the other hand, it is a method of detecting radioactivity on chromatograms, a method that can often be replaced by chromatogram-scanning devices of one sort or another. It seems to me that the characteristics which make nuclear emulsions so valuable for the detection and measurement of radioactivity in single cells also open up very interesting possibilities in the field of microchemical determinations.

Most existing techniques of detecting and measuring compounds of biological interest require the presence of 10^{14} or more molecules in the sample. But, to the cytologist interested in the physiology of a single cell or group of cells, this may be an impossibly high lower limit. Autoradiographic measurements have already been made on biological structures in the range of 10^6–10^7 molecules[42,43]. There is no reason whatever why the same measurements could not be made on purified chemical samples.

Specimens giving as little as 1 disintegration per day can be detected in ideal circumstances by autoradiography, provided only that they are sufficiently small for their effects to be registered by a minute volume of emulsion. Separative techniques that produce a spacing of 20 μm or more between bands that are less than 5 μm wide should give suitable specimens for autoradiography. This is not too ridiculous when one considers the elegant microchemical techniques of Edström[44], who has been able to separate the RNA bases from single nerve cells by electrophoresis along a cellulose fibre.

Many possibilities await exploration in this particular field. Ficq[45] has suggested that nuclear emulsion itself might be the matrix in which separation of related compounds could be effected, and Lambiotte[46] has described a technique for the electrophoretic separation of compounds in an X-ray emulsion.

Prophecy is always a risky business. It seems fair to guess, however, that the high resolution and sensitivity of nuclear emulsions for isotopes that emit β particles, the long exposures and low background levels that can be achieved, and the precision with which absolute measurements of radioactivity can be made, will make autoradiography an attractive prospect to the biochemist interested in developing techniques for the separation and measurement of compounds from single cells or subcellular structures.

DESCRIPTION OF TECHNIQUES

(a) *Impregnation by dipping in Ilford G5 emulsion*

Ilford G5 emulsion in gel form should be stored at 4°C. Like other emulsions in gel form, it should never be frozen. Its shelf-life is at least one month. The safelight should be a 15 W bulb behind an Ilford S 902 filter, at a distance of at least 3 feet.

(i) *Equipment needed.* Thermostatically controlled water-bath at 43°C. Two 50-ml graduated cylinders. One dipping dish, the size and shape dependent on specimen — large and shallow for paper chromatograms, smaller and deeper for small irregular specimens. One pair of plastic print forceps. One glass stirring rod.

(ii) *Procedure.* Following the general procedure on p. 368 for liquid emulsion work, prepare a molten, diluted emulsion in the dipping dish. The final volume will depend on the size of the specimen to be impregnated. The emulsion should be diluted 1 part to 3 parts distilled water, to which glycerol should be added to make 1% by volume of the final solution.

Dip the specimen gently in the diluted emulsion, and allow it to dry with as little handling as possible. Drying should be slow and gentle. The darkroom temperature should be 18–21°C, and the relative humidity around 50%. Circulation of air should be slow, and a direct blast from a fan is not recommended.

After 1.5–2 hours the emulsion should be dry, and the specimen can be placed in a lightproof box for exposure. Dried silica-gel can be placed in the box during the first 24 hours of exposure, provided the specimen does not produce stress artefacts in the emulsion. Over-enthusiastic drying can lead to emulsion separating from the specimen.

(iii) *Development.* Place the specimen in Ilford Phen-X developer, diluted with an equal volume of distilled water, at 20°C for 5 minutes, without agitation. Transfer to a 1% acetic acid stopbath for 2 minutes. Fix in 30% sodium thiosulphate for 15 minutes. Wash in gently running tapwater for 20 minutes and allow to dry.

If difficulties are experienced with emulsion separating from the specimen during exposure or processing, it may help to dip the specimen first in a dilute solution of gelatin (p. 135), which is allowed to dry completely before dipping the specimen in emulsion. A thicker emulsion layer is best achieved by dipping the specimen a second time, after about 30 minutes drying: attempts to increase the viscosity of the emulsion by reducing the concentration of distilled water in the dipping bath usually make for very uneven final coating with emulsion.

(b) *Impregnation by spraying with Ilford G5 emulsion*

(i) *Equipment needed.* The equipment needed is the same as for impregnation by dipping, except that the dipping dish is replaced by an all glass chromatogram spray with large orifice, and a 500-ml conical flask with ground glass joint to fit the spray. A small nitrogen cylinder is also needed, connected to the spray.

(ii) *Procedure.* The molten, diluted emulsion is this time prepared in the conical flask. The emulsion should be diluted 1 part to 7 parts distilled water, with glycerol added to 1% of the final volume.

When the emulsion is prepared, allow it to cool to room temperature in the conical flask before spraying, swirling the flask at frequent intervals to keep the emulsion well mixed. When the emulsion is cool, attach the spray head, apply gentle pressure from the nitrogen cylinder, and spray the specimen with steady, even strokes from a distance of about 9 inches.

Drying should be complete in about 1.5 hours at 18–21°C and 50% relative humidity.

Exposure and development are as for impregnation by dipping in G5 emulsion.

(c) *Removal of emulsion from one side of a processed X-ray film*

Take a large, flat dish big enough to hold the X-ray film. Place 6 layers of filter paper on the bottom of the dish, and pour on enough 10% sodium hydroxide to wet the paper thoroughly without producing puddles over it. Place the film carefully on the filter paper, unwanted emulsion layer downwards, taking care to avoid trapping air bubbles between film and paper. After a few minutes at room temperature, the emulsion layer will disintegrate: this can be easily recognised by the appearance of a reticulated pattern beneath the film. Remove the film, and wash it thoroughly in rapidly running tapwater for 20 minutes.

At no stage should the sodium hydroxide be allowed to wet the upper surface of the film.

REFERENCES

1 R.G. Tsuk, T. Castro, L. Laufer and D.R. Schwartz, in J. Sirchis (Ed.), *Proc. Conf. Methods of Preparing and Storing Marked Molecules, Euratom (EUR 1625 e)*, Brussels, 1964.
2 M.J. Daniels and D.G. Wild, *Anal. Biochem.*, 35 (1970) 544.
3 Eastman Kodak Company Research Laboratories, personal communication.
4 G.S. Richardson, I. Weliky, W. Batchelder, M. Griffith and L.L. Engel, *J. Chromatog.*, 12 (1963) 115.
5 S.A.M. Cross, A.D. Groves and T. Hesselbo, *Int. J. Appl. Radiation Isotopes*, 25 (1974) 381.
6 R. Lim, J.J. Huang and G.A. Davis, *Anal. Biochem.*, 29 (1969) 48.
7 I. Haikal and O. Bobleter, *J. Chromatog.*, 76 (1973) 191.
8 J. Chamberlain, A. Hughes, A.W. Rogers and G.H. Thomas, *Nature*, 201 (1964) 774.
9 D.D. Jackson and M. Kahn, *Int. J. Appl. Radiation Isotopes*, 20 (1969) 742.
10 C.O. Tio and S.F. Sisenwine, *J. Chromatog.*, 48 (1970) 555.
11 B. Nelson and B. Pritchard, *Beitrag. Path. Bd.*, 149 (1973) 206.
12 K.M. Prescott and G.S. Davis, *Anal. Biochem.*, 57 (1974) 232.
13 V. Brisgunov, B. Rebentish, I. Gordon and V. Debabov, *Anal. Biochem*, 66 (1975) 100.
14 R.H. Liss, D.B. Lindsay and C.E. Catchpole, *Nature*, 242 (1973) 523.
15 Bibliography of Chromatography, *J. Chromatog.* Suppl. (1968) 700.
16 A.W. Rogers, *Nature*, 184 (1961) 721.
17 B. Markman, *J. Chromatog.*, 11 (1963) 118.
18 A.T. Wilson, *Biochim. Biophys. Acta*, 40 (1960) 522.
19 E.V. Parups, I. Hoffman and H.R. Jackson, *Talanta*, 5 (1960) 75.
20 U. Lüthi and P.G. Waser, *Nature*, 205 (1965) 1190.
21 K. Randerath, *Anal. Chem.*, 41 (1969) 991.
22 S. Prydz, T.B. Melö, J.F. Koren and E.L. Eriksen, *Anal. Chem.*, 42 (1970) 156.
23 J.F. Koren, T.B. Melö and S. Prydz, *J. Chromatog.*, 46 (1970) 129.
24 L.H. Landmark, A.G. Hognestad and S. Prydz, *J. Chromatog.*, 46 (1970) 267.
25 S. Prydz, T.B. Melö, E.L. Eriksen and J.F. Koren, *J. Chromatog.*, 47 (1970) 157.
26 R.A. Laskey and A.D. Mills, *Eur. J. Biochem.*, 56 (1975) 335.
27 B. Larsson and S. Ullberg, *Science Tools*, Special Issue, 1977.
28 P.P. Amaldi, in D.M. Prescott (Ed.), *Methods in Cell Biology, Vol. 10*, Academic Press, New York, 1975.
29 E. Gruenstein and T.W. Smith, *Anal. Biochem.*, 61 (1974) 429.

334

30 S. Prydz, T.B. Melö and J.F. Koren, *J. Chromatog.*, 59 (1971) 99.
31 E.B. Chain, A.E. Lowe and K.R.L. Mansford, *J. Chromatog.*, 53 (1970) 293.
32 S. Ullberg, *Acta Radiol.*, Suppl. 118 (1954).
33 W.J. Waddell and G.C. Marlowe, *Perinatal Pharmacol. Ther.*, 3 (1976) 119.
34 E.N. Cohen and N. Hood, *Anaesthesiol.*, 30 (1969) 306.
35 S. Ullberg, *Science Tools*, Special Issue, 1977.
36 M. Berlin and S. Ullberg, *Arch. Environ. Health*, 6 (1963) 589.
37 G.W.W. Stevens, *J. Microscopy*, 106 (1976) 285.
38 D.A. Farebrother and N.C. Woods, *J. Microscopy*, 97 (1973) 373.
39 D.A. Farebrother and D.M. Creasy, *J. Microscopy*, 108 (1976) 195.
40 E.N. Cohen and N. Hood, *Anaesthesiol.*, 31 (1969) 61.
41 E.N. Cohen and N. Hood, *Anaesthesiol.*, 31 (1969) 553.
42 A.W. Rogers, Z. Darżynkiewicz, K. Ostrowski, E.A. Barnard and M.M. Salpeter, *J. Cell Biol.*, 41 (1969) 665.
43 Z. Darżynkiewicz, A.W. Rogers and E.A. Barnard, *J. Histochem. Cytochem.*, 14 (1966) 379.
44 J.-E. Edström, *Biochim. Biophys. Acta*, 22 (1956) 378.
45 A. Ficq, in J. Brachet and A.E. Mirsky (Eds.), *The Cell*, Vol. 1, Academic Press, New York, 1959.
46 M. Lambiotte, *Compt. Rend.*, 260 (1965) 1799.

CHAPTER 15

The Stripping-Film Technique

The stripping-film technique as used at present was developed by Berriman, Herz and Stevens[1], of Kodak Ltd., England in close association with Doniach and Pelc[2]. The product, Kodak AR-10 stripping film, and the technique that they evolved have been very widely used ever since, and have probably been the basis for more investigations at the histological level than any other autoradiographic method.

The technique is described by Pelc and Doniach[2,3], and is also summarised in two descriptive leaflets issued by Kodak, Ltd. (Nos. Pl. 1157 and SC-10).

Briefly, AR-10 stripping-film consists of a thin layer of nuclear emulsion carried on a layer of plain gelatin. These arrive from the factory mounted on a glass support, with the gelatin layer in contact with the glass. When it is required for use, an appropriate area of emulsion plus gelatin is cut around with a scalpel or sharp knife, and gently stripped off the glass support on to the surface of water. Here it floats for a minute or two, emulsion side downwards, while it imbibes water, spreading as it does so. When it has reached its fullest dimensions, the stripped film is picked up on the specimen — usually histological material on a microscope slide — by dipping the slide in the water under the film and lifting it out with the film draped over it. As the film dries, it makes very close contact with the specimen. The layer of plain gelatin makes it easier to handle the very thin layer of emulsion in the stripping stage, and protects the emulsion from scratches during drying and exposure. Both emulsion and gelatin remain in contact with the specimen throughout development and preparation for microscopy.

EMULSIONS AVAILABLE

Kodak AR-10 stripping-film is available from Kodak Ltd. (U.K.), and is obtainable in the U.S.A. through Eastman Kodak. There was a recent suggestion that AR-10 would no longer be manufactured. Fortunately, the protests of users were given considerate attention, and Kodak have agreed to continue manufacture. Although the use of this emulsion has declined in recent years, it still plays a vital part in quantitative autoradiography with isotopes of the energy of carbon-14 or higher, and its withdrawal would have left a gap that would have been difficult to fill by liquid emulsion techniques.

Kodak AR-10 arrives from the factory as a 5 μm layer of emulsion on a 10 μm gelatin base. The silver halide crystals have a mean diameter of about 0.2 μm. The emulsion is not fully sensitive to electrons at minimum ionisation: in other words, it would record the passage of β particles of say, 200 keV and over as an occasional developed grain rather than a complete track, if it were to be used in thick layers. On the Ilford scale of sensitivity (see Appendix, p. 420), this is roughly equivalent to the third level of sensitisation. This will be the only stripping film discussed in detail.

Kodak has also provided two other products in the past, which have been described in the literature, but which are now no longer available. Kodak Experimental Stripping Plates V1062 had the same basic emulsion as AR-10 in a layer 4 μm thick, unsupported by the layer of gelatin. They were originally produced for two-emulsion autoradiography for material containing both tritium and carbon-14 (see p. 307). If two separate layers of AR-10 stripping-film are used in this technique, the total thickness of the emulsion and gelatin becomes sufficiently great to make staining, mounting and viewing the specimen rather difficult. The intention was to use V1062 for one of the two emulsion layers, as described by Dawson, Field and Stevens[4]. Some autoradiographers found it a useful product in place of AR-10 for experiments with one isotope alone: the advantage lies in the better staining and visibility of the specimen. The film was more fragile and difficult to handle than AR-10, however.

Kodak's Special Autoradiographic Stripping Plates were designed to give better resolution than AR-10. They had the same emulsion as AR-10, but in a layer only 1–2 μm thick; this was on a gelatin base 5 μm thick. They have been used, for instance, in high-resolution studies of chromosome replication.

Ilford, Ltd., have in the past provided their emulsions as stripping-films. Their emulsions are no longer available in this form. Apart from the Russian Niikhimfoto Type MSM[5], therefore, Kodak AR-10 is now the only stripping-film that is commercially available.

Several papers have appeared recently describing methods for making a sort of stripping-film in the laboratory[6,7]. These techniques are really designed for electron microscopic autoradiography, and will be discussed in more detail in that context. The rest of this chapter will deal with Kodak AR-10.

THE ADVANTAGE OF THE STRIPPING-FILM TECHNIQUE

The relative merits of stripping-film and of liquid emulsion techniques have been discussed in some detail in Chapter 13, p. 291. To summarise the main advantage of stripping-film briefly, it is an excellent compromise designed to meet all the main requirements of autoradiography at the light microscope level adequately. The emulsion layer and the crystal size give reasonably good resolution in the range of 0.3–5 μm, depending on the energy spectrum of the isotope used. It can therefore give satisfactory results in any study of the distribution of radioactivity at the cellular level, and will also, in favourable conditions, distinguish between major cell compartments, such as nucleus and cytoplasm. The uniformity of the emulsion layer, which is claimed by Kodak to be within the range of $\pm 10\%$ of the stated 5 μm, is better than can be obtained by liquid emulsion techniques. Valid comparisons of the radioactivity of different structures can be based on the grain counts observed over them in controlled conditions, even with isotopes of high maximum energy.

The number of experiments that are carried out with isotopes other than tritium is remarkably small. This is in a sense unfortunate, since the problems of making reproducible, quantitative measurements of radioactivity are so severe with tritium. Self-absorption is so significant and measurements of specimen thickness and density are so difficult that there are real advantages to be gained from moving to isotopes of higher energies. The recent work of Dörmer[8] provides an excellent illustration. He has chosen to work with thymidine labelled with carbon-14, rather than with tritium, and has built up a technique of great quantitative precision, based on grain density autoradiographs with AR-10. His measurements of the rates of synthesis of DNA in bone marrow cells would have been very difficult to make by any other method. He and his co-workers have recently investigated the quantitative aspects of autoradiography with iodine-125 also[9,10]. Their work shows well the one situation in which AR-10 holds a significant advantage over other methods at the light microscope level.

In addition to this one clear advantage, stripping-film can give acceptable results in the great majority of other autoradiographic experiments. The technique is not difficult, and the shelf-life of the film is of the order of 6 months under reasonable conditions. The laboratory in which an autoradiographic experiment is only needed occasionally would clearly do well to standardise its procedures around this reproducible technique, for which a great deal of data is already available in the literature.

THE LIMITATIONS OF THE STRIPPING-FILM TECHNIQUE

A single compromise technique like stripping-film autoradiography, which gives satisfactory results in a number of experimental situations, will often be inferior to a more specialised technique, designed for one particular application. If, for instance, the highest possible resolving power in the light microscope is required, a very thin layer of a fine-grained emulsion such as Ilford L4 will be superior to the thicker emulsion and larger crystal size of AR-10 stripping-film. If high efficiency is particularly important, with very low levels of labelling or a very short isotopic half-life, a thicker layer of a more sensitive emulsion, such as Ilford K5, will give better results. If quantitative precision is important, the more specialised techniques of β-track autoradiography will be preferable. Inevitably, the single sensitivity, the one diameter of halide crystal, and the one emulsion thickness available, place stripping-film at a disadvantage with respect to the wide range of products and of techniques possible with liquid emulsions.

In addition, there are a number of drawbacks inherent in the stripping-film technique itself which should be mentioned. The process of stripping the film from the glass support inevitably involves more handling, with the chance of creating a higher initial background, than the relatively mild methods of liquid emulsion work. In inappropriate conditions of temperature and humidity in the darkroom, stripping can cause visible flashes of static discharge between film and glass support, resulting in a very high background. It is, in addition, almost impossible to keep the water surface of the stripping bath completely free of dust, and this is the very surface that becomes trapped between emulsion and specimen. The time required for the emulsion to swell on the surface of the water places a limit on the speed of preparation of autoradiographs. If batches of 30–50 slides have to be covered with emulsion, dipping in liquid emulsion is quicker than using stripping-film.

When one considers development, fixation and washing, stripping-film adheres to slide and specimen less firmly than a layer of liquid emulsion applied by dipping. It is possible for the film to slip relative to the specimen, particularly in the final stage of washing after fixation: it may even be lost altogether if it is carelessly handled. These problems are far more acute if the specimen has been covered with an inert layer of, for instance, polyvinyl chloride (p. 146) before applying the film. In these conditions, it is difficult to keep the film in place, even if the PVC membrane has been coated with dilute gelatin. This tendency of the film to move in processing can be minimised. The film is permitted to spread fully before applying it to the slide by floating on water at 25°C for a full 3 minutes, while the temperature of all processing solutions is kept to 18°C or below. The film can be wrapped around 3 sides of the slide when it is picked up from the stripping bath. Finally, the very considerable swelling of the emulsion that occurs in the usual stopbath of distilled water can be reduced by controlling the salt concentration of the stopbath, following a suggestion by Stevens[11].

In preparing stripping-film autoradiographs for microscopy, it sometimes happens that small pockets form under the film in and around the specimen, with a different refractive index to the rest of the preparation. These are probably due to a failure of the mounting medium to penetrate the film completely. Many remedies have been suggested for this: the use of polyvinyl alcohol is quite effective[12], and will be described later (p. 347).

Finally, the presence of the layer of gelatin above the emulsion may have, as noted above, an unfortunate effect on the appearance of the specimen under the microscope. Many stains applied through the emulsion after development and fixation are taken up to some extent by gelatin. This will be a much more noticeable effect in the presence of the relatively thick layer of plain gelatin of stripping-film than with a single layer of liquid emulsion. Nor is this disadvantage purely cosmetic. It may make difficult the use of dark-field incident illumination for viewing and photography, and produce a high background in incident light photometry from light scattered by the stain in the gelatin layer. Dörmer has shown that this problem can be adequately controlled, however, and has generated large series of photometric measurements of grain density from stained smears of bone marrow cells[8-10].

Many of these limitations of stripping-film autoradiography can be minimised by appropriate techniques. They are all, however, relevant to the selection of the most suitable method for any proposed autoradiographic experiment. In spite of its drawbacks, stripping-film remains the only basic technique that is applicable to almost the entire range of observations that one might wish to make at the light microscope level.

USEFUL FACTS AND FIGURES ABOUT AR-10

In a series of tests with stripping-film on the surface of water at 25°C, the area of the film was found to increase by a factor of 1.38. This seemed to be reached in 2–3 minutes, and no further increase followed over 15–20 minutes. This implies that the thickness of the emulsion layer during exposure is approximately 3.6 μm.

The uniformity of the emulsion layer appears to be well within the tolerance quoted by Kodak, Ltd., in a series of experiments in which

AR-10 was applied to a plane surface on a microscope slide. Emulsion thickness probably varies slightly more when the film is applied to a specimen with an irregular upper profile, such as a paraffin-embedded section after dewaxing.

The grain yield of stripping-film, i.e. the number of developed grains per incident β particle, has been calculated by a number of authors for different isotopes[8]. The values range from around 1 for tritium down to about 0.7 for phosphorus-32. Naturally, the efficiency of stripping-film, defined as the number of developed grains per disintegration in the source, varies widely with the nature of the specimen. The efficiency is a function of the autoradiograph as a whole, and includes factors such as self-absorption. Hughes, Bond, Brecher, Cronkite, Painter, Quastler and Sherman[13] give a figure of 1 developed grain per 20 disintegrations in smears of white blood cells labelled with tritium. The figure of 1 silver grain per 100–200 disintegrations is sometimes quoted for the efficiency of a 3 μm section labelled with tritium[14]. There are technical grounds for doubting the validity of this measurement[15], which is inconsistent with several other values in the literature similar to that of Hughes and co-workers quoted above.

Stripping-film, as used in the normal way, is clearly unsuited to the autoradiography of soluble materials. Attempts have been made to float the stripped film on the surface of mercury, to avoid the necessity of immersing the slide in water while picking up the film. This technique has little to recommend it. It is possible to use stripping-film in the Appleton[16] or Stumpf and Roth[17] techniques for diffusible material by covering the slide with film, gelatin layer in contact with the glass, and drying the film thoroughly, before placing the section in contact with the emulsion layer (see Chapter 8).

Stubblefield[18] has exploited the relatively poor adhesion between stripping-film and slide in an ingenious technique to improve the statistics of grain counting over small sources. If the grain densities over such sources are kept low to avoid a high probability of adjacent crystals receiving double or multiple hits, it will never be possible in a single exposure to collect more than 1 or 2 grains per source. In his technique, a relatively short exposure is made, the autoradiographs are developed and photographed, and the film floated off the specimen. A second layer of film is then put on the slide, which is exposed for the second time. By serial short exposures in this way it is possible to build up quite high grain counts over sources as small as mammalian chromosomes.

O'Callaghan, Stevens and Wood[19] have drawn attention to the leaching of bromide ions out of the stripping-film while it is floating on the water prior to picking up, and to the increased rate of build-up of background that may result during long exposures. They have shown that floating the film on a solution of potassium bromide and glucose can significantly improve the signal-to-noise ratio and the reproducibility of autoradiographs. This step should be incorporated in the technique whenever exposure times of more than 2 or 3 weeks are anticipated. It is described in detail later (p. 343).

Some papers have given the unfortunate impression that latent image fading and negative chemography are unlikely to occur with AR-10 stripping-film (see, for instance, Pelc, Appleton and Welton[20]). There is no reason to expect AR-10 to be exempt from effects which are basic to the nature of the photographic process itself, and no excuse for neglecting the controls which should be part of every autoradiographic experiment. Dörmer[8] illustrates negative chemography found over bone marrow cells with AR-10, and I have also several times encountered similar problems with this emulsion.

ATTEMPTS TO INCREASE THE EFFICIENCY OF AR-10

There have been several attempts to increase the efficiency of AR-10. Herz[21] suggested hypersensitisation with triethanolamine prior to exposure, but I know of no experiments which have used this approach.

Latensification with gold salts prior to development should produce a significant increase in efficiency. The methods are described later (p. 409), and should be transferrable to AR-10 with little modification. Since latent image intensification increases the probability of development of classes of silver halide crystal with relatively small deposits of silver at sensitivity specks, great care must be taken in all steps of preparing and developing the emulsion to keep to a minimum all the hazards which can cause background grains. Exposure to safelighting should be minimised strictly, and drying must be very slow and gentle. Otherwise, the main effect of gold latensification will be a very considerable increase in background, swamping the improvement in efficiency.

Papers by Platkowska[22], Sawicki, Ostrowski and Platkowska[23], and Mukherjee and Chatterjee[24] have claimed very significant increases in efficiency from exposure in scintillation fluid at $-70°C$. From first principles, it is unlikely that such great improvements could be achieved. The major loss of efficiency with tritium comes from self-absorption in the section or smear, and it is difficult to believe that a brief immersion in scintillation fluid will effectively impregnate the section with scintillator through the overlying film. Photons have a longer range than β particles, and produce one-atom nuclei of silver in hit crystals instead of the multi-atom nuclei characteristic of hits by charged particles. Combining these two factors, one would predict a rather different distribution of hit crystals over a source if effective energy conversion to photons is taking place within the specimen. There should be a number of crystals over the source with relatively large deposits of latent image silver in them, which have been hit by many photons. At increasing distances from the source, there will be crystals with smaller and smaller deposits. If methods of development evolved for the visualisation of the relatively large silver deposits produced by β particles are used unchanged on such an emulsion, only the crystals over the source with large deposits from multiple hits will be developed. This will give no great loss in resolution, but also no great increase in efficiency. More significant increases in efficiency should follow if carefully adjusted development, with gold latensification, results in the visualisation of more and more crystals with relatively small silver deposits in them, but this can only occur with a corresponding loss of resolution. It is inconceivable that much higher efficiencies could be achieved with no loss of resolution, by routine development procedures. Yet this is what is claimed.

Before scintillation autoradiography can become an accepted method at the light microscope level, it must be clearly shown that its benefits are reproducible and reliable. Many attempts to reproduce the results claimed in the literature have failed. It may be that quenching of the scintillation process by reactive groups present in the specimen is partly to blame. If the system becomes reliable, which it is not at present, autoradiographers will still have to face the problems of a non-linear emulsion response to photons — a problem clearly recognised in the scintillation autoradiography of chromatograms[25] and whole-body sections[26].

DETAILED DESCRIPTION OF TECHNIQUE

The shelf-life of AR-10 is about 6 months. It is best kept at about 4°C, in the plastic bags inside the box in which it comes.

The conditions within the darkroom are very important to successful autoradiography. The temperature should be 18–21°C, the relative humidity ideally between 60 and 65%. If it is too dry, spontaneous stripping of the film away from its glass support may occur, and static discharge between film and glass will give high levels of background. If it is too humid, it may be difficult to strip the film at all. There should be no violent air currents, such as are created by a powerful exhaust fan. The emulsion should be handled and processed at least 1.1 m (4 feet) from a safelight fitted with a 25 W bulb and a Kodak safelight filter No. 1 (red). The best position for the safelight is above bench height, so that the light can be seen reflected in the waterbath while one is standing in the spot most convenient for working. As always, the darkroom should be clean: a dusty room will make it impossible to keep the surface of the water free from specks, which will be trapped between emulsion and specimen.

The stripping-film should be brought out of the refrigerator an hour to two before it is needed, to allow it to come to room temperature,

(i) *Equipment needed.* A dish containing distilled water at 25°C will be needed; for exposures which may last more than 2–3 weeks, the water should be replaced with a solution of potassium bromide (10 mg/litre) with glucose (5%)[19]. The size of the dish is not very critical, but a bigger dish will hold more pieces of film while they are imbibing water and spreading. The water should be at least 1.5 inches deep, and its surface should be near the top of the dish, so that there is no awkward rim around the water to interfere with picking up the film. It is easier to see the film clearly if one is working against a black background: I use a glass dish standing on a piece of black paper.

A clean, sharp knife or scalpel and a pair of forceps (not rat-toothed) will also be required. It is useful to have a pair of straight scissors available. The surface of the slide must, of course, be gelatinised to get good adhesion between emulsion and slide, unless a blood smear or similar specimen is present. The smear is easier to make on a clean slide without gelatin, and the protein film produced by the dried smear provides sufficiently good adhesion.

(ii) *Packaging of the emulsion.* Stripping-film plates come in boxes of 12: each box contains 3 plastic bags of 4 plates each. Fig. 86 illustrates the way in which the 4 plates are stacked: the innermost pair of plates are in

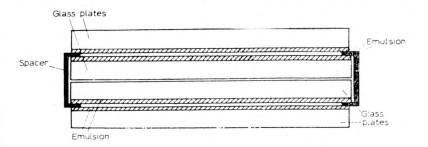

Fig. 86. A diagram to illustrate the packing of Kodak stripping-film plates, with cardboard spacers to prevent the emulsion surfaces from coming into contact.

direct contact, back to back, and their emulsion-coated surfaces face the emulsion of the outer two plates, from which they are separated by small cardboard spacers. It is important to prevent the emulsion from being scratched when returning unused plates to their plastic containers. It is often a good idea to reconstitute the stack of 4 plates just as they were received, before putting them away again, even if one or two of the plates have had all their emulsion removed, as these used plates can still protect the emulsion on the other ones.

(iii) *Stripping the emulsion from its glass support.* If there is any doubt which side of the glass is covered with film, gentle scratching with a fingernail at the very corner of the plate will at once give the answer. The film is firmly attached to the glass for a distance of about 0.5 cm from the edge of the plate. With the knife, cut around an oblong area of film at one corner of the plate, starting about 1 cm in from the edge: the size of the piece of film depends on the specimen to be covered. Remembering that the film will spread on the surface of the water, it should be big enough to lap over 3 sides of the microscope slide, or whatever the specimen is mounted on, and to cover the specimen with film extending at least 1 cm from it on all sides. Given this minimum size, nothing is gained by cutting pieces of film any bigger.

Usually, the cut piece of film will lift slightly from the glass at its edges, so that it can be held gently at one corner with the forceps. If it does not do this, one corner can be detached from the glass with the edge of the knife or scalpel. The piece of film should be grasped in the forceps by the corner nearest the centre of the plate, and stripped from the glass with a steady, slow movement. Always strip away from the film that is left on the plate, to avoid scratching it (Fig. 87). If the film crackles on stripping, or minute flashes of light due to static electricity are seen, it often helps to breathe gently on it.

During stripping, hold the plate vertically, near the surface of the water, and strip the piece of film off downwards towards the water. In this way, as the film separates from the glass, it will be facing the water surface, emulsion side downwards. As the film parts company with the glass, it should be placed on the surface of the water. It is a good idea to practice this in the light before attempting it for the first time. With very little effort, one gets the knack of floating the film in an unwrinkled condition on the water. If the film is very wrinkled or curled up, remove it from the waterbath and strip another piece: it is a waste of time to try and straighten it out.

Difficulties with this stage of stripping are usually due to incorrect conditions of temperature and humidity in the darkroom. Occasionally,

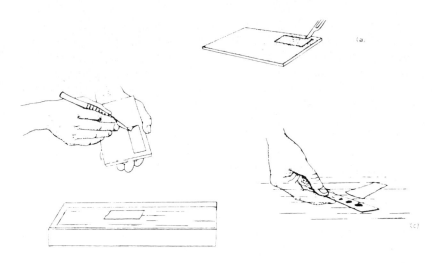

Fig. 87. Three diagrams to illustrate the preparation of stripping-film autoradiographs. (a), An area of film is cut with a sharp scalpel, taking care to leave a margin of 0.5 cm around the edge of the plate. (b), Holding the plate vertical near the surface of the water, a corner of the cut square of film is taken with forceps, and the square stripped off the plate on to the water, emulsion surface downwards. Care should be taken to strip away from the rest of the film on the plate. (c), After the cut square of film has imbibed water for 3 min it is picked up on the surface of the slide that has the specimens on it. The emulsion thus achieves very close contact with the specimen.

large areas of film strip spontaneously from the glass, particularly after a few pieces have already been taken off. It is almost impossible to cut this peeling film with a knife or scalpel, but this can be done with scissors, and suitable bits of film salvaged for use. Avoid holding the plate up at face level during stripping: it will probably be too near the safelight at this height, and the shorter the distance the stripped film has to travel to the water surface, the less chance there is of it curling up in transit.

Experts in the use of stripping-film claim that its appearance changes while it is on the water, and that it becomes "granulated" when spreading is complete and it is ready for picking up. I must admit I have never been convinced of this change, and I prefer to leave the film floating for 3 minutes by the darkroom clock before going on to the next stage.

The forceps should be carefully dried before stripping the next piece of film, otherwise it will stick to them instead of floating free on the water.

(iv) *Picking up the film.* Holding the slide at the end furthest from the specimen, place it in the water and move it under the piece of film, at an angle of about 30° to the horizontal. Then lift the film out of the water on the slide. One edge of the film should attach itself to the slide first, the rest folding neatly around the slide as the latter comes out of the water. If there are wrinkles, or the film is not covering the specimen properly, do not attempt to move the film along the slide while it is out of the water, but try to refloat the film and pick it up again. If things get too complicated at this stage, reject the piece of film and start again.

Around the edge of the strip of film, the background may be higher, due to the act of cutting and the grasp of the forceps. It is important to get a reasonable margin of film around the specimen, so that it lies under an area of low background.

(v) *Drying and exposure.* The slide with film on it should stand vertically to dry in a gentle stream of cool air for 20–30 minutes. Sawicki and Pawinska[27] have drawn attention to the lower background levels that result from very slow, and possibly even incomplete, drying of the emulsion. A certain amount of moisture left in the film would favour fading of the latent

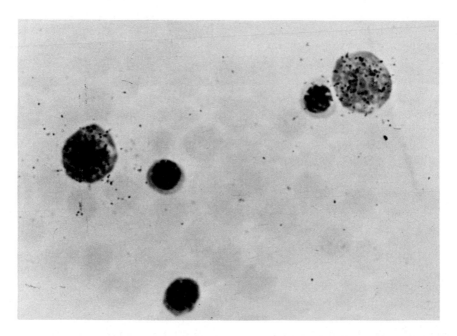

Fig. 88. An autoradiograph of a smear of human bone marrow cells from a patient with spherocytosis. The cells have been incubated in vitro with [^{14}C]thymidine and exposed to AR-10 stripping-film for 25 days. An amidol developer was used and staining with Giemsa. ($\times$ 1200). (Picture kindly provided by Dr. P. Dörmer)

image. Since an appreciable background is usually present at the start of exposure from the process of stripping, this fading would result in an initial decrease in background (p. 124).

Yet the drying must be complete if latent image fading is to be avoided. A suitable compromise is to allow the slides to remain in the darkroom for a further 1 hour, and then to transfer them to a desiccator at room temperature over dried silica-gel for 18–24 hours. After this, they can be put away in lightproof boxes to expose. This drying routine appears to be somewhat more reproducible than exposure in the presence of a few crystals of drying agent, which may on occasions become saturated. If long exposures or latent image regression from any cause are anticipated, exposure should take place in an atmosphere of carbon dioxide or an inert gas[21], at a temperature of $-20°C$ or lower. Otherwise, exposure in air at 4°C, or even at room temperature[28], is usually satisfactory.

Very low background levels can be achieved with stripping-film with experience (Fig. 88).

(vi) *Development and fixation.* These stages should all be carried out at a temperature of 18°C. At this temperature, the emulsion will not swell to the same extent as it did at 25°C, thus reducing the chances that it will part company with the slide.

The recommended developer is Kodak developer D19

"Elon"	2.0 g
Sodium sulphite (anhydrous)	40.0 g
Hydroquinone	9.0 g
Sodium carbonate (anhydrous)	38.0 g
Citric acid	0.7 g
Potassium metasulphite	1.5 g
Distilled water to	1000 ml

The chemicals should be dissolved in the order listed, and the developer used without further dilution.

Development should take about 5 min at 18°C.

Rinse for 30–60 seconds in distilled water.

Fix in 30% w/v solution of sodium thiosulphate for 10 minutes. Alternatively, Kodak Acid Fixer may be used.

Wash in gently running tapwater for 5 minutes.

After staining, the slides can be allowed to dry in air, and viewed dry or with immersion oil applied directly. Alternatively, they can be dehydrated, cleared in xylene and mounted under a coverglass for microscopy in the usual way. If the latter process produces opaque areas around the specimen under the film, the slides can be treated with polyvinyl alcohol during dehydration[12]. After rinsing the slide in 50% alcohol, it is transferred to a

2% solution of polyvinyl alcohol ("Elvanol", Dupont 51-05) in 50% alcohol for 1–2 hours: it is advisable for the slide to be lying horizontal during this time. After drying in the air, the slide can be dipped in absolute alcohol, followed by xylene, and mounted under a coverglass in one of the conventional histological mounting media.

With plant tissues the variations in section profile may be very considerable (Fig. 51), with cellulose structures remaining at the original section thickness while other elements shrink on fixation and dehydration. Jona and Goren[29] have found the Elvanol method unsuccessful with plant material, and recommend impregnation with 30% glycerol, or a gelatin–glycerol mixture, before mounting under a coverslip.

Swelling of the emulsion in the stopbath, with possible displacement of the image, can be minimised by substituting a rinse in Kodak SB-4 stopbath for the more usual distilled water[11]. This stopbath has the following formula:

Potassium chrome alum	30 g
Sodium sulphate (anhydrous)	60 g
(Alternatively, crystalline sodium sulphate	140 g)
Distilled water to	1 litre

Agitate the slides for 30–45 seconds in the stopbath, and then leave them for 3 minutes. Discard the stopbath after use.

REFERENCES

1 R.W. Berriman, R.H. Herz and G.W.W. Stevens, *Brit. J. Radiol.*, 23 (1950) 472.

2 I. Doniach and S.R. Pelc, *Brit. J. Radiol.*, 23 (1950) 184.

3 S.R. Pelc, *Int. J. Appl. Radiation Isotopes*, 1 (1956) 172.

4 K.B. Dawson, E.O. Field and G.W.W. Stevens, *Nature*, 195 (1962) 510.

5 M.F. Merkulov, T.N. Garrilova, I.F. Razorenova and V.A. Myl'tseva, *Byull. Eksp. Biol. Med.*, 24 (1972) 120.

6 J.R. Williamson and H. van den Bosch, *J. Histochem. Cytochem.*, 19 (1971) 304.

7 R.W. Burry and R.S. Lasher, *J. Microscopy*, 104 (1975) 307.

8 P. Dörmer, in V. Neuhoff (Ed.), *Molecular Biology, Biochemistry and Biophysics, Vol. 14*, Springer-Verlag, Berlin, 1973.

9 E. Thiel, P. Dörmer, H. Rodt and S. Thierfelder, *J. Immunol. Meth.*, 6 (1975) 317.

10 E. Thiel, P. Dörmer, W. Ruppelt and S. Thierfelder, *J. Immunol. Meth.*, 12 (1976) 237.

11 G.W.W. Stevens, *Microphotography*, 2nd ed., Chapman and Hall, London, 1968.

12 M.J. Schlesinger, H. Levi and R. Weyant, *Rev. Sci. Instruments*, 27 (1956) 969.

13 W.L. Hughes, V.P. Bond, G. Brecher, E.P. Cronkite, R.B. Painter, H. Quastler and F.G. Sherman, *Proc. Nat. Acad. Sci. (U.S.A.)*, 44 (1958) 476.

14 W.E. Kisieleski, R. Baserga and J. Vanpotic, *Radiation Res.*, 15 (1961) 341.

15 H.K. Oja, S.S. Oja and J. Hasan, *Exptl. Cell Res.*, 45 (1967) 1.

16 T.C. Appleton, *J. Roy. Microscop. Soc.*, 83 (1964) 277.

17 W.E. Stumpf and L.J. Roth, *Stain Technol.*, 39 (1964) 219.

18 E. Stubblefield, *Fed. Proc.*, 23 (1964) 332.

19 C. O'Callaghan, G.W.W. Stevens and J.F. Wood, *Brit. J. Radiol.*, 42 (1969) 862.

20 S.R. Pelc, T.C. Appleton and M.E. Welton, in C.P. Leblond and K.B. Warren (Eds.), *The Use of Radioautography in Investigating Protein Synthesis*, Academic Press, New York, 1965.

21 R.H. Herz, *Lab. Invest.*, 8 (1959) 71.

22 E. Platkowska, *Bull. Pol. Acad. Sci.*, 24 (1976) 701.

23 W. Sawicki, K. Ostrowski and E. Platkowska, *Histochemistry*, 52 (1977) 341.

24 A.S. Mukherjee and R.N. Chatterjee, *Histochemistry*, 52 (1977) 73.

25 R.A. Laskey and A.D. Mills, *Eur. J. Biochem.*, 56 (1975) 335.

26 D.A. Farebrother and D.M. Creasy, *J. Microscopy*, 108 (1976) 195.

27 W. Sawicki and M. Pawinska, *Stain Technol.*, 40 (1965) 67.

28 W. Sawicki, K. Ostrowski and J. Rowinski, *Stain Technol.*, 43 (1968) 35.

29 R. Jona and R. Goren, *Stain Technol.*, 46 (1971) 156.

CHAPTER 16

Liquid Emulsion Techniques for
Grain Density Autoradiographs

Belanger and Leblond[1] pioneered the use of liquid emulsions in order to achieve very close apposition between specimen and emulsion. Their original technique involved melting the emulsion, and applying it to the specimen by means of a paintbrush. In 1955, Joftes and Warren[2] published a method in which the slide, with specimen on it, was dipped bodily in molten emulsion. This technique was described again, in slightly modified form, by Joftes[3]. Messier and Leblond[4] adapted this basic procedure of dipping the slide in liquid emulsion to their own requirements, and further modifications to the technique were described by Kopriwa and Leblond[5]. This process of dipping in liquid emulsion is the most widely used autoradiographic method in biological laboratories at the present time.

Before going further into descriptions of techniques it would be well, at the risk of repeating statements made in Chapter 13, to discuss briefly the value and the very real limitations of autoradiographs prepared in this way.

THE ADVANTAGES AND LIMITATIONS OF
DIPPING TECHNIQUES

After dipping, the biological specimen is covered, and even to some extent perhaps impregnated, with emulsion. No other method of applying the emulsion gives such close contact between emulsion and specimen. It is possible, by regulating the dilution of the molten emulsion, to produce a very thin layer over the specimen. Emulsions are available for autoradiography with crystal diameters significantly smaller than that of Kodak AR-10 stripping-film. If one takes advantage of all these possibilities, the resulting autoradiograph will have a significantly better resolution with many isotopes than can be achieved with stripping-film. If high resolving

351

power is needed in an autoradiograph for the light microscope, a technique based on dipping in liquid emulsion offers the best possibilities.

The dipping technique has other advantages. It is very quick and simple to prepare the autoradiographs, and the layer of gelatin that remains after processing is so thin that staining, mounting under a coverglass, viewing and photomicrography are all improved relative to stripping-film. The emulsion layer adheres to specimen and slide much more firmly than with stripping-film, making it simpler to modify the basic technique for use with impermeable layers of plastic (see p. 146). There is a wide range of emulsions available, differing in sensitivity and in crystal diameters.

The one great limitation to this type of autoradiograph is that it is all but impossible to produce an emulsion layer of constant and reproducible thickness. Leblond, Kopriwa and Messier[6] have investigated this in considerable detail. Even with their experience of this technique, the best they can achieve is a fairly uniform thickness over an area of perhaps one-half of the microscope slide. They present no data on the variations of emulsion thickness from one slide to the next.

If variations of emulsion thickness can occur within one series of autoradiographs, comparisons of the radioactivity within different structures become very difficult. It must be shown that the observed variations in grain density are not consequences of the differences in emulsion thickness. It is in this situation, where comparisons of grain density must be made, that stripping-film has greater precision.

There are special circumstances, however, in which comparisons of grain density are valid with the dipping technique. With tritium and several isotopes that emit extra-nuclear electrons, the energies of the emitted particles are so low that few, if any, might be expected to travel more than 2 μm through emulsion. If the emulsion is nowhere less than 2 μm thick, the grain densities due to these isotopes will be independent of variations in thickness. The thickness of the recording layer is in effect determined by the maximum range of the particle in this case. It is quite acceptable to use a dipping technique for quantitative work with such isotopes, provided the emulsion layer is sufficiently thick.

It is sometimes possible to base comparisons of radioactivity on grain densities in liquid emulsion autoradiographs, even with isotopes of higher energy. Variations in emulsion thickness occur over relatively large distances, and it is unlikely that significant differences will be found over two sources separated by only a few microns. In experiments with sulphur-35 and emulsion layers about 3 μm thick, paired observations were made over experimental and control areas about 5 μm apart. Although it was clear from the grain counts that considerable differences in emulsion thickness were occurring from one section to another, the statistical treatment of these paired observations produced valid and useful results, which were independent of this variation[7].

To sum up, liquid emulsion techniques can provide grain density autoradiographs of high resolution. Comparisons of radioactivity between one site and another are possible, provided it can be shown that variations in emulsion thickness are not affecting the grain counts. With a thin emulsion layer and very small developed grains, the best possible conditions for viewing the underlying specimen can be realised.

THE CHOICE OF EMULSION

Some of the emulsions available commercially are listed in the Appendix with a note of their major characteristics. Most of the published work has involved the Eastman Kodak NTB-2 and NTB-3, or the Ilford emulsions, so that detailed discussion will be limited to these. Agfa-Gevaert, who used to manufacture NUC-715 and NUC-307, have ceased to make nuclear emulsions. Fuji and Sakura emulsions are manufactured on a limited scale in Japan, but are not available to overseas purchasers. The Russian nuclear emulsions are likewise not sold abroad.

It is seldom appreciated that the Eastman Kodak and the Ilford emulsions differ considerably in many respects. To take only one example, a slide dipped in Kodak NTB-3 may be covered by an emulsion layer 4 μm thick while, using identical conditions, Ilford G5 will give a layer 25–30 μm thick. Not only does the physical consistency of the emulsions from the two sources vary, but also the sensitivity to latent image fading, the safelighting requirements, and the extent of formation of background fog on extreme drying.

The techniques that have been described in the literature for Eastman Kodak emulsions do not give optimal results with Ilford products, and vice versa. It follows that one must select a technique appropriate to the source from which emulsions will be normally obtained.

(a) Eastman Kodak emulsions

The two emulsions generally used for grain density autoradiographs in light microscopy are NTB-2 and NTB-3. The former is less sensitive, being suitable for tritium, iodine-125, and perhaps even carbon-14 and sulphur-35; the latter should be used for isotopes of higher energy. The less sensitive NTB-2 has a slightly lower background, as would be expected.

Both emulsions have the same physical characteristics, differing only, so far as one can judge, in the degree of sensitisation. It follows that procedures devised for the one are fully applicable to the other. Both are reproducible in use, and have a low level of background.

NTB emulsions can be stored for up to two months, preferably at about 4°C. Nuclear emulsions in bulk should never be frozen.

The emulsion is usually melted in the glass or plastic container in which it was received. The same batch of emulsion may be melted and used several times without deterioration. It is usually not diluted, although dilution with distilled water is possible without altering the properties of the dried emulsion. From the data presented by Leblond and his co-workers[4,6], it appears that NTB emulsions are very liable to latent image fading. Careful drying, together with exposure in the presence of a drying agent, are recommended. These emulsions require strict safelighting conditions.

(b) Ilford emulsions

There are three series of emulsions produced by Ilford, designated G, K, and L, respectively. They differ from each other in the crystal diameter, and in other characteristics (see Appendix, p. 420). The development routine that is optimal, the liability to latent image fading, and so on, are not identical in the three series.

Within each series, emulsions of different sensitivities may be obtained. The level of sensitisation is indicated by a number after the series letter: thus level 5 is the highest normally produced, and G5 and K5, while differing in crystal diameter, have the same sensitivity. The emulsions available are G5 and K5, for isotopes of high energy, and K2, which is excellent for tritium and iodine-125. In the L series, the only level of sensitisation available is L4.

Ilford emulsions should not be stored longer than two months: 4°C is the optimal temperature for the bulk emulsion. As with the NTB emulsions, the bulk emulsion should never be frozen.

The Ilford emulsions should not be heated more than once, as this produces a high background. On each occasion that slides are to be dipped, a known amount of emulsion should be taken from the stock jar: this may be done by weighing the emulsion or, more conveniently, by measuring it in a graduated cylinder. The emulsion is then melted, and added to a known volume of distilled water. This dilution is essential to obtain an emulsion layer thin enough for grain density autoradiography. The precise dilution required will be discussed later.

Rigorous drying, such as that recommended by Messier and Leblond[4] for NTB emulsions, is unnecessary and even harmful in the case of the Ilford emulsions. The Ilford Research Laboratory recommend exposure at 0–4°C, and a relative humidity of 40–50% and, in these conditions, latent image fading is not serious. Attempts to dry the emulsion too much will result in a high background, due to stressing of the silver halide crystals by the gelatin. This tendency to the formation of stress background can be considerably reduced by incorporating glycerol in the molten diluted emulsion, to a concentration of 1% of the final volume.

The Ilford emulsions can be used in lighter safelight conditions than the NTB emulsions without producing a significantly increased background. This has considerable advantages in special circumstances, such as the autoradiography of soluble isotopes, where complicated procedures have to be carried out under safelighting, but probably makes no great difference with the very simple routine of dipping slides.

FACTORS AFFECTING THE THICKNESS OF THE EMULSION LAYER

A number of factors influence the thickness of the layer that covers the specimen when a slide is dipped in liquid emulsion. Some idea of their relative importance is essential if the full potentialities of the technique are to be realised. Some of these factors can be varied deliberately to give variations in emulsion thickness; others require stating only so that they may be controlled in the interests of consistency and reproducibility.

(a) *The dilution of the emulsion*

It is clear that dipping a slide in undiluted emulsion will result in a thicker layer than dipping in a mixture of 1 part emulsion to 3 parts distilled water. It is quite reasonable to dilute bulk emulsions in gel form in this way. During drying of the slide, the excess water evaporates, and the composition of the emulsion during exposure returns to that of undiluted emulsion. There is no reason to expect changes in the composition or behaviour of the emulsion with dilutions up to 1 part in 10 parts of distilled water, which is far more dilute than will be normally required. The diluent should be glass-distilled or ion-free water. Small traces of metallic ions may cause fogging of the emulsion.

This is the variable that will normally be altered, if the emulsion thickness needs to be changed. A known and easily controllable change in dilution will produce a predictable change in emulsion thickness, if all the other factors affecting thickness remain constant.

(b) *The temperature of the emulsion*

The hotter the molten emulsion is, the less viscous it will be, and the thinner the layer that results when a slide is dipped and allowed to drain dry.

This parameter is only of limited usefulness, however. At temperatures above 50°C, emulsions tend to develop high levels of background, while at the lower end of the temperature range at which emulsions are molten, a very slight change in temperature produces a large change in viscosity.

Caro and van Tubergen[8] have suggested the use of liquid emulsion at a temperature near the gel point, to prevent redistribution of silver halide crystals during the process of drying. It is certainly true that a very thin emulsion layer at a temperature of around 50°C will dry extremely fast, with a rather uneven distribution of silver grains. With thicker emulsion layers this redistribution phenomenon becomes less troublesome. It is, in any case, possible to avoid this artefact by careful attention to the conditions of drying. In the techniques that will be described, their suggestion is not adopted, as it is felt to be more convenient and reproducible to hold the emulsion throughout at a temperature of 42–43°C.

(c) *The temperature and wetness of the slide*

If the slide to be dipped is warmed to 40°C, the emulsion picked up will drain off more rapidly than off a cold slide, resulting in a thinner emulsion layer. Similarly, a thinner layer will result if the slide is thoroughly wetted before dipping. These are not particularly useful parameters to vary in order to achieve a given emulsion thickness, but they should be controlled in the interests of reproducibility. It is useful to establish a set routine for the handling of the slides prior to dipping, so that they always reach the emulsion in the same condition. If they require dewaxing before autoradiography, it may for instance be convenient to keep them in the darkroom in a dish of distilled water until a few minutes before dipping, and then to allow excess water to drain off them.

(d) *The temperature and humidity during drying*

Ideally, the molten emulsion should gel before it begins to dry, to prevent the redistribution of silver halide crystals mentioned above. This is most conveniently achieved by placing the slides on a cool surface as soon as possible after dipping. But the emulsion will dry more rapidly and thoroughly at a higher temperature, and this is not harmful once the emulsion has gelled. I have a metal plate, which is cooled beforehand by placing containers filled with ice on it, on which the slides are placed lying flat after dipping (Fig. 89). After 15–20 minutes on this plate, the slides are placed, still lying flat, on the bench top for a further hour or so; the darkroom conditions are about 20°C and 45–50% R.H.

If the temperature or humidity of the darkroom are very variable, and no attempt is made to cool the slides immediately after dipping, an uncontrol-

Fig. 89. A levelled metal plate to facilitate the drying of autoradiographs: there are removable containers in each end compartment, which can be filled with ice or with water at any desired temperature. The closely fitting black plastic lid permits drying in an atmosphere of dried CO_2 if required.

led variable will be introduced which may affect the draining of emulsion from the slides, and thus its final thickness.

The effects of several combinations of temperature and humidity during drying on the thickness and uniformity of the emulsion layer have been studied by Leblond, Kopriwa and Messier[6]. They have suggested that high temperatures during drying help to give uniform layers, since excess emulsion can drain off the slide without gelling. But, as we have seen, high temperatures during drying result in very rapid drying, with high stress backgrounds in the emulsion. This can be prevented by maintaining a high relative humidity at the same time, which effectively slows down the rate of drying, and helps to keep the emulsion fluid while draining from the slide. They found that a temperature of 28°C and a relative humidity between 60 and 80% gave reasonably uniform emulsion layers with a low background. The emulsion was "dry" in 30 minutes in these conditions, in the sense that the slides were ready for exposure to begin. Exposure took place in the presence of a drying agent, so it is certain that drying continued during the early stages of exposure. This is an interesting suggestion, and worth following up if the uniformity and reproducibility of the emulsion layer become of critical importance.

(e) *The technique of dipping by hand*

Even if all the factors mentioned above are kept constant, it is still possible to introduce differences into the thickness of the emulsion layer by varying the sequence and timing of events on withdrawing the slide from the emulsion. The slide may be withdrawn slowly or quickly; it may be kept vertical, standing up against a support to drain, or it may be held immediately in a horizontal position, and laid flat for drying to start. In order to get a uniform emulsion coat, the slide must be withdrawn steadily from the dipping jar. Each autoradiographer tends to develop his or her own technique of dipping, so that the slides produced by one person should have a fairly reproducible emulsion thickness. But different people, working with apparently the same conditions of dilution, temperature, and so on, can have very different emulsion layers on their slides.

My own procedure is to keep the dipping jar vertical, and to withdraw the slide fairly slowly, holding it upright to drain for 1–2 seconds. The back of the slide is then wiped clean with a paper tissue, and it is placed on the cold metal plate, described above, lying emulsion side upwards. The thicknesses quoted below have resulted from this procedure in my hands, and will only provide an approximate guide to the emulsion layers others will produce if they try to follow this procedure.

(f) *Attempts to produce more uniform emulsion layers*

Apart from the experiments of Leblond, Kopriwa and Messier[6] into the temperature and humidity of drying, there have been several attempts to get better reproducibility of emulsion thickness by controlling more accurately the application of the emulsion. Kopriwa[9] described a small, motor-driven winch for removing the slides from the emulsion at constant speeds. This approach undoubtedly does give better control of thickness. More complicated machines have been described for the same task, such as the one by Marshall and Faulkner[10]. Fig. 90 gives the working drawings for a very simple dipping machine, designed by Dilys Parry. The basis of this design is that a weight with holes in it falls under gravity down a cylinder filled with oil of standard viscosity. The movement of the weight pulls the slide out of the dipping jar. The speed of descent of the weight can be controlled by varying the sizes of the holes in the weight.

All three designs do the same job of making reproducible the withdrawal of the slides from the emulsion. If all the other sources of variability in

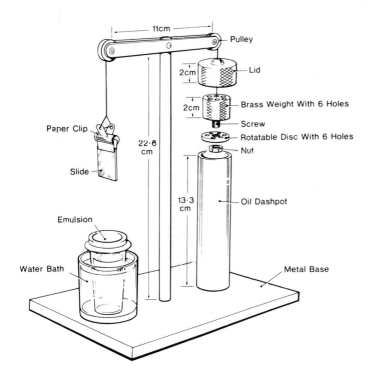

Fig. 90. Drawings for a simple dipping machine. The slide is lowered manually into the emulsion, and is then drawn out at a controlled speed by the brass weight, which falls through the oil in the dashpot under gravity. The speed if withdrawal is controlled by the viscosity of the oil in the dashpot, and by the degree of alignment between the holes in the weight and those in the rotatable disc, which can be varied as desired. (Designed by Miss D. Parry, who kindly provided the drawings)

emulsion thickness are well controlled, it is possible to produce layers comparable in thickness and reproducibility to stripping-film.

Sanderson[11] has described another approach, involving the dispensing of emulsion on to carefully levelled slides at constant temperature. This method can also give reasonably reproducible emulsion layers, though it is rather more fiddly than simple dipping techniques.

SELECTING AN APPROPRIATE EMULSION THICKNESS

Generally speaking, there are two situations in which one wishes to dip slides in molten emulsion. The first is to produce an autoradiograph with the highest possible resolving power for viewing in the light microscope. In this case, the emulsion layer should be as thin as practicable, let us say between 1 and 2 μm during exposure. The second situation involves the preparation of autoradiographs for quantitative work with tritium or isotopes emitting low energy electrons when, as we have seen (p. 292), the emulsion layer must never be less than about 2.5 μm during exposure, implying a mean thickness of between 3 and 4 μm to be on the safe side.

How can one accurately estimate the emulsion thickness during exposure? One very simple method is to take advantage of the limited range of β particles from tritium through nuclear emulsion. One can take as test objects slides with sections labelled with, for instance, tritiated thymidine. After development and fixation, they should be mounted in an aqueous mounting medium such as glycerin jelly, which leaves the emulsion in a swollen state. If labelled nuclei are then examined under the highest power objective available, it should be clear whether or not the silver grains over them extend all the way up to the upper surface of the emulsion, or are separated from it by a layer of emulsion that contains only a few scattered background grains (Fig. 91). If the emulsion is consistently thicker than the maximum range of the β particles, suitable conditions have been established for quantitative work with tritium. If a layer between 1 and 2 μm thick has been obtained, it will be quite clear that the grains over labelled nuclei extend to the upper surface of the emulsion.

In making a thin emulsion layer for high resolution work, it is very easy to go too far, producing a layer that is too thin. Generally speaking, it is not advisable to try to dip to less than about 1.5 μm, unless steps are taken to prevent the emulsion drying very rapidly, with consequent chances for redistribution of silver grains during drying, and a higher stress background. A very thin layer also has a considerably lower efficiency as a recording medium, resulting in longer exposure periods for the same grain density. It is useful to take a coated slide into the light after drying, and to examine the emulsion under the microscope. Over the section, there should be no gaps and no areas where the emulsion appears to be only one or two silver halide crystals deep. The emulsion layer should look reasonably uniform, and perhaps 5 crystals deep, though this may be difficult to determine accurately.

It is always a useful step to bring one slide of each batch out into the light, to check that gross variations in emulsion thickness are not occurring.

In view of what has been said in the previous section, even if the techniques outlined at the end of this chapter are carefully followed, it is necessary to check that the emulsion layers are of the correct thickness. It should not be a surprise to find that they are not, as slight differences in the timing and sequence of events after dipping can introduce relatively large

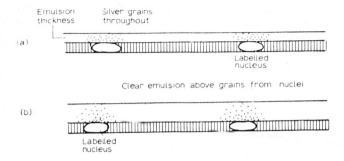

Fig. 91. Two diagrams to illustrate a simple method of checking that a thin emulsion layer is sufficiently thick for the quantitative autoradiography of tritium. Heavily labelled structures, such as cell nuclei that have incorporated tritiated thymidine, are examined with an emulsion layer that has been swollen prior to microscopy. (a), The silver grains over the nuclei extend right up to the upper surface of the emulsion. This emulsion is too thin for quantitative work: slight variations in emulsion thickness will alter the autoradiographic efficiency significantly. (b), The emulsion layer is thicker than the maximum range of β particle. Variations in thickness at this level will not affect the efficiency of the system.

variations into the emulsion thickness. The dilution of the emulsion should then be adjusted on future occasions so that, with the routine that is most convenient to the autoradiographer concerned, the desired thickness will be obtained.

CONDITIONS OF DRYING AND EXPOSURE

These have been mentioned already, insofar as they influence the thickness of the emulsion layer. In general, the emulsion should gel before significant drying has time to take place. Once it has gelled, drying should be slow and gentle.

Placing the slides on a cool metal plate immediately after dipping helps the emulsion to gel fast. It also helps to ensure that the initial rate of drying will be fairly slow.

The work of Leblond and his collaborators[5,6], which has already been referred to, illustrates the way in which the temperature and humidity of the atmosphere in which the slides are dried can be varied. Their results show that the lowest background levels are produced by very slow drying, which may even extend into the early stages of exposure itself. Sawicki and Pawinska[12] have similarly demonstrated lower background levels from very slow drying with Kodak AR-10 stripping-film.

Low background levels can also be produced by failure to dry the emulsion adequately before the start of exposure (p. 265). If fading of latent images from this cause continues throughout exposure, the low background will also be accompanied by low grain yields over labelled structures. But if incompletely dried emulsion is exposed in the presence of a drying agent, such as dried silica-gel, latent image fading can be limited to the first part of exposure. This can result in low background levels, since all the latent images present in the emulsion before exposure will have been lost, while the grain yields can remain high, because the emulsion is fully dry and free from fading for most of the exposure time. It should always be remembered that prolongation of drying into exposure in this way makes it very difficult to estimate the true duration of exposure accurately. This may not matter, if the distribution of radioactivity within a specimen is being studied. A more reproducible drying technique is preferable in quantitative work, however. The routine we use at present is to cool the slides for 15–20 minutes on the metal plate after dipping, then place them on the bench in the darkroom still lying flat, for a further one hour or so, at about 20°C and 45–50% R.H. After this, the slides are placed in their exposure boxes, with the lids off, in a desiccator over dried silica-gel overnight at room temperature. The next morning, the boxes are closed, and placed in the refrigerator to expose. This method gives slow but complete drying, with no evidence of latent image fading in exposures of up to two months. Drying has been more reproducible in this way than with silica gel in the exposure box. There have been occasions when the silica gel has been found saturated at the end of exposure, and the emulsion not completely dry: perhaps the slides were wetter than usual on putting them into the boxes, or the amount or dryness of the silica gel itself less than adequate.

From what has been said above, it should be quite clear that directing a blast of hot air from a hairdryer on to the slides immediately after dipping them is not an ideal method of drying (Fig. 38). Gentle but thorough drying is the ideal to aim at.

Rechenmann and Wittendorp[13,14] have drawn attention to the increase in efficiency that can result from latent image intensification with gold salts before development. This procedure is discussed at greater length on p. 34. Since latensification increases the probability of development of crystals with only small deposits of silver at latent image specks, it can increase the background very significantly, unless great care is taken in preparing the slides. Sanderson[11] has described a drying device, based on a suggestion of Rechenmann's, which helps to assure controlled drying of emulsions before exposure. It is essentially a box in which the temperature of the floor and the roof can be separately regulated. The slides lie on the floor, and their temperature and their rate of drying can be accurately controlled:

the drying rate depends on the temperature difference between floor and roof.

If latent image fading must be prevented — as, for instance, in the case of a very low level of labelling requiring an exposure of many weeks to give a trace — it may be necessary to expose the slides in an atmosphere that does not contain oxygen[15]. This can be done with nitrogen, argon, or any inert gas, but is probably most convenient with carbon dioxide. When the slides are placed in the desiccator over dried silica-gel, a small lump of solid carbon dioxide is put at the bottom, where it sublimes, filling the desiccator by displacement. When the exposure boxes are closed next morning, care is taken to put their lids on under the dry carbon dioxide layer in the desiccator. If the boxes are impermeable, such as the plastic slide boxes usually used for this purpose, the slides will remain in this atmosphere throughout exposure. If permeable boxes made of cardboard are used, it will be necessary to store them during exposure in a larger vessel with the same filling gas.

Caro and van Tubergen[8] found it satisfactory to expose very thin layers of Ilford L4 at room temperature. This is not the usual practice, however, and other workers who have examined the conditions of exposure agree with the recommendations of the manufacturers that exposure should normally take place at 0–4°C. Exposure can be made at considerably lower temperatures without significant loss of sensitivity, but this is seldom necessary or convenient, except in the autoradiography of soluble material (see Chapter 8), and as a measure to reduce the severity of chemography (p. 120).

ESTIMATES OF RESOLUTION AND EFFICIENCY

There are very few published observations on the resolution and efficiency to be expected from liquid emulsions applied in thin layers. The difficulty of achieving a uniform and reproducible emulsion thickness makes detailed measurements rather pointless. The best that can be done is to equate the performance of these emulsions to that of Kodak AR-10 in rather general terms, making certain assumptions as to the emulsion thickness.

If one looks first at the emulsion layers of 1.5–2 μm that are desirable for high resolving power, it is clear on theoretical grounds (see Chapter 4), that a higher resolution should be obtained for isotopes of the energy of carbon-14 or higher than with the 3.8 μm layer of AR-10 during exposure. The smaller the crystal size of the emulsion, the greater this improvement is likely to be. Fig. 24 shows that an overlapping monolayer of L4 will have an HD for carbon-14 of 0.8 μm, in conditions where AR-10 will have 2.0 μm[16]. If the specimen is a section 5 μm thick, there is little point in juggling the emulsion variables to get the highest resolution: reducing the section thickness will be a more effective approach.

With tritium, the position is rather different. The short range of the β particle is the most important factor in determining the resolution. As can be seen from Fig. 24, there is little difference in HD between AR-10 and an overlapping monolayer of L4 for this isotope.

When one considers efficiency, the situation is rather complex. Clearly, a thinner emulsion layer is likely to have a lower efficiency for isotopes of the energy of carbon-14, sulphur-35, or higher. In the cases of Ilford G5, K5 and L4, and of Eastman Kodak NTB3, this effect is offset by the higher sensitivities of these emulsions relative to AR-10 stripping-film. The overall efficiency of a 1 μm layer of K5 should be much the same as that of AR-10 for phosphorus-32, for instance. The same emulsion, applied as a layer 3.5–4 μm thick would give similar resolution to AR-10, but a significantly higher efficiency.

With tritium, efficiency is unaffected by changes in emulsion thickness over 2 μm. There is little gain in efficiency from using emulsions of higher sensitivity but similar crystal size — K5 instead of K2, for instance. Changing to an emulsion of smaller crystal size — L4 instead of K2 — should give a significant increase in efficiency, however, from the greater number of crystals per unit volume of emulsion. An emulsion such as L4 has the added advantage of being able to record a higher grain density over labelled sources before corrections for crystals hit by more than one particle become significant (p. 102). Leblond[17] has reported briefly on some results of Kopriwa on the relative sensitivities of Eastman Kodak NTB-2, NTB, and Kodak AR-10, to tritium in tissue sections. Under comparable conditions of exposure, the relative grain densities found were 1.0, 0.4 and 0.5.

By the selection of a suitable emulsion, and careful control of its thickness, it is possible to achieve optimal conditions for nearly every type of autoradiographic experiment.

ATTEMPTS TO IMPROVE THE EFFICIENCY OF LIQUID EMULSIONS

These fall into two categories: latent image intensification before development, and scintillation autoradiography.

Latensification is usually taken to mean treatment of the emulsion with gold salts just prior to development. Other techniques do exist, such as exposure of the emulsion to a light flash of very low intensity[18], or treatment of the emulsion with polyethylene glycol[19]. Gold latensification is certainly the most widely used and reproducible method, however. Rechenmann and Wittendorp[13,14] have pioneered its use at the light microscope level, and described in detail the increases in efficiency that can be had with various emulsions and developers. They have also emphasised the improvement in resolution that will result, since crystals with stable latent sub-images are more likely early in the track of a β particle than near its end. Gold treatment is far less effective with some emulsions than with others. Eastman Kodak recommend that emulsion Type 129-01 should not be treated with gold, as apparently this has already been done in the process of sensitisation. In much the same way, it appears that gold treatment is part of the routine method of raising Ilford emulsions above level 2 of sensitivity. Although Rechenmann demonstrates a slight increase in efficiency on gold latensification of Ilford K5, it is clear that much greater improvements result from similar treatment of K2 (Fig. 6).

Attention has already been drawn to the need for great care in handling emulsions that will have gold latensification (p. 34). Exposure to safelight must be minimal, and all possible stages should be carried out in darkness. Drying must be very gentle and slow.

The solutions for gold latensification and their use are described on p. 409.

Scintillation autoradiography at the light microscope level with liquid emulsions has been described by Durie and Salmon[20] and by Panayi and Neill[21]. Both papers claim considerable increases in efficiency from exposure in scintillation fluid. The similar claims for scintillation autoradiography with AR-10 stripping-film have already been discussed, and much of what was written on p. 341 applies equally to the liquid emulsion work. In a series of experiments[22], I tried to achieve similar improvements with Ilford K2 emulsion, exposed to sections of [³H]methacrylate. The results, in Fig. 92, showed a reduction in efficiency from exposure in scintillation fluid at $-70°C$, rather than an improvement. It could be argued that my test system was biased, since the scintillation fluid could not penetrate the section, and the chief loss of efficiency with tritium is self-absorption. But wax sections of small intestine labelled with [³H]thymidine were also included in these experiments, and showed an even more dramatic loss of efficiency: the group exposed in scintillation fluid at $-70°C$ had so few grains over the tissue that it was impossible to identify labelled cells. Several other laboratories have informed me of failure to achieve increases in efficiency by this technique.

In addition to the factors discussed on p. 341, it should be realised that nuclear emulsions are relatively insensitive to light. The problem remains

366

Exposure conditions	Photometric grain density ($\pm$ S.D)
No scintillator — room temperature	302 ± 16
No scintillator — $-79°C$	266 ± 15
Dioxane only — $-79°C$	228 ± 26
Scintillator — $-70°C$	162 ± 47

Fig. 92. The results of an experiment at light microscope level into scintillation autoradiography. Sections of [³H]methyl methacrylate were covered with thin layers of Ilford K2 emulsion and thoroughly dried. They were then allocated to 4 groups and exposed for 24 hours in the conditions indicated. Following development together in a developing tank (p. 219), grain densities were measured photometrically in incident light. The values listed are the means ($\pm$ SD) of 20 readings from each of several sections in each group after subtraction of background.

that some groups of workers are claiming very significant advantages from this technique. The answer appears to lie in their experimental design. Conventional autoradiographs, exposed for several days or weeks, are compared to scintillation autoradiographs exposed for hours. If the drying of the emulsions in the conventional autoradiographs were incomplete, latent image fading would give a spuriously low grain count in this group. Certainly the drying described by Durie and Salmon[20] is inadequate in the light of the work of Messier and Leblond[4]. Immersion in dioxane, followed by scintillation fluid, would in contrast be a rapid and effective method of drying.

One can only speculate on the explanation for the claimed successes of this technique. It is certain that its advantages are not reproducibly obtainable. In view of this, and the disadvantages of poorer resolution and non-linearity of emulsion response, scintillation autoradiography in this form is best avoided at present.

DETAILED DESCRIPTION OF TECHNIQUES

In view of the many differences between Eastman Kodak and Ilford emulsions, their use will be described separately, starting with the Eastman Kodak NTB series. Detailed descriptions of techniques of applying these emulsions have been given by Joftes and Warren[2,3], and by Leblond and his co-workers[4-6].

(1) *Emulsions*: Eastman Kodak NTB-2 and NTB-3.
This technique should give emulsion layers of 3–4 μm.

Safelight: Wratten No. 2.

Darkroom conditions: Temperature, 18–20°C; relative humidity, 40–50%.

Equipment required: Thermostatically controlled waterbath at 43°C; a suitable dipping jar, such as a 50-ml graduated cylinder cut short at the 40-ml mark, or a 100-ml beaker, depending on the size of the slides; a levelled metal plate with provision for placing containers of ice at each end (Fig. 89).

Preparing the autoradiographs: The NTB emulsion arrives in a plastic jar. Under safelighting, the jar should be placed in the waterbath at 43°C, and allowed to stand there for 30 minutes (this 30 minute period can be in absolute darkness). By this time it will have become molten, and most of the bubbles in the emulsion will have risen to the surface.

Pour the emulsion gently into the dipping jar to the required level. Stand the dipping jar once more in the waterbath.

Dip a clean slide into the emulsion, and take it up to the safelight to check that the emulsion is uniform and free from bubbles. If many bubbles are present, wait 2 minutes and dip another clean slide.

Take the experimental slides, and dip them individually into the emulsion. Keep the slide vertical in the emulsion, and withdraw it slowly and steadily. Holding the slide in a vertical position for several seconds, allow excess emulsion to drain into a paper tissue.

Wipe the back of the slide with a paper tissue, and place the slide, face up, on the cooled metal plate.

When all the slides have been dipped, leave them on the metal plate, in complete darkness, for 15–20 minutes. Transfer them, still lying flat, to the bench top for a further one hour. Then place them in their exposure boxes, with lids off, in a desiccator over dried silica-gel overnight at room temperature. In the morning, close the boxes, and place them in a refrigerator at 4°C to expose. A teaspoonful of dried silica gel wrapped in a paper tissue may be included in the exposure box with the slides if desired.

Processing: Under safelighting, transfer the slides to stainless steel or glass slide racks. All solutions should be at the same temperature: the development times stated should be suitable for a developer temperature of 17°C, and for subsequent viewing of the slides by transmitted light.

Eastman Kodak Dektol developer, 1 part stock solution diluted with 2 parts distilled water, for 2 minutes.

Rinse in distilled water.

30% sodium thiosulphate solution for 8 minutes.

Wash in running tapwater for 15 minutes.

For emulsion layers 1–2 μm thick, for high resolution studies, dilute the emulsion in the dipping jar 2 parts to 1 part of distilled water.

If uniformity of the emulsion layer is not good enough by this technique, it may be possible to improve the method in this respect by varying the temperature and humidity at which the slides are initially dried[5,6] (p. 117). Alternatively, semi-automatic withdrawal of the slides from the emulsion should be tried (p. 358).

If latent image fading occurs during exposure, in spite of the presence of a drying agent, it may be necessary to expose the slides in an atmosphere of dry carbon dioxide, or of nitrogen (p. 266).

The development conditions are only approximate, and will require examination in the light of the needs of the particular experiment.

Eastman Kodak NTB emulsions should be stored at 4°C. It is possible to melt the same batch of emulsion several times without significant increase in background levels.

(2) *Emulsion*: Ilford G5, K2, K5 or L4.

This technique should give an emulsion layer of 3–4 μm.

Safelight: Ilford 'F904' or Wratten 'OC'.

Darkroom conditions: Temperature, 18–20°C; relative humidity, 40–50%.

Equipment required: Thermostatically controlled waterbath at 43°C, one 50-ml graduated measuring cylinder; one dipping jar; one 25-ml graduated measuring cylinder; one pair plastic print forceps; one glass stirring rod; a levelled metal plate with provision for placing containers of ice at each end (Fig. 89).

Preparing the autoradiographs. With a black grease pencil, make a mark on the 25-ml measuring cylinder at 12 ml. Place 11.76 ml distilled water and 0.24 ml glycerol in the dipping jar. Stand the dipping jar and the two measuring cylinders in the waterbath at 43°C.

Under safelighting, transfer emulsion from the stock bottle to the 50-ml cylinder, using the print forceps. Shake the emulsion down gently, and continue until it reaches 20–25 ml. Stand this cylinder again in the waterbath for 10 minutes, stirring the emulsion very gently with the glass rod. Vigorous stirring will whip up a froth of bubbles: one revolution per second is adequate.

Take the 50-ml and 25-ml measuring cylinders out of the waterbath, wiping their sides with a paper tissue to remove water. Pour molten emulsion from the large cylinder to the small one, until the level reaches the black grease pencil mark at 12 ml.

Pour the molten emulsion from the 25-ml cylinder into the dipping jar. With the dipping jar standing in the waterbath, stir the emulsion gently for a minute or so to ensure complete mixing. Then leave the emulsion standing in the waterbath a further 2 minutes to allow bubbles to disperse.

Dip a clean slide into the emulsion, and take it up to the safelight to check that the emulsion is uniform and free from bubbles. If the emulsion is not uniform, it will appear thicker at the bottom of the slide: gentle stirring with a clean slide, moving it in a rotary, vertical direction will help. If bubbles are present, leave the emulsion a further 2 minutes, then dip another clean slide.

Take the experimental slides, and dip them individually in the emulsion. Keep the slide vertical in the emulsion, and withdraw it slowly and steadily. Holding the slide vertical for 1–2 seconds, allow excess emulsion to drain into a paper tissue.

Wipe the back of the slide with a paper tissue, and place the slide, face up, on the cooled metal plate, from which the ice should now be removed.

When all the slides have been dipped, leave them on the metal plate, in complete darkness, for 15–20 minutes. Transfer them, still lying flat, to the bench top for a further one hour. Then place them in their exposure boxes, with the lids off, in a desiccator over dried silica-gel overnight at room temperature. In the morning, close the boxes and put them in a refrigerator at 4°C to expose.

Processing. Under safelighting, transfer the slides to stainless steel or glass slide racks. All solutions should be at the same temperature: the times stated should be suitable for the K2 emulsion, for a developer temperature of 20°C, and for subsequent viewing of the slides with transmitted light.

Ilford Phen-X developer, diluted with an equal volume of distilled water for 8 minutes.

Rinse in distilled water.

30% sodium thiosulphate solution for 8 minutes.

Wash in running tapwater for 15 minutes.

For emulsion layers 1–2 μm thick, for high resolution studies, dilute the emulsion in the dipping jar 3 parts to 5 parts distilled water.

The description given here will produce 25 ml of molten, diluted emulsion, which is sufficient to cover about two-thirds of a 3 in. × 1 in. microscope slide when placed in a suitable dipping jar. It is simple to scale the volume of water, glycerol and emulsion up or down according to the needs of the experiment.

If latent image fading is occurring, exposure can take place in an atmosphere of dry carbon dioxide or nitrogen (p. 266).

Once again, the development conditions are only approximate, and should be altered, if necessary, to meet the needs of the experiment. Ilford G5 emulsion in general requires more gentle development than either the K or the L series: it may be convenient with G5 to dilute the developer with 2 parts of distilled water, instead of an equal volume.

Ilford emulsions should be stored at 4°C. They should never be heated a

370

second time. Emulsion left in the dipping jar should therefore be discarded.

REFERENCES

1 L.F. Belanger and C.P. Leblond, *Endocrinology*, 39 (1946) 8.
2 D.L. Joftes and S. Warren, *J. Biol. Phot. Assoc.*, 23 (1955) 145.
3 D.L. Joftes, *Lab. Invest.*, 8 (1959) 131.
4 B. Messier and C.P. Leblond, *Proc. Soc. Exptl. Biol. Med.*, 96 (1957) 7.
5 B.M. Kopriwa and C.P. Leblond, *J. Histochem. Cytochem.*, 10 (1962) 269.
6 C.P. Leblond, B.M. Kopriwa and B. Messier, in R. Wegmann (Ed.), *Histochemistry and Cytochemistry*, Pergamon, London, 1963.
7 D. Darlington and A.W. Rogers, *J. Anat.*, 100 (1966) 813.
8 L.G. Caro and R.P. van Tubergen, *J. Cell Biol.*, 15 (1962) 173.
9 B.M. Kopriwa, *J. Histochem. Cytochem.*, 14 (1967) 923.
10 J. Marshall and D. Faulkner, *J. Microscopy*, 109 (1977) 227.
11 J. Sanderson, *J. Microscopy*, 104 (1975) 179.
12 W. Sawicki and M. Pawinska, *Stain Technol.*, 40 (1965) 67.
13 R.V. Rechenmann and E. Wittendorp, *J. Microscopy*, 96 (1972) 227.
14 R.V. Rechenmann and E. Wittendorp, *J. Microscop. Biol. Cell.*, 27 (1976) 91.
15 R.H. Herz, *Lab. Invest.*, 8 (1959) 71.
16 M.M. Salpeter, G.C. Budd and S. Mattimoe, *J. Histochem. Cytochem.*, 22 (1974) 217.
17 C.P. Leblond, in R.J.C. Harris (Ed.,) *The Use of Autoradiography in the Investigation of Protein Synthesis*, Academic Press, New York, 1965, p. 21.
18 R.A. Laskey and A.D. Mills, *Eur. J. Biochem.*, 56 (1975) 335.
19 N.R. Novikova and G.V. Kaliamina, *Zh. Nanch. Piece. Prikl. Fotogr. Kinematogr.*, 18 (1973) 372.
20 B.G.M. Durie and S.E. Salmon, *Science*, 190 (1975) 1093.
21 G.S. Panayi and W.A. Neill, *J. Immunol. Meth.*, 2 (1972) 115.
22 A.W. Rogers, *Acta Pharmacol. Toxicol.*, 41, Suppl. 1 (1977) 70.

CHAPTER 17

Liquid Emulsion Techniques for Track Autoradiography

Nuclear emulsions owe their existence to the physicists who wished to study the behaviour of ionising particles, and a great deal of information on the interaction of these particles with matter has been obtained from studying the characteristics of their tracks in thick layers of emulsion.

It is surprising how seldom biological research has been carried out with thick emulsion layers giving track records, when one remembers this historical evolution of the nuclear emulsions and the techniques of their use. In the early days of autoradiography, the emphasis lay heavily on identifying the site at which radioactivity was localised within a tissue, and the techniques based on thick emulsion layers were difficult and time-consuming by comparison with the grain density methods which relied on thin layers of liquid emulsion or stripping film. Certainly, track recording techniques are more difficult, but, as has been pointed out in Chapter 12 (p. 277), the advantages of track autoradiography are substantial when measurements of radioactivity are needed. The information on grain spacing and scattering angles required from a track record by a particle physicist represents a higher level of technical achievement than the straightforward recording of recognisable tracks, which is sufficient for quantitative autoradiography. There is thus no real doubt that the emulsions and methods developed for the former task should be quite adequate for the latter.

The recording and recognition of the tracks of α particles present no great problem. Quantiative studies based on counting α tracks have been described by several workers[1-3]. The short range and high rate of energy loss characteristic of the α particle combine to make a thin layer of a relatively insensitive emulsion the method of choice. In fact, either Kodak AR-10 stripping-film or the technique employing Ilford K2 emulsion for work with tritium (described in Chapters 15 and 16 respectively) are suitable.

The recording of the tracks of β particles is made difficult by their long range and relatively slow rate of energy loss, as well as by their very irregular trajectory. The emulsion layers used must be considerably thicker, and the emulsion must have a sufficiently high sensitivity to give a close spacing of developed silver grains, if these tortuous tracks are to be recognised clearly. The technical problems of β-track autoradiography all stem from the need for a thick layer of emulsion of high sensitivity.

Hilde Levi, of Copenhagen, has been associated with much of the basic work on β-track autoradiography[4-6]. In 1954, she published a careful description of the technique developed in her laboratory for preparing 60 μm layers of Ilford G5 emulsion[7]. Levi, Rogers, Bentzon and Nielsen[8] presented further detailed descriptions of techniques for handling 60- and 120-μm layers of Ilford G5, together with the correlations between initial particle energy, track length, grain number and spacing and track radius, for β particles in the energy range of 20–400 keV.

Ficq[9-11], at Brussels, has also made extensive use of β-track recording, though this has usually been for purposes of localisation rather than quantitation. Her techniques are described in detail in a chapter in The Cell (1961), edited by Brachet and Mirsky[12].

The first attempt to realise the quantitative potentialities of this approach was the work of Levinthal[13] (1956) and Levinthal and Thomas[14] (1957), who measured the absolute disintegration rate of bacteriophage virus labelled with phosphorus-32. Their techniques are described in the latter paper. Phage particles were suspended in thick layers of Ilford G5 emulsion, and their position subsequently recognised by the star of β-tracks radiating out from a common origin. The number of tracks per star formed the basic data for calculating the disintegration rate, and thus the number of atoms of phosphorus per virus.

Similar techniques have been applied to measurements of the phosphorylation of acetylcholinesterase by diisopropylfluorophosphate (DFP) in motor endplates[15], and in megakaryocytes[16]. From these values it has been possible to calculate, in absolute terms, the number of molecules of the enzyme acetylcholinesterase in these two sites. This work is interesting for the correlation obtained between track counts from DFP labelled with phosphorus-32, and liquid scintillation counting from DFP labelled with tritium[15,17].

A considerable advance in the methods of processing thick emulsion layers has come from the work of Rechenmann and Wittendorp[18,19]. Their careful control of development, following latensification with gold salts, has resulted in a high density of grains within the tracks combined with a low background of individual grains. The track autoradiographs made by their methods in several laboratories have had an excellent signal-to-noise ratio.

BASIC CHARACTERISTICS OF β-TRACK AUTORADIOGRAPHS

Many facets of β-track autoradiography have been dealt with already in earlier chapters of this book. Here, the important facts will only be briefly summarised.

The use of a thick layer of emulsion to record the passage of β particles as tracks of developed silver grains is really only justified if accurate determinations of the disintegration rates of labelled sources are required. The localisation of radioactivity is more conveniently studied with thin emulsion layers and, in most cases, the relative concentrations of isotope in different structures can also be compared more simply by comparing grain densities in thin emulsion layers. The measurement of disintegration rates in absolute terms, however, is usually carried out or calibrated[20] by means of track autoradiography.

It is usually convenient, when preparing a track autoradiograph, to mount the biological material on a microscope slide, and cover it on one side only with nuclear emulsion. The technique of suspending the source in the emulsion, while it provides a record of tracks over the full space angle, is not essential for quantitative studies.

The basic concepts of autoradiography, such as efficiency, resolution, and background, require restating in terms unfamiliar to those accustomed to working with grain-density autoradiographs. They will therefore be briefly recalled.

(a) *Efficiency*

This is discussed in detail in Chapter 5 (p. 106). In a track autoradiograph, the efficiency may be defined as the percentage of disintegrations taking place in the source during exposure which give rise to recognisable tracks. With no self-absorption, high initial particle energies, and emulsion completely surrounding the source, the efficiency of a track autoradiograph should be 100%: this is the sort of situation Levinthal and Thomas[13,14] achieved, with viruses labelled with phosphorus-32 and suspended in emulsion.

Efficiency may be reduced by three factors. The first is self-absorption, which becomes increasingly important at lower particle energies, and with increasing dimensions and mass of the source. For tissue sections of 5 μm or less, and for smears of cell suspensions, self-absorption can be ignored with phosphorus-32, but may be significant with carbon-14. The second factor concerns the ability of particles of low initial energy to give rise to a recognisable track — usually defined as 4 or more developed grains. With carbon-14, for instance, 14% of the total particles emitted will not produce as many as 4 grains in their trajectory[8].

The third factor tending to reduce the efficiency of track autoradiographs from the theoretical maximum of 100% is the geometrical relationship between source and emulsion. If the source is mounted on a glass slide, and covered by emulsion on one side only, the β particles that leave to enter the slide will not produce a recognisable track. In this situation (see Fig. 30, p. 84), it is reasonable to assume that 50% of the particles leaving the source will enter the emulsion directly. Some of these will subsequently be scattered back into the glass, and some that enter the glass initially will be scattered into the emulsion, but, at a first approximation, the tracks that enter the emulsion directly over a source of cellular dimensions represent half the tracks that would have been observed if the source were completely suspended in emulsion.

If one assumes reasonable care in preparing the autoradiograph, so that factors like fading of the latent image and loss of activity in histological processing may be ignored, these three factors are all that need to be considered in relating the observed track count to the disintegration rate within the source.

(b) *Resolution*

This is discussed in Chapter 4 (p. 83). Briefly, one cannot specify the point at which a β particle is emitted, unless one is dealing with a small source suspended in emulsion: one can only observe the point at which the particle enters the recording emulsion. With a tissue section mounted on a glass slide (see Fig. 30), we have seen that 50% of the emitted particles enter the emulsion directly from the source. The number of particles that are scattered subsequently across the glass–emulsion interface is difficult to determine accurately, but seems to be in the order of 10% of the total β flux from the source, from preliminary observations made in this laboratory with phosphorus-32.

If one defines the resolution of a track autoradiograph as the minimum radius around a point source that contains the points of entry of 50% of all the tracks produced by particles from that source, a resolution of about 2 μm should be achieved, even with phosphorus-32.

(c) *Background*

One of the great advantages of track autoradiographs is their low background. In Chapter 6 (p. 128), we saw that many factors apart from ionising particles can produce developed grains in nuclear emulsions — heat, light, pressure, and chemical agents, for example. One cannot distinguish between these silver grains and the ones produced by radiation in a thin emulsion layer. None of these agents gives rise to a β track, however, so that all the causes of random background grains can be discriminated against, in a track autoradiograph.

Background β tracks do occur. They arise from potassium-40 in the glassware, carbon-14 in the gelatin of the emulsion, and secondary electrons caused by cosmic rays. They should be rare, and their nature will often be obvious. A track of less than 70 grains, starting in the emulsion itself without any obvious event or structure as its origin, is probably carbon-14 in the gelatin; while secondary electrons usually start from the track of some other charged particle through the emulsion.

Of course, if the single grain background rises above a certain level, it gets increasingly difficult to recognise β-tracks. Such high backgrounds are rare if the emulsions are handled and developed carefully.

(d) *Latent image fading*

Fading of the latent image has been discussed in Chapter 12 (p. 266). Particularly in the presence of oxidising agents and of moisture, latent images formed early in exposure are liable to regress. This shows itself in track autoradiographs in a decrease in the number of grains per unit length of track and, if this process goes far enough, it will be impossible to determine the trajectory of the β particle from the few grains that remain.

Levi, Rogers, Bentzon and Nielsen[8] have investigated the grain spacing in β tracks up to about 400 keV in Ilford G5, and this seldom drops below 10 grains per 25 μm, even at the start of tracks at the upper end of the energy range they examined. Even for phosphorus-32, it is unlikely to fall below 7 grains per 25 μm.

If, therefore, tracks are observed which have a grain spacing below that predicted, latent image fading should be suspected. Such "faded" tracks will co-exist with tracks with a normal grain spacing, which were produced late in exposure. The presence of a normal track thus cannot be taken as evidence for lack of fading: the latter can only be assumed if no "faded" tracks are seen. Normal and "faded" tracks are illustrated in Fig. 37.

It is possible for lesser degrees of fading to occur without altering the observed track count, if all the "faded" tracks are still recognisable; but, if clear evidence for the existence of fading is found, it is dangerous to assume that tracks have not been lost, without additional supporting evidence.

One related problem in quantitative work comes from the relatively long drying period needed for thick emulsions, and the difficulty in deciding when exactly effective exposure starts. Since exposure times are often 48 hours or less, an uncertainty of 2 hours at the start of exposure may represent a significant proportion of the total counts. During the final stages of drying, "faded" tracks will be created which may well survive to development.

(e) *Problems of track recognition*

These have already been discussed in some detail (p. 208). In material that is technically good, with a suitable density of β-tracks, a low background of random silver grains, and correct conditions for microscopy, it is very easy to recognise and to count β-tracks. On several occasions, I have had track counts checked by microscopists who have had no previous experience of track autoradiography, and statistical analysis has failed to show a difference between the two sets of counts. Even in good autoradiographs there will always be patterns of silver grains that can be interpreted in more than one way, but these should form only a small percentage of the total.

Levi and Hogben[21] showed that the variance of track counts increased with increasing track density, illustrating how the percentage of patterns capable of several interpretations increases as the tracks cross with greater frequency. As a rough guide, the exposure time and specific activity should be adjusted so that no more than 8–10 tracks per 500 μm^2 for carbon-14 or sulphur-35, and 12–16 tracks per 500 μm^2 for phosphorus-32, are produced.

The process of re-swelling the emulsion prior to microscopy greatly assists in the interpretation of track patterns. After photographic fixation, dehydration, and mounting in one of the routine non-aqueous histological media, the emulsion layer is much thinner than it was during exposure. Tracks which in fact crossed at different levels in the emulsion appear to run into one another. Swelling the emulsion back to its original thickness restores the distance separating the tracks, and simplifies their recognition[8].

In any thick emulsion layer, the upper few microns will be rather unpleasant after processing, with a higher density of random background grains, and occasional scratches, bits of dust, and crystals of various sorts. It is an additional merit of re-swelling prior to microscopy that this layer which is so full of artefacts is lifted away from the biological material deep in the emulsion, making observation easier.

(f) *The choice of emulsion*

The emulsion selected for β-track autoradiography must have a high sensitivity in order to give sufficient grains along the trajectory of the β particle to form a recognisible track. In the Ilford range of nuclear emulsion, only G5, K5, and L4 come into this category. Eastman Kodak NTB-3 is also sensitive enough.

It is impossible to view emulsion layers more than about 20 μm thick with dark-field incident lighting, as light scattering in the emulsion interferes with the reflection of the incident beam by silver grains more than a few microns below the surface. The developed silver grains must, therefore, be large enough to be seen clearly by transmitted light, in conditions which are not ideal for microscopy. A large grain size is therefore preferable.

The descriptions of techniques in the literature are almost without exception based on Ilford G5 emulsion, and the comprehensive data assembled by Levi, Rogers, Bentzon and Nielsen[8] on the characteristics of β-tracks are again based on this emulsion. Therefore, while it is clearly possible to make track autoradiographs with K5 or NTB-3, G5 would seem to be the emulsion of choice.

While there are valid reasons for working with G5, it must be admitted that some very good autoradiographs of carbon-14 have been made recently with K2, an emulsion not usually regarded as sufficiently sensitive for this job. With gold latensification and careful development, however, the grain density in carbon-14 tracks is quite acceptable, and the random grain background unusually low.

THE PREPARATION OF THICK EMULSION LAYERS

It will be seen, in the discussion on the processing of thick emulsion layers, that the difficulties of track autoradiography increase rapidly with increasing emulsion thickness. There is every incentive to keep the emulsion as thin as possible, without sacrificing the basic advantages of the technique.

The precise thickness of the emulsion layer will vary somewhat with the experimental material, and with the energy of isotope. As an example, if the material is a section or smear mounted on a glass slide, and labelled with carbon-14 or sulphur-35, the emulsion should be 15–20 μm thick. This is sufficient to contain all but a small percentage of the tracks. For phosphorus-32 in a similar geometrical situation, 60 μm is thick enough. If one wishes to suspend the sources in emulsion, as, for instance, one might with labelled bacteria or algae, 20 μm of emulsion should lie above and below the highest and lowest sources respectively with carbon-14 or sulphur-35, and 60 μm with phosphorus-32.

A layer of emulsion 20 μm thick may be very simply applied by a modified dipping technique[22]. For the thicker layers, it is more convenient to pipette the warmed, diluted emulsion on to the microscope slide. If suspended sources are to be studied, the first layer of emulsion is applied and allowed to dry. Then a few drops of liquid emulsion, in which the labelled sources have been previously suspended, are placed in the centre of the slide, and allowed to dry. Finally, a second thick layer of emulsion is pipetted on to the slide, making a kind of sandwich, with the sources in the thin middle layer.

The drying of these thick emulsions is of crucial importance. Unless it is complete and thorough, latent image fading will be severe in the deeper levels, precisely where the biological material lies. Unless it is slow and gentle, a high background of random silver grains may be produced by stress within the gelatin. Emulsions intended for gold latensification before development require particularly slow drying and minimal exposure to safelighting, if a high background of random silver grains is to be avoided.

Exposure is usually very short by comparison with grain density autoradiographs. Seldom will more than 48 hours be needed.

THE PROCESSING OF THICK EMULSION LAYERS

With thick emulsion layers, it may take an appreciable time for solutions to diffuse in as far as the lower levels of the emulsion. Similarly, the removal of the products of fixation will take much longer than with thin layers. In the case of the development of a 60 μm layer, for instance, if it were to be immersed in developer at 20°C in the usual way for thin emulsion layers, the upper surface would be completely developed before developer had even penetrated to the lower surface. For emulsions not more than about 30 μm thick, it is usually sufficient to increase the time of development relative to the time of penetration, by lowering the temperature of the developer. For thicker layers, it is better to allow developer to penetrate completely into the emulsion at 5°C, at which temperature very little development takes place, and then to warm the emulsion to permit development to proceed[23]. If this warming up is done by transferring the slides to developer at, say, 20°C, there may still be an appreciable gradient of development, as the upper layers warm up faster than those next to the glass slide. Hauser[24] has suggested a useful method of avoiding this, by diluting the developer used at 20°C. Thus, in effect, the upper surface will have had a longer period of development than the lower surface, but in a more dilute solution. Developers based on Amidol are usually recommended for thicker emulsion layers, as their rate of penetration into the emulsion appears to be higher than with other developers. Amidol has the added advantage of acting in neutral or even slightly acid media, which helps to limit the swelling and softening of the emulsion which may occur with more alkaline developers. Rechenmann and Wittendorp[18,19] have worked very successfully with developers based on ferrous/ferric oxalate.

It would take a long time to wash the developer out of the emulsion with a distilled water rinse, as is common in thin emulsion work. It is preferable to use an acid stopbath to halt development.

Fixation presents problems also. At a rough approximation, the time needed for fixation increases as the square of the emulsion thickness. Doubling the emulsion thickness in the technique for working with phosphorus-32 on p. 382, for instance, would mean taking about 16 hours over fixation instead of the stated 4 hours. Unfortunately, acid fixers, which act more rapidly than plain thiosulphate, should not be used with thick emulsion layers. Their presence in the emulsion for long periods etches and finally dissolves away many of the developed silver grains. Plain sodium thiosulphate is probably the best fixative. Its concentration is not critical, as it has a broad peak of efficiency between 25 and 35%. A large volume of fixative which is mechanically stirred, and replaced with fresh if necessary, will give the best results.

Washing after fixation must also be long and thorough.

It is a good idea to carry out all the stages of processing with the slides horizontal, and to keep the temperatures of all solutions below about 22°C. If these precautions are not observed, lateral distortions of the emulsion may take place, particularly in the fixation and subsequent washing.

THE MICROSCOPY OF THICK EMULSION LAYERS

The value of re-swelling the processed emulsion has already been mentioned. It is necessary to use transmitted light and relatively high magnifications to view these thick emulsions. Unfortunately, the objectives designed for conventional microscopy seldom have a sufficiently long working distance to permit the scanning of the lower layers of re-swollen emulsions. Most manufacturers of microscopes make special nuclear-track objectives for this purpose, such as the Leitz KS X53 and X100, both immersion lenses.

Many artefacts will be found at the surface of the emulsion, which often has small cracks and fissures produced during drying as well. These should not interfere with the examination of the rest of the emulsion, particularly if re-swelling has been carried out.

The photography of β-tracks is extremely difficult. A high magnification is usually needed to see the grains clearly, but the three-dimensional nature of the tracks makes it difficult to get more than a few adjacent grains in focus at any one time. It may be necessary to scan many slides before a convincing track pattern is found, lying in one focal plane. It may be a good idea to expose a slide deliberately for longer than the optimum period for track counting, and to dehydrate and mount it histologically in the conventional way, without re-swelling the emulsion, to increase the probability of seeing several tracks at the one focal plane. Figs. 58 and 59 illustrate the sort of picture that can be obtained with sulphur-35 and with phosphorus-32.

The problems of recognition of β-tracks have already been discussed (p. 208).

DETAILED DESCRIPTION OF TECHNIQUES

The preparation and processing of track autoradiographs for carbon-14 and sulphur-35.

(1) *Preparing the autoradiographs*
Emulsion: Ilford G5 in gel form
Safelight: Ilford "F904" or Eastman-Kodak "OC"
Darkroom conditions: Temperature 16–22°C; relative humidity, 45–50%.
Material: Sections or smears on gelatinised slides.
Equipment needed: a thermostatically controlled waterbath; one 50-ml graduated measuring cylinder, cut short at the 40-ml mark (the dipping jar); one 50-ml graduated cylinder; one 25-ml graduated cylinder; one glass rod; one pair of plastic print forceps; a levelled metal plate with provision for cooling by placing ice on it (Fig. 89).

(i) Make a mark on the 25-ml cylinder at 15 ml with a black marking pencil.

(ii) Measure into the dipping jar 4.8 ml distilled water and 0.2 ml glycerol.

(iii) Stand these two measuring cylinders in the waterbath at 43°C.

(iv) Under safelighting, transfer G5 emulsion from the stock bottle to the 50-ml cylinder, using the print forceps, until, after gentle shaking down, the emulsion fills it to 30 ml.

(v) With the 50-ml cylinder now in the waterbath, allow the emulsion to melt for 10 minutes, stirring gently with the glass rod. This stirring must be slow and gentle: one revolution per second is quite fast enough.

(vi) Holding the 25-ml cylinder so that the black mark at 15 ml can be clearly seen, pour molten emulsion into it from the 50-ml cylinder, up to the black mark.

(vii) Pour this 15 ml of molten emulsion into the dipping jar which already contains the diluting water and glycerol.

(viii) Return the dipping jar to the waterbath for 2 minutes, stirring gently to ensure complete mixing.

(ix) Place the dipping jar in a beaker of distilled water at room temperature. Dip into the emulsion a clean slide, and place it close to the safelight to check that mixing has been complete, and there are no bubbles. If the emulsion contains many bubbles, leave it to stand for 1 minute, and dip in another clean slide.

(x) When the emulsion appears satisfactorily mixed and free from bubbles, dip in the slides with biological material on them. On withdrawing each one from the dipping jar, hold it horizontal, with the section facing upwards, wipe the emulsion off the lower surface with a paper tissue, and place the slide flat on the cooled metal plate to dry. When all the slides are on the plate, and their emulsion has gelled, remove the ice, allowing the plate to come slowly up to room temperature.

(xi) Ideal drying conditions are provided by an ambient temperature of 18–20°C, a relative humidity of 40–45%, and a gentle current of air. The emulsion should be hard, and present a shiny surface, after about 45 minutes. During drying, the safelight should be off.

(xii) Continue drying for a further period of at least 4 hours in a desiccator over dried silica gel, at room temperature.

(xiii) Expose in lightproof boxes at 4°C.

(2) *Processing the autoradiographs*

(i) Make up the Amidol developer, following the procedure given on page 385. Dilute it 1 part with 2 parts distilled water, and place it in a developing dish in a thermostatically controlled waterbath at 17°C.

(ii) Under safelighting, place the slides, preferably lying horizontally, in the developer for 12–15 minutes. They should be shielded from safelight while in the developer.

(iii) Transfer the slides to a 1% solution of acetic acid for 2 minutes.

(iv) Transfer the slides to a 30% solution of sodium thiosulphate for 20 minutes, with mechanical stirring. They should clear completely at this stage.

(v) Transfer the slides to a 10% solution of sodium thiosulphate for 20 minutes, with mechanical stirring.

(vi) Wash the slides in gently running tap water for 40–60 minutes.

Staining of the biological material may then be carried out.

After staining, soak the slides in a 20% solution of glycerol in a petri dish for 20 minutes. After removing excess fluid from the surface of the emulsion, mount in glycerine jelly or in Farrant's medium.

To prepare autoradiographs of suspended sources labelled with carbon-14 or sulphur-35, prepare an emulsion layer on a gelatinised slide, as outlined above, and, after drying, pipette a few drops of a suspension of the sources in dilute, molten emulsion on to the slide. When this layer has dried, pipette a further 0.5 ml of dilute, molten emulsion on top, following a modified version of the procedure outlined below. Processing should then follow the routine outlined below for emulsion layers 60 μm thick.

For K2 emulsion which will be latensified with gold salts prior to development, the same methods of preparing emulsion layers can be followed. The recipe for the gold solution is given on p. 409. First find the optimum development time in Amidol alone, without gold treatment, and then try the effects of various times in the gold solution, followed by the time in Amidol that has been already determined. Rechenmann and Wittendorp[18,19] describe a rather more complex method of finding the optimum development routine, based on measurements of optical density. Though more complicated to describe, it should not take very long, and it identifies with more precision the best signal-to-noise ratio obtainable.

If Eastman Kodak NTB-3 emulsion is to be used instead of Ilford G5, this technique will have to be modified. Dipping in molten but undiluted NTB-3 does not appear to produce a thick enough layer of emulsion during exposure. The method of choice is a modification of the one that follows, for 60 μm thick layers of G5. The slides should lie flat on the level plate, and have molten but undiluted emulsion pipetted on them. If 4 slides 3 in. $\times$ 1 in. are uniformly coated over most of their surface with a total of 2.5–3.0 ml emulsion, this should give an emulsion layer during exposure comparable to the method above for G5, suitable for tracks from carbon-14 or sulphur-35. Drying, exposure and development should follow the routine given above.

The preparation and processing of track autoradiographs for phosphorus-32

(1) *Preparing the autoradiographs*
  Emulsion: Ilford G5 in gel form.
  Safelight: Ilford "F 904" or Eastman-Kodak "OC".
  Material: Sections or smears on gelatinised slides.
  Darkroom conditions: Temperature, 16–22°C; relative humidity, 45–50%.
  Equipment needed: A thermostatically controlled waterbath; one 50-ml graduated cylinder, cut short at the 40-ml mark (the dipping jar); one 50-ml graduated cylinder; one 25-ml graduated cylinder; one glass rod; one pair of plastic print forceps; two 250-ml beakers; several Pasteur pipettes; one fine paintbrush; about 18 inches of rubber tubing, with a short length of glass tubing at one end; one large desiccator.

In addition some device will be needed to provide a current of dry carbon dioxide, and a level surface the temperature of which can be varied between about 5 and 30°C. A suitable device is illustrated in Fig. 89. It consists of a metal plate, fitted with levelling screws, and measuring 24 in. × 7 in. At each end, a small container of rigid plastic is attached. The remaining, central portion of the plate has a removable cover, also of rigid plastic, with a small tubular opening at one end, and a small vent, situated near the roof of the cover at the other end. This cover can be connected by tubing to a drying flask filled with dry silica-gel, and thence to a closed flask of solid carbon dioxide. The cover should fit well on the plate, so that when solid carbon dioxide is allowed to sublime in the first flask, it will flow through the drying flask into the cover, filling it, and providing an appreciable current of carbon dioxide at the outlet vent.

(i) Level the drying plate with a spirit level, and place the slides on it, with the sections or smears facing upwards. Warm the plate and slides by placing copper troughs filled with hot water in the plastic containers at each end of the plate; 25–30°C is sufficient.

(ii) With a black marking pencil, make a mark at 1 ml on each of the Pasteur pipettes, and attach one of them to the length of tubing. Fill the beakers with distilled water, place the pipettes in the first beaker, and put both beakers into the waterbath at 43°C.

(iii) The emulsion should be diluted 3 parts to 1 part of water for use, with glycerol making up 1% of the final mixture. The total needed will depend on the number of slides to be covered; allow 1 ml diluted emulsion per slide, with 5 ml in addition. The figures given would be suitable for 15 slides.

(iv) Follow steps (i)–(viii) of the technique for track autoradiographs for carbon-14 and sulphur-35 described above, with the volumes adjusted if necessary, as in step (iii) above.

(v) Leaving the dipping jar with molten diluted emulsion in it, in the waterbath, take the Pasteur pipette with the tube and mouthpiece attached, and fill it to the 1-ml mark with emulsion by gently sucking at the mouthpiece.

(vi) Allow the pipette to empty on the first slide. It is important to have warmed the pipette with distilled water at 43°C before filling it, and not to delay too long before covering the slide, or the emulsion may gel in the pipette. Once the pipette is running freely, pressure should not be applied to empty it, except with great caution, otherwise bubbles will be blown over the slide. With a little practice, it is possible to pipette emulsion very accurately in this way, without making bubbles.

(vii) With the paint brush, spread the emulsion to cover the whole slide. If any bubbles have been produced, they can often be guided away from the smear or section. If only half the slide is covered with emulsion, it will be far too thick, making the fixation times unreasonably long.

(viii) Rinse the pipette several times in the second beaker of distilled water in the waterbath. This cleans it and warms it before it is filled with emulsion again to cover the next slide.

(ix) When all the slides have been covered, remove the copper troughs of hot water from the plastic containers, and replace them filled with ice. Allow the slides to stand on the cooled plate for at least 20 minutes, for the emulsion to gel.

(x) Replace the troughs of ice with warm water once more, so that the temperature of the plate rises again to room temperature. Allow the slides to dry in a gentle current of air. Ideal conditions are provided by an ambient temperature of 18–20°C and a relative humidity of 40–45%. The emulsion should be hard, and present a shiny surface, after 1.5–2 hours. During drying, the safelight should be off.

(xi) When the emulsion appears to be dry, place the cover on the plate, and allow a gentle current of dry carbon dioxide to flow over the slides for a further 1 hour.

(xii) Place the slides in lightproof boxes for exposure, and put the boxes open, with their lids beside them, in a desiccator with dry silica gel and a few pieces of solid carbon dioxide. After 2–3 hours put the lids on the boxes without removing them from the desiccator, thus filling them with dry carbon dioxide.

(xiii) Expose at 4°C.

(2) *Processing the autoradiographs*

(i) Make up the Amidol developer, following the procedure given on p. 385.

(ii) Place a shallow dish full of concentrated, fresh developer in the refrigerator in the darkroom, so that it reaches a temperature of 6°C or less.

(iii) Dilute the rest of the developer 1 part to 2 parts of distilled water, and place it in a developing dish in the waterbath at 20°C.

(iv) Under safelighting, place the slides, lying horizontally, in the dish of concentrated developer in the refrigerator for 15 minutes.

(v) Transfer the slides to the dish of developer at 20°C for 20 minutes. The slides should be shielded from safelighting while in developer.

(vi) Transfer the slides to a 1% solution of acetic acid for 15 minutes. While here, wipe the surface of each slide gently several times with a paper tissue soaked in the acetic acid.

(vii) Transfer the slides to a 30% solution of sodium thiosulphate for 1 hour, with mechanical stirring.

(viii) Transfer the slides to a second change of 30% sodium thiosulphate for 1.5 hours, with mechanical stirring. They should clear completely during this time.

(ix) Transfer the slides to a 10% solution of sodium thiosulphate for 1.5 hours, with mechanical stirring.

(x) Wash the slides in gently running tapwater for 2 hours. Staining, re-swelling in 20% glycerol solution, and mounting follow the procedure outlined on p. 381 above.

Relatively large volumes of solution are necessary in steps (v)–(ix) above: 1 litre per 6 slides would be reasonable. The slides should be kept lying horizontally throughout development, fixation, and washing.

Mechanical stirring during fixation should be gentle, and care must be taken not to raise the temperature of the solution above 25°C.

To prepare autoradiographs of suspended sources labelled with phosphorus-32, pipette 1 ml of emulsion on to a gelatinised slide, as outlined above. When this has dried, pipette on a few drops of a suspension of the labelled sources in molten, diluted emulsion. Subsequently, cover this by pipetting on a further 1 ml of emulsion. The drying plate will have to be kept cool during the pipetting of the second and third layers.

Processing should follow the procedure outlined above, except that step (vi) should last for 20 minutes, and considerably longer times will be needed in fixer. Either larger volumes of fixer in steps (vii) or (viii), or two changes at each step, should be allowed. Steps (vii) and (viii) together should take 6–6.5 hours. Step (ix) should take 2–3 hours, and step (x), 4 hours.

Even with such thick layers, there is no need to extend the period in developer. With G5 and the Amidol developer recommended here, there is a long plateau of development, so that times in stage (v) above can be varied from 15 to 40 minutes, without a significant change in the grain densities in the β-tracks. The size of the developed grains in the tracks, and the density of random background grains, will increase with longer times; the development time should be selected on the basis of the ease of recognition of the tracks under the conditions of the experiment.

If Eastman Kodak NTB-3 emulsion is to be used, follow the same technique, but pipette 1–2 ml of molten, diluted emulsion on each slide. Drying, exposure and processing should follow the same routine outlined above, except that the 1% acetic acid stopbath should be replaced by a hardening stopbath (p. 348), otherwise the emulsion layer will be lost in the stopbath and subsequent fix.

Sometimes the processed emulsion contains very many fine, dust-like particles which interfere with microscopy. This can be avoided by buffering the 30% sodium thiosulphate used for fixation. The sodium sulphite: sodium hydrogen sulphite buffer used in making up the Amidol developer (see below), is satisfactory for this purpose.

Amidol developer

Dissolve 2.2 g sodium sulphite (7 H_2O) in 100 ml water. To a further 210

386

ml water, add 0.46 ml of a solution of sodium hydrogen sulphite (specific gravity 1.34). Mix the two solutions. Add 1 g Amidol, filter, and use immediately. Amidol is photosensitive and should be stored in darkness.

REFERENCES

1 H. Levi, *Biochim. Biophys. Acta*, 7 (1951) 198.
2 B.L. Miller and F.E. Hoecker, *Nucleonics*, 8 (1951) No. 5, 44.
3 T.F. Dougherty (Ed.), *Some Aspects of Internal Irradiation*, Pergamon, London, 1962.
4 G.A. Boyd and H. Levi, *Science*, 111 (1950) 58.
5 H. Levi, *Exptl. Cell Res.*, Suppl. 4 (1957) 207.
6 H. Levi and A. Nielsen, *Lab. Invest.*, 8 (1959) 82.
7 H. Levi, *Exptl. Cell Res.*, 7 (1954) 44.
8 H. Levi, A.W. Rogers, M.W. Bentzon and A. Nielsen, *Kgl. Danske Videnskab. Selskab, Mat.-Fys. Medd.*, 33 (1963) No. 11.
9 A. Ficq, *Exptl. Cell Res.*, 9 (1955) 286.
10 A. Ficq, *Lab. Invest.*, 8 (1959) 237.
11 J. Brachet and A. Ficq, *Exptl. Cell Res.*, 38 (1965) 153.
12 A. Ficq, in J. Brachet and A.E. Mirksy (Eds.), *The Cell, Vol. 1*, Academic Press, New York, 1959.
13 C. Levinthal, *Proc. Nat. Acad. Sci. (U.S.A.)*, 42 (1956) 394.
14 C. Levinthal and C.A. Thomas, *Biochim. Biophys Acta*, 23 (1957) 453.
15 A.W. Rogers, Z. Darzynkiewicz, K. Ostrowski, E.A. Barnard and M.M. Salpeter, *J. Cell Biol.*, 41 (1969) 665.
16 Z. Darzynkiewicz, A.W. Rogers and E.A. Barnard, *J. Histochem. Cytochem.*, 14 (1967) 915.
17 A.W. Rogers and E.A. Barnard, *J. Cell Biol.*, 41 (1969) 686.
18 R.V. Rechenmann and E. Wittendorp, *J. Microscopy*, 96 (1972) 227.
19 R.V. Rechenmann and E. Wittendorp, *J. Microscop. Biol. Cell.*, 27 (1976) 91.
20 P. Dörmer and W. Brinkmann, *Histochemie*, 29 (1972) 248.
21 H. Levi and A.S. Hogben, *Kgl. Danske Videnskab. Selskab. Mat-Fys. Medd.*, 30 (1955) No. 1.
22 D. Darlington and A.W. Rogers, *J. Anat.*, 100 (1966) 813.
23 C.C. Dilworth, C.P.S. Occhialini and R.M. Payne, *Nature*, 162 (1948) 102.
24 J. Hauser, *Photographie Corpusculaire*, 2 (1959) 207.

CHAPTER 18

Autoradiography with the Electron Microscope

The development of electron microscopy has greatly widened the scope of cytology, bringing into view a range of structures below the cellular level right down to individual molecules in favourable circumstances. Inevitably, as with every other technique that produces resolution in space between biological structures, autoradiography has been used to study the distributions of particular labelled compounds against this backcloth of sub-microscopic structures.

Unfortunately, there is a considerable mismatch between the sizes of structures that can be resolved by transmission electron microscopy and the resolution of the autoradiographic method. This mismatch is inherent in the autoradiographic process, and it places limits on the usefulness of the marriage between the two techniques. While the resolving power of the present-day transmission electron microscope is 0.1–0.3 nm, the most favourable conditions of autoradiography will only permit an HD of 50–70 nm. The β particles have finite ranges in matter, and the nuclear emulsion is made up of silver halide crystals which must have a minimum diameter to be effective detectors. Given these starting points, it is difficult to see how electron microscopic autoradiography can ever approach the resolution of the microscope itself.

Two important consequences follow from this mismatch. First, the techniques of autoradiography are more demanding than at the macro-scopic or light microscopic levels. To make the autoradiographic resolution as good as possible, one is working at the limit of the technique. We shall see later that this involves a specimen as thin as possible, and a monolayer of the smallest available silver halide crystals. Since both these steps act to reduce the final grain density, prevention of latent image loss and every conceivable trick to increase the efficiency of development are necessary parts of most experiments. To take another example, the standard method of whole-body autoradiography will detect freely diffusible radioactive material, and has been modified to cope with volatile anaesthetics such as

387

chloroform. At the light microscope level, special techniques permit diffusible materials to be localised, with care and effort. At the electron microscopic level, in spite of many years of dedicated work in a number of laboratories, no really acceptable technique exists, though the first few autoradiographs which claim to show the distribution of labelled materials are beginning to appear in the literature.

The second consequence of the gulf between the resolutions of electron microscopy and of autoradiography is that the analysis of the finished autoradiographs is far more difficult than at other levels of magnification. With an HD of 70 nm, which is near the limit, half the silver grains produced by a point source will lie further than 120 nm from it, some of them as far away as 500 nm. In cytoplasm, with identifiable organelles and structures sometimes very closely packed, it is clearly difficult to allocate silver grains to their real sites of origin.

Why, then, bother with electron microscopic autoradiography? The answer is that, difficult and time-consuming and limited as it is, it remains the only way to solve many of the problems challenging biologists. An autoradiographic experiment at this level requires quite a commitment in time and effort, of which a significant proportion will be spent on analysis. It is not a quick, easy way to look at the distribution of radioactivity, and it should be reserved for situations in which the maximum information has already been obtained by simpler methods, such as scintillation counting and light microscope autoradiography. The fact remains, however, that it is the only method we have for studying the distribution of labelled molecules at the ultrastructural level. Methods of X-ray microanalysis may develop, in conjunction with the tagging of molecules with atoms of particular elements, which will be able to compete with autoradiography; but it is likely that the range of molecules that can be labelled in this way without interfering with their biological activity will be fairly limited.

The very first electron microscopic autoradiograph was not a biological specimen. Comer and Skipper, in 1954, produced some autoradiographs of α-tracks from specimens of minerals, and used electron microscopy to examine the fine particles from which the tracks came[1]. The earliest biological attempt was by Liquier-Milward[2], who published pictures of β-tracks in 1956.

Since that start, which is surprisingly recent, progress has been fast. New techniques of preparing autoradiographs evolved. In 1961, Pelc, Coombes and Budd[3] described mounting ultrathin sections on a formvar film covering a hole in a plastic slide. A thin layer of nuclear emulsion was then pipetted over the sections. After exposure and development, the film was cut out and mounted on a grid for viewing. In 1962, Caro and van Tubergen[4] published a method in which a thin film or bubble of emulsion was picked up on a wire loop. After gelling, this was placed over the sections which were already on their grids. Koehler, Muhlethaler and Frey Wyssling[5] suggested the use of a centrifuge in order to obtain suitable monolayers of emulsion; and, in the same year, 1964, Salpeter and Bachmann[6] described their flat substrate technique. In this, the sections were mounted on a microscope slide, over a layer of collodion, and the emulsion layer was applied by dipping the slide in liquid emulsion.

As these techniques evolved, new emulsions were produced more suited to the demands of electron microscopic autoradiography. Ilford's L4 emulsion, with a mean crystal diameter of 140 nm, has remained the standard product throughout this period, while a number of emulsions with smaller crystals appeared. Agfa-Gevaert NUC 307, with a mean crystal diameter of 70 nm, was described in use by Granboulan in 1963[7]. Salpeter and Bachmann[6] reported on Eastman Kodak NTE, with crystal diameters between 30 and 50 nm, the following year. Fine-grain emulsions suitable for this type of work have also been manufactured by Sakura in Japan and Niikhimfoto in the U.S.S.R. NUC 307 is no longer produced, and NTE is now being replaced by Eastman Kodak 129–01, whose characteristics were described by Salpeter and Szabo[8] in 1976.

Progress in the analysis of electron microscopic autoradiographs lagged behind the development of techniques. The first careful discussion of the factors affecting resolution came from Caro[9], in 1962. Salpeter and her co-workers[10-12] contributed a series of papers on the distribution of silver grains around standard sources of various sizes and shapes, on which our present techniques of analysis are firmly based, and introduced the concepts of the HD and HR (p. 66). The problems of applying this information to the analysis of autoradiographs complicated by crossfire between labelled structures was tackled by Williams[13] in 1969. A considerable breakthrough came in 1973, with the method proposed by Blackett and Parry[14], which has recently been improved by them[15], and by Salpeter, McHenry and Salpeter[16].

Only 12 years ago, it was a considerable achievement to have made a good series of autoradiographs at the electron microscopic level: analysis had progressed little further than recording the structures immediately under each grain. Now, powerful techniques exist which enable one to specify the most probable distribution of radioactivity for any observed grain distribution.

Of necessity, this is a very incomplete list of the many contributions to the rapid growth of electron microscope autoradiography. For further reviews, the reader is referred to Neumann[17], and to Jacob[18]. For an excellent and recent review of current techniques in autoradiography at the electron microscope level, see Williams[57].

RESOLUTION AND EFFICIENCY

The extensive work on the resolution of electron microscope autoradiography is discussed at some length in Chapter 4. It will not be re-examined here, but some of the principal conclusions will be restated briefly.

A number of factors affect resolution, of which some of the more important are the thickness of the specimen and the emulsion layer, the thickness of any materials separating specimen from emulsion, the mean crystal diameter of the emulsion and the density of the specimen. Overshadowing all these is the initial energy of the particles emitted by the particular isotope used. This factor not only contributes decisively to the final resolution, it determines the relative importance of the other factors[19].

With carbon-14, sulphur-35 and isotopes of higher energy still, the HD increases linearly with increasing specimen thickness, and nearly linearly with increasing emulsion thickness. Any intervening layer has a serious effect on the HD. So the specimen should be as thin as possible, as should any intervening layer, and the emulsion should be a packed monolayer of crystals. Other factors, including crystal diameter, are of relatively little importance.

With the much lower energies of tritium, increasing the specimen thickness is not quite so dramatic in its effects. There is virtually no difference in HD between monolayers and double layers of crystals. The thickness of the intervening layer is also less important. In contrast, a significant improvement in HD now results if the crystal diameter is reduced from 140 nm to 50 nm, and a further, very slight, improvement results from the use of fine-grained development techniques.

With iodine-125, the autoradiographic image relies mainly on extra-nuclear electrons of 3–4 keV. At this energy, there is little effect of specimen or emulsion thickness on HD, but a substantial improvement from reducing the crystal diameter. With this isotope, heavy metal staining of the section can improve the HD by about 20%, and a further improvement can be expected from fine-grained development.

In practice, then, the factors that need particular care in order to achieve the best HD most economically can be identified. Examples of the values to be expected in a number of experimental situations are given in Fig. 29.

In much the same way, the various factors affecting efficiency (which are discussed in detail in Chapter 5) vary in importance with the energy of particle.

With carbon-14 and isotopes of higher energy, increasing the thickness of the source does not affect efficiency, though clearly affecting the total radioactivity of the specimen. At this level, the ultrathin section is at infinite thinness. The efficiency is linearly related to the emulsion thickness, however. It is virtually unaffected by the thickness of any likely intervening layer or by heavy metal staining of the specimen. It depends crucially on the sensitivity of the emulsion and the efficiency of development.

The β particles of tritium are of considerably lower energy, and self-absorption begins to reduce efficiency at about 100 nm section thickness. Intervening layers of 5 nm do not significantly affect efficiency: heavy metal staining will reduce efficiency by barely detectable amounts. Emulsion sensitivity is less critical. Going from a monolayer to a double layer of crystals increases efficiency with Ilford L4 by 10–20%; with the smaller crystals of NTE the effect is bigger, of the order of 80%. The development routine still has a considerable effect on efficiency.

With iodine-125, self-absorption begins to reduce efficiency at thicknesses above about 40 nm, and heavy metal staining reduces efficiency still further by about 10%. No increase in efficiency results from doubling the monolayer of L4: the increase with NTE is about 50%.

Once again, as with resolution, it is not possible to give simple instructions on how to get the best efficiency: it all depends on the isotope. The values that can be expected in a number of different situations are given in Chapter 5 and Fig. 93.

It will be seen at once that the conditions that give the best resolution are not always those that produce the highest efficiencies. Not only that, but the requirement for a very thin specimen reduces the total radioactivity that it is often possible to get into the section. So that when it comes to choosing a technique, one frequently has to compromise, settling for a poorer resolution in order to get a sufficient number of silver grains per grid square to make a sensible analysis possible.

As a rough guide, when working with tritium and sections 100 nm thick, the lower limit for useful detection with a monolayer of L4 is 1–1.5 disintegrations per cubic micron per 10 days.

The disintegration rate can be found either by light microscope autoradiography or by scintillation counting. With NTE, the lower limit is about 6 disintegrations per cubic micron per 10 days; it is likely to be about half that figure with the new 129-01 emulsion.

Electron microscopic autoradiography would seem to be an ideal place to try techniques which would increase efficiency, such as scintillation autoradiography, since developed silver grains are at such a premium. Unlike light microscopic and whole-body work, however, there is not a large, untapped reservoir of β particles locked within the specimen by self-absorption, so that the potential gains are smaller. The difficulties of analysis are already great at the electron microscope level, and they are likely to be made impossibly greater by superimposing photons, with different HD and efficiency values, on the existing distribution of β

Authors	Emulsion	Developer	Efficiency %
Bachmann and	L4 monolayer	Microdol X	10
Salpeter, 1967	NTE monolayer	Dektol	4–5
		Gold-EAS	12–13
	NTE double layer	Dektol	8–9
Kopriwa, 1967	L4 thick	D19b	9
	monolayer	Microdol X	4
	NTE thick	D19b	3–4
	monolayer	Microdol X	0.2
Wisse and	L4 monolayer	Microdol X	10
Tates, 1968		Gold–EAS	55
Vrensen, 1969	L4 thick	D19b	36
	monolayer	Told–EAS	33
Salpeter and	L4 thick	Microdol X	10–20
Szabo, 1972	monolayer	D-19	20–30
		Gold–EAS	22–28
Fertuck and	L4 thick	Gold–EAS	26
Salpeter, 1974	monolayer	Microdol X	17
		p-Phenylenediamine	10
	NTE monolayer	Gold–EAS	7

Fig. 93. A summary of efficiency values for tritium in electron microscope autoradiography taken from the literature. Note that conditions of exposure and development are not directly comparable, even when the same developer was used: this is particularly true of gold–EA, where the developer may be made up to different formulae. As emphasized by Salpeter and Szabo (1972) efficiency varies with the final grain density. All these values must be regarded as approximate only.

particles. Two attempts have been reported. Fischer, Korr, Thiele and Werner[20] incorporated PPO into the embedding medium and claimed increases by a factor of ×3 in sections of tritium-labelled material autoradiographed with L4. In Bouteille's laboratory[21,58], sections have been mounted on scintillation plastic, to get some benefit from the β particles going away from the emulsion, and an increase of efficiency of × 1.8 found.

THE ANALYSIS OF AUTORADIOGRAPHS

The theoretical basis for this is dealt with at some length in Chapter 11. It is most unlikely that any worthwhile experiment will give useful results without a fairly rigorous statistical analysis. It is a good idea to plan the details of the experiment with the analytical procedure in mind.

The analytical methods available fall into two main groups: those for sources which are so widely spaced that there is little or no crossfire, and those for situations in which labelled structures are so closely packed that crossfire is inevitable.

Both types of analysis require you to know the HD value for the autoradiographs. If the autoradiographic system you choose is one of the standard ones, it is quite acceptable to take an HD value from the literature. A number of these values is given in Fig. 29. If, however, there is any reason to expect that your material might have a different resolution, you will need to prepare your own hot line source and determine the HD for your own system. Methods for preparing hot lines are clearly described by Salpeter and Bachmann[12].

It is usually sufficient to find the relative radioactivities of the various structures in the specimen. If, however, you wish to convert your results to disintegration rates within the specimen, the efficiency of the autoradiographic system must be found. This can be done from the hot line autoradiographs, providing the absolute radioactivity of the material making up the hot line is known.

In determining the HD and the efficiency, it is important to count silver grains out to at least 5 times the expected HD from the hot line. Counting only with a narrower band will give spuriously low values for both factors.

Having decided on an appropriate analytical method, it is vitally important to base the analysis on a really representative set of micrographs. Work out a method of sampling the autoradiographs that eliminates as far as possible decision-making while you can actually see the specimen in the microscope. If you leave it to the whim of the moment whether or not to include a particular field in the series, subjective bias becomes a distinct possibility. It is a pity to expend all the time and effort on autoradiography and analysis only to end up with results based on an unrepresentative set of micrographs. In particular, decide clearly whether or not suspected sources without silver grains over them are to be included in the series: whether comparisons of radioactivity or absolute measurements are the objective, fields without silver grains must contribute to the estimate, and it is such fields in particular that tend to be left out if your micrographs are taken in a subjectively random manner.

SELECTING A SUITABLE TECHNIQUE

At first reading, the literature seems to contain a bewildering variety of techniques. In fact these simplify down to three basic methods. Two of

394

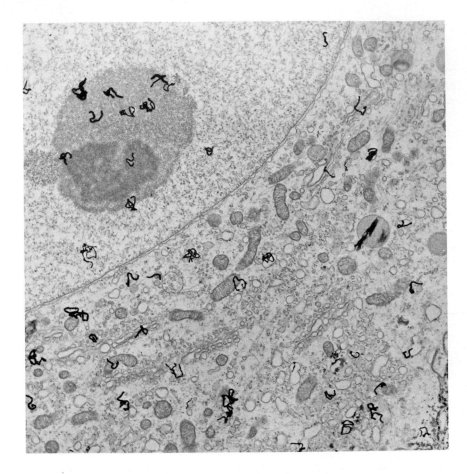

Fig. 94. An electron microscope autoradiograph of a neurone from a sensory ganglion of a newt, after incorporation of tritiated histidine. The autoradiograph was prepared on a collodion-coated slide, which was coated with a monolayer of Ilford L4 emulsion by dipping. Development was in Microdol-X. The collodion membrane was stripped off the slide after development, and attached to a grid for viewing. (× 10,800). (Material prepared by Dr. M.M. Salpeter)

these are based on the use of Ilford L4 emulsion, which is by far the most widely used one at the electron microscope level (Fig. 94). Its advantages are reliability, reasonably good sensitivity, and relative freedom from chemography and latent image fading. The same emulsion can be used at the light microscope level on semithin sections. Its principal disadvantage is the crystal diameter of 140 nm, which limits the resolution obtainable.

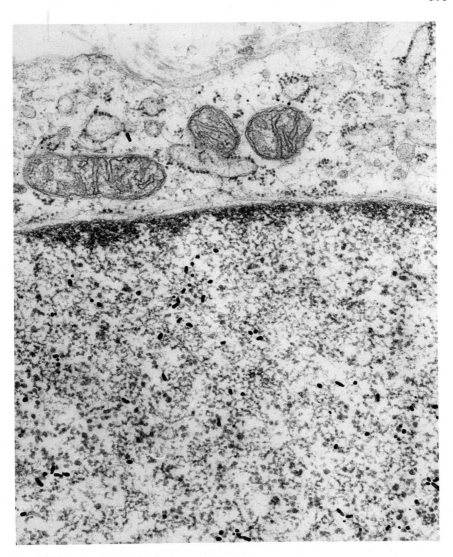

Fig. 95. An electron microscope autoradiograph of part of a mesenchymatous cell from a newt, injected with [³H]thymidine: it was prepared with a monolayer of NTE emulsion and developed for the highest resolution with gold latensification and Elon–ascorbic acid. (× 36,000). (From Bachmann and Salpeter, 1965)

The third technique involves the new Eastman Kodak product, 129-01, the successor to NTE (Fig. 95). With this emulsion the much smaller crystal diameter (30–50 nm) makes possible better resolution, particularly with tritium, iodine-125 and similar isotopes. It is early yet to know how this emulsion will perform, but the preliminary reports[8] suggest that it is a distinct improvement on NTE. Even so, the very small crystal diameter places it at the limits of emulsion technology. One would predict that it will be less sensitive than L4, less reproducible in sensitivity and more liable to latent image fading and chemography.

Eastman Kodak 129-01 is not the place to start electron microscope autoradiography for the first time. It could be argued that L4 is the emulsion of choice for each experiment initially, with 129-01 reserved for experiments which have been shown to be feasible with L4, and in which the higher resolution obtainable with 129-01 is likely to make the analysis significantly easier.

Each of the three techniques described below has evolved in a number of laboratories. While it is, of course, possible to mix them, grafting parts of one technique on to another, there are usually good reasons why each particular sequence has evolved, and such mixing will often prove disappointing.

(a) *The loop method with Ilford L4*

The loop method was first proposed by Caro and van Tubergen[4,22]. In one or another form, it is the most widely used technique at the present time, to judge from the literature. Essentially it is an attempt to prepare and expose the emulsion layer on the grid. If the specimen is already mounted on a grid, it is impossible to produce an even monolayer by dipping or pipetting emulsion over it. The only alternative is to form a suitable monolayer and allow it to gel before placing it over the specimen. The various methods that have evolved generally rely on forming this layer in a wire loop which is dipped into liquid L4 held just above its gelling temperature. Telford and Matsumura[23] proposed an expandable loop as a way of producing more uniform monolayers: this idea has been carried further by Maraldi, Biagini, Simoni and Laschi[24].

While the loop has been the standard way of making an emulsion monolayer, there is no reason why the recently devised methods of producing monolayers of L4 as home-made stripping-films[25,26] should not also be used.

The advantage of exposure on the grid is that the conventional techniques of transmission electron microscopy require very little modification. Grids are covered with a support film of collodion over which a 5–6 nm layer of carbon is evaporated. The sections are place on the grid in the usual way and dried. The grid is then attached to some sort of holder and, in the darkroom, a gelled emulsion layer is placed over it. After photographic processing and staining, the grid is immediately ready for viewing.

The method, while simple, has several weak spots. Accurate determination of section thickness by interference microscopy is not possible with the section on a grid. With the older wire loop of fixed size, it was possible to produce layers of emulsion with quite variable thickness, visible as gross swirling patterns in the emulsion layer. Even with the support film present, there is a tendency for the specimen to sag between the grid bars, with a possibility of background grains clustering around the grid bars in the finished autoradiograph. The expandable slide developed by Maraldi and co-workers seems to improve the consistency of thickness of the emulsion layer very considerably.

A number of clear descriptions of this technique have recently appeared, notably those by Fischer and Werner,[27] by Henrickson[28] and by Bouteille[29].

(b) *The dipping technique with Ilford L4*

This method was developed by Salpeter and Bachmann[6], and is often called the flat substrate method (Fig. 96). The essential feature is that the specimen is placed on a microscope slide initially rather than a grid. This provides a flat preparation which can then be covered with emulsion by dipping in liquid emulsion. The thickness and uniformity of the emulsion layer are controlled by observing the interference colours in control slides taken out of the darkroom, and by the monochrome bands produced by the same interference phenomena in emulsion layers viewed by safelighting (Fig. 97).

The flat substrate provides a number of advantages over the loop method. It is possible to check section thickness by interference microscopy before coating with emulsion, rejecting sections that lie outside a selected range. It is possible to make emulsion layers of reproducible thickness over the specimen, without any of the problems that arise from having a specimen with an uneven upper profile. The disadvantage is that the completed autoradiograph must be removed from the microscope slide and mounted on a grid. Some laboratories had great difficulty with this final stage in the early years of this technique. The conditions which favour successful removal of support film, section and processed emulsion from the slide are a bit better understood now, and there should not be significant loss of material at this stage.

Detailed recent descriptions of this method have been published by Salpeter and Bachmann[12] and by Kopriwa.[30]

(c) *Autoradiography with Eastman Kodak 129-01*

The NTE emulsion from which 129-01 has been developed, had a number of idiosyncrasies[6]. The ratio of gelatin to silver bromide was high,

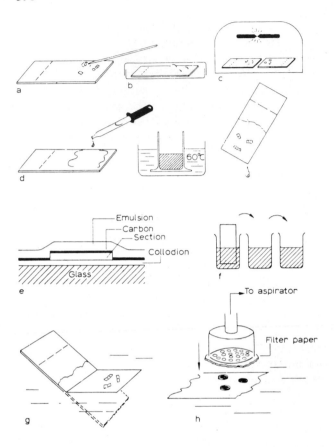

Fig. 96. The preparation of electron microscope autoradiographs by the flat substrate method. (a), The ribbons of sections are mounted on a collodion-coated slide. (b), The slides are dried in a dust-free atmosphere. (c), After staining, a layer of carbon, 5–6 nm, is applied over the sections. (d), The slide is either dipped into the molten emulsion (L4), or diluted emulsion is pipetted on to the slide and drained off again (129-01). (e), A diagram to show the various layers on the slide during exposure. (f), After exposure, the emulsion is developed and fixed. (g), The collodion membrane, carrying sections, carbon and emulsion is stripped off the slide on to the surface of distilled water. (h), Grids are placed over the sections, and the membrane with sections and grids on it is picked up from the water surface and dried, ready for examination in the electron microscope. (From Salpeter and Bachmann, 1965)

which necessitated the centrifugation of the molten emulsion to concentrate the crystals before use. The emulsion did not easily form a satisfactory layer in the loop technique, although Maraldi, Biagini, Simoni and Laschi[31] succeeded in making satisfactory layers with a specially designed modifica-

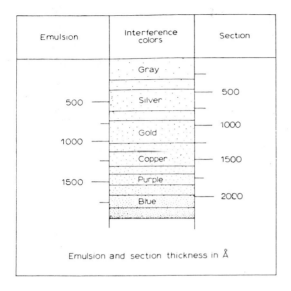

Fig. 97. A table indicating the thickness in Ångstrom units that corresponds to various interference colours of section and emulsion. A monolayer of Eastman-Kodak NTE emulsion has a silver to pale gold colour: a monolayer of Ilford L4 is purple. (From Bachmann and Salpeter, 1965)

tion of their expanding loop device. It has therefore been customary to use the flat substrate method. But because the emulsion was fairly expensive, and it was difficult to get large volumes after centrifugation, it was usually applied by pipetting a drop on to the sections and draining it off again rapidly, leaving a very thin layer over the sections.

This technique had the disadvantage of the flat substrate method — that the final autoradiograph had to be stripped off the microscope slide to be put on a grid — plus the uncertainty that the emulsion layer was reproducibly a monolayer of crystals. Hayashi and Trelstad[59] proposed a routine of preparing an even layer of collodion on a grid, and pipetting the molten, centrifuged NTE on to that. While this avoids the former problem, the latter one, of uncertain emulsion thickness, remains.

Product 129-01[8] has only just appeared on the market at the time of publication, and it is early to assess the technique that will prove the most successful. It differs in several respects from NTE, though the crystal size remains the same. The gelatin to silver bromide ratio has been changed, and the emulsion no longer requires concentration by centrifugation before use. The sensitisation process includes treatment with gold salts, so that

Eastman Kodak do not recommend gold latensification for 129-01. The technique that will be described at the end of this chapter will be based on the methods evolved for NTE, modified in the light of Salpeter and Szabo's[8] description of the new product. Clearly, other methods are likely to evolve with increasing use.

SPECIALISED TECHNIQUES

The three methods described above are suitable for studying radioactivity that is satisfactorily retained through conventional processing for transmission electron microscopy. A number of more specialised techniques have emerged in recent years for particular applications of electron microscopy.

First, what is the situation with labelled compounds that are not quantitatively retained in the tissue through fixation, embedding and sectioning? This problem is discussed at greater length in Chapter 8. Briefly, there are two lines of attack. The first is an extension to the electron microscope level of the methods of Stirling and Kinter[32]. The tissue is rapidly frozen, freeze-dried and vapour fixed. It is then impregnated with monomer which is polymerised, and the resulting block sectioned. Considerable losses of diffusible materials may occur as the section floats on the water surface behind the knife. The answer is to cut on a dry knife, a process which is easier at low temperatures. A description of the method, with illustrations of its application to the autoradiography of labelled steroids in the testis, is given by Frederick and Klepper[33]. The alternative approach is to section fresh frozen material, freeze-dry the sections, mount them on a coated grid and evaporate carbon over them, and apply an emulsion layer as a gelled film on a loop. This approach has been described by Baker and Appleton[34,35], amongst others. There are many problems with each approach, and neither is yet at the stage of providing an acceptable method for general use.

The techniques discussed in the previous section assume that the specimen is a conventional section for transmission microscopy. Techniques have been described by Maraldi, Biagini, Simoni and Laschi[31] and by Haworth and Chapman[36] for autoradiographing non-sectioned material, such as molecules or organelles. These may be negatively stained, if necessary.

Autoradiography has been combined with scanning electron microscopy by Hodges and Muir[37]. Imaging by means of the back-scattered electrons gives an interesting effect similar to that achieved by dark-field methods with incident light at the light microscope level. The same authors[38] have pointed out that X-ray spectroscopy can be used to give the mass of developed silver per unit area, acting as an expensive but effective grain counter. It is also possible to combine autoradiography with freeze-fracture techniques. Fisher and Branton[39] describe a way of achieving this marriage and comment critically on its possibilities. One of the chief problems in all these techniques is the use of a solid specimen of infinite thickness with respect to the isotopes being studied. The radioactivity from an appreciable depth within the specimen is recorded by the emulsion, but only the surface structures are visualised. Fisher and Branton[39] worked with a system that labelled only the cell membranes of isolated cells, so that radioactivity was effectively limited to the surface of the specimen.

With the difficulties of producing good contrast in autoradiographs for viewing in the transmission electron microscope, there may be real advantages in working with the scanning transmission microscope.

STAGES IN PREPARING AN ELECTRON MICROSCOPIC AUTORADIOGRAPH

A number of the steps involved in preparing an autoradiograph present problems which are common to all three techniques. These will be discussed now, before the detailed descriptions of the techniques.

(a) *Preparing a suitable section*
The choice of section thickness depends on many factors, including the radioactivity of the block and the isotope involved. The thickness and reproducibility of the sections is of considerable importance if any sensible analysis of the grain distribution is needed. It is usual to assess the thickness by observing the interference colour of the section as it floats on the water surface behind the knife edge[40,41] (Fig. 97). Williams and Meek[42] have drawn attention to the variations in section thickness that can be obtained by this method which obviously relies on the subjective judgement of the observer. If the conditions of observation and illumination vary, the same observer may award a section a different colour. Some error can also occur as the section is transferred to and flattened on the grid or microscope slide, particularly if chloroform vapour is used to spread the section, with sections of identical initial thickness ending up covering rather different areas. For really accurate work, the section thickness should be measured by interference microscopy when it is on its support. The simplest way of doing this is by incident light: the upper surface of the section reflects sufficient light for the difference in path length between this surface and the support to be accurately measured. The method involved is

described fully by Salpeter and Bachmann[12]. It is also possible, though less convenient, to measure section thickness by transmitted light interferometry. As pointed out earlier, these measurements can only be made if the flat substrate method is used.

Williams[13] has drawn attention to variations in thickness within individual sections, which can be visualised by shadowing techniques. Red blood cells, bacterial cells and bundles of collagen fibres are often raised above the general section level. Not only can the section thickness vary, but also its density, particularly after fixation in osmium and staining with heavy metals. The effects of heavy metal staining on resolution and efficiency have been discussed elsewhere (pp. 70, 93).

For many experiments, it is important that the section should be as thin as possible. It will be viewed together with a support film and a layer of processed emulsion. The combination is not ideal for micrographs of high contrast. Staining is a vital part of the process of preparing a specimen for viewing, and it is often rather difficult. There are two approaches: staining before applying the emulsion or after photographic processing. These have been discussed on pp. 141–146.

Briefly, to review them, prestaining risks introducing groups into the specimen that may react with the emulsion. Staining with uranium salts has the added hazard of introducing radioactivity into the section[43]. The standard stains of lead citrate and uranyl nitrate are to some extent removed from the section by photographic processing. Chemography and stain removal are minimised by coating the stained section with a layer of 5–6 nm of evaporated carbon. Salpeter and Bachmann[12] discuss these problems. If, at the end of all the steps of preparing the autoradiograph, the contrast is too poor, it is very difficult to introduce more stain into the section, which is lying on a support film one side and coated with carbon on the other. It is possible sometimes to dissolve away the support film and stain the section again from that side, but this is not to be recommended as a routine. If prestaining is to be used, 2–5 minutes in 2% aqueous uranyl acetate followed by 8–20 minutes in Reynold's lead citrate is suitable; alternatively, 2–3 hours in uranyl acetate alone. Coating with carbon should always follow prestaining.

Post-staining carries a fairly high risk of stain deposits in the gelatin. It cannot be very easily carried out through a carbon layer. It is a good idea to evaporate carbon on to the support film before placing the section on it, if post-staining is to be attempted. This will leave the section in direct contact with the emulsion, with a slightly higher risk of chemical interactions between them. It is particularly important to run controls against positive and negative chemography if this sequence is followed. Post-staining is usually limited to uranyl acetate. The rather alkaline lead stains may remove the gelatin of the emulsion. In fact, Revel and Hay[44] suggested this as a means of improving the contrast, but the dangers of removing or displacing silver grains at the same time have deterred most people.

Cleanliness is absolutely essential, particularly with post-staining. Stains should be freshly made up with boiled, distilled water, centrifuged and filtered several times through Millipore filters. The stains should be pipetted on to the specimen while they are lying in a petri dish: the dish should be covered to prevent evaporation. Stains should be flushed off after staining with several changes of distilled water. If, in spite of all precautions, stain deposits still appear in the gelatin while post-staining, the problem usually lies in the emulsion rather than the stain. After fixation and washing, the gelatin may contain residual thiosulphate ions, some of them complexed with silver: it may also be at a pH quite inappropriate for the stain. Try washing the emulsion thoroughly after fixation, then soaking it briefly in a solution of potassium iodide (0.1 g/litre) in distilled water. Wash several times in distilled water, then soak the emulsion for 1 minute in a 0.02 M buffer at the pH of the stain to be used.

The grids in general use in electron microscopy are copper, and this metal is well known for affecting nuclear emulsions. Since the grid is usually separated from the emulsion by a support film, a section and a layer of evaporated carbon, this may not matter. A number of workers use other grids, such as silver[43] or platinum, for all their autoradiographs, nevertheless.

The viewing of an autoradiograph in the transmission electron microscope presents few specific problems. Care should be taken to work at low beam currents, as a torn support film due to uneven heating of the specimen becomes a major tragedy at the end of the long process of preparing an autoradiograph. It is possible to evaporate off the silver grains if one is over-enthusiastic.

(b) *The emulsion layer*

The choice lies between a monolayer of emulsion and a double layer. There is some evidence that crystals may redistribute themselves on immersing an emulsion layer that has significant gaps in it in water[10]; this would add considerably to the difficulties of analysis if it occurred at the end of exposure, on placing the emulsion in developer. It is, anyhow, clear that an emulsion layer with gaps in it will not record with uniform, high efficiency. So it is important to make sure that a monolayer is a reasonably well packed one, as in Fig. 98. With L4, a minimum figure should be 45–50 crystals per square micron; with 129-01, about 400 per square micron. Vrensen[45] has drawn attention to the increase in efficiency obtained from using a slightly thicker emulsion layer, with little loss of resolution. Values of 70–80 crystals per square micron with L4 are often used in place of a simple monolayer.

Whatever thickness is selected, it is important to check that the layers of emulsion produced are in fact the desired thickness. Examine test emulsion layers in the electron microscope periodically.

As with any other emulsion, latent image fading is favoured by the presence of water and of oxidising agents. With a monolayer, every crystal is in contact with the gas in which exposure takes place. Though L4 can be exposed in dry air for up to 2 months without latent image fading[10], longer exposures are best carried out in an inert gas such as nitrogen or argon.

Fig. 98. A packed monolayer of Eastman-Kodak NTE crystals viewed in the electron microscope. Such a layer gives a silver to pale gold interference colour on the slide, and is the ideal to aim at in high resolution studies. ($\times$15,000). (Material prepared by Dr. M.M. Salpeter)

With Eastman Kodak 129-01 drying and the exclusion of oxygen are needed to prevent fading, even with shorter exposures.

A recent paper has drawn attention to the efficiency of exposure in vacuo in preventing fading and chemical desensitisation[46]. It may be that this step could be profitably used in electron microscope autoradiography, where efficiency is at such a premium and it matters to accumulate every possible developed grain.

Since chemical interaction between stained section and emulsion is known to occur, and fading of latent images is favoured by a reduction in crystal diameter, it makes sense to expose control sections of non-radioactive material alongside experimental sections, in the same way as for light microscope autoradiographs. Any increase in grain density above background levels over the control section indicates positive chemography; any reduction of grain density over the section by comparison with the emulsion away from the section, in a specimen given a hefty dose of β radiation before exposure, indicates negative chemography; any drop in overall grain density with increasing exposure time in material given β radiation indicates generalised fading of latent images.

(c) Development

A wide range of developers has been used in electron microscope autoradiography. What factors should influence one's choice?

Chemical development with D-19 or similar Metol-hydroquinone developers gives large, curly silver grains. These are perhaps a good idea if one is starting the technique for the first time, as they provide an unmistakeable signal, not easy to confuse with stain deposits or dirt. They have their disadvantages also. Being large, they tend to obscure detail in the underlying specimen: they make slightly worse the resolution obtainable (p. 72): they may make contact with adjacent crystals, producing "infectious" development of neighbours that lack latent images[47]; finally, they become very difficult to disentangle and count at high densities.

Microdol-X was a widely used developer at one time, since it tended to produce smaller grains than D-19 (Fig. 94). It has lower efficiencies than D-19, however, and is particularly liable to dose dependence, a phenomenon described by Salpeter and Szabo[48] (p. 36). It is not much used now.

Amongst the other chemical developers, Elon-ascorbic acid produces quite small grains, but also low efficiencies. Used after gold latensification, however, the efficiencies are comparable with those of D-19 (Fig. 95). Dektol is another chemical developer capable of producing small grains: it is the one recommended by Salpeter and Szabo[8] for use with 129-01, for which gold latensification is not advisable.

Physical development was first suggested for electron microscope autoradiographs by Caro and van Tubergen[4], who used para-phenylenediamine. This can give very small grains indeed and, in their hands, had an efficiency comparable to that of Microdol-X. It has proved rather disappointing in general use, unfortunately, acquiring a reputation for being fickle and not reproducible.

A number of other developing agents have been used. Fisher and Werner[27] use Kodak D76. Bouteille[29] uses phenidone after gold latensification. Heremans[49] proposed a fine-grain physical developer based on Metol. Agfa 8/66 developer[50] can be used to produce very small, dense round grains if long development times at 0°C are employed.

Kopriwa[51] reviewed the efficiency, background and grain morphology resulting from a number of emulsion–developer combinations. She has more recently[47] looked at a further series of developers used after gold latensification.

Developers that give very small silver deposits at the site of latent images can produce spuriously high efficiency values, since a crystal that has been hit by a β particle may contain several latent images, resulting in a cluster of 3 or 4 small "grains" within a small radius (Fig. 5). Such clusters should be treated as a single grain in the analysis of autoradiographs, and in calculating efficiencies[6,48,51].

DETAILED DESCRIPTIONS OF TECHNIQUES

Autoradiography of tritium with Ilford L4 emulsion by the two techniques described below should give a resolution with a HD value of about 1450 Å for a section 500 Å thick, or about 1650 Å for a section 1200 Å thick[11]. The efficiency should be between 20 and 30% (Figs. 26 and 93). This emulsion has a shelf-life of at least 3 months. The silver halide crystals have diameters for the most part between 1200 and 1400 Å. The appropriate safelight is Ilford "F 904" or Wratten "OC".

Suggested darkroom conditions are a temperature of 18–20°C and a relative humidity of 45–50%.

(a) *The loop technique*[4,22]

Take a grid which has been coated with a collodion film, over which a thin layer (50–60 Å) of carbon has been evaporated. The specimen is mounted on this film, and carefully dried. The grid is then attached by a small piece of double-coated masking tape (Scotch tape No. 400) near one end of a microscope slide: up to 4 grids can be mounted on each slide.

The day before it is used for the first time, take the 50 ml bottle of L4 into the darkroom and stand it in a waterbath at 45°C until it is molten; take it out to allow it to gel again.

When the specimens are ready for coating with emulsion, take them into the darkroom. The emulsion can be weighed out for melting and dilution, or measured out volumetrically using the procedures outlined on p. 368. If the emulsion is to be weighed, take 15 g and melt and mix it with 15 ml water in a 300-ml beaker, standing in a waterbath at 43–45°C. If it is to be measured volumetrically, take about 8 ml of gelled emulsion into a 25-ml measuring cylinder, and stand it in the waterbath until it melts: this will take about 10 minutes. Pour from this 4 ml of molten emulsion into a second measuring cylinder, and transfer this to a 300-ml beaker that holds 15 ml distilled water.

When the molten, diluted emulsion is homogeneous and smooth, transfer the beaker to an ice-bath for 1–5 minutes, then into a waterbath at 20°C. The emulsion should have the correct consistency for use from 10–20 minutes later, i.e. it should gel immediately on being picked up in the platinum loop.

The loop, which should be of thin platinum wire, should have a diameter of about 4 cm, and must be clean. Dip it in the emulsion and withdraw it slowly, picking up a thin film of emulsion over the loop, which should gel at once, giving an emulsion film which looks uniform. Take one of the slides with grids mounted on it, and touch the end nearest the grids against the film, which should fall from the loop to cover the grids. With experience, the gelling of the emulsion film can be recognised, and the transfer of the emulsion between the two waterbaths and the icebath continued to keep it at the correct consistency.

Place the emulsion-coated slides in an open slide-box in a desiccator over dried silica gel overnight. Next day, the boxes can be closed and taped, and placed in a refrigerator at 4°C to expose. If long exposures of over 3 months are anticipated, or latent image fading is a problem, the boxes should be filled with dry nitrogen, argon or carbon dioxide; in addition, a small packet of drying agent can be added before closing the box.

All processing solutions should be freshly made up and as clean as humanly possible. For a simple method of development giving large developed grains, use D-19 developer, full strength, for 2 minutes at 20°C, in a petri dish, and place the slide in the dish, emulsion up, lying horizontal. Then transfer the slide to distilled water for 30 seconds, to 1% acetic acid for 10 seconds, and to fresh distilled water for 30 seconds. Fix in Kodak rapid fixer or in buffered 30% sodium thiosulphate (p. 218) for 3 minutes. Finally, wash in 3 changes of distilled water, for at least 1 minute each.

To stain the preparations for viewing, prepare a 1% solution of uranyl acetate in distilled water: mix 70 ml of this solution with 30 ml absolute ethanol just before use. Immerse the slide in this solution for 10–45 minutes, washing the stain off afterwards with distilled water. Allow the slide to dry in a clean atmosphere.

Finally, cut the emulsion around the grids with a fine scalpel. The grids are now ready for viewing.

Reproducible emulsion layers may be more readily obtained with the expandable loop device of Telford and Matsumura[23], or the sliding device of Maraldi, Biagini, Simoni and Laschi[24]. A number of authors have described special holders for the grids, instead of microscope slides[27,53,54].

(b) *The dipping technique*[6,12] (*see Fig. 96*)

3×1 in. microscope slides should be carefully cleaned with alcohol and dried. They are then coated with a thin layer of collodion by dipping once in a 0.5% solution in amyl acetate. A ribbon of sections is floated out on the surface of a drop of water on the slide, taking care not to damage the collodion membrane in doing so. The sections should be one-third of the way from one end of the slide. The drop of water is then drained off. The section thickness can be estimated from the interference colour (Fig. 97), while the sections are still floating on water in the microtome trough.

The sections are next stained. A suggested routine is immersion for 2–5 minutes in 2% aqueous uranyl acetae, followed by 8–10 minutes in lead citrate[9]. Alternatively, 2–3 hours in uranyl acetate alone may be used. Staining is performed by placing a few drops of stain over the section, and flushing it off with distilled water after the stated period. Care must be taken to prevent evaporation of the staining solution from the slide. The slide with stained sections on it is coated with 50–60 Å of carbon. It is then ready for autoradiography.

The steps in preparing the L4 emulsion for dipping should follow the sequence outlined on p. 368. The initial dilution of the emulsion should be 1 part emulsion to 4 parts water. The diluted emulsion, in the dipping jar, should be cooled to 25°C and held at that temperature.

Taking a blank slide, dip it once in the diluted emulsion, removing it slowly and evenly, and holding it vertical. Allow the slide to dry in the vertical position, and take it out of the darkroom to examine the emulsion layer. An area of the slide, approximately one-third of its length in the bottom half of the slide, should be uniform in appearance. From the interference colour of the emulsion layer, its thickness can be estimated (Fig. 97). It will probably be thicker than the 1500 Å (purple interference colour) which represents a monolayer. The emulsion should be further diluted, if this is the case, until test slides show the correct thickness of emulsion has been achieved.

When the correct dilution of emulsion is ready, dip the slides with sections on them in identical fashion, leaving them vertical to dry.

Drying and exposure conditions have been described already on p. 407. Development can follow the steps outlined on the same page, using D-19 developer; alternatively, the following procedure can be used. This gold–Elon–ascorbic acid recipe will produce fairly large developed grains, without visualising individual latent images within each grain. It may give slightly more reliable efficiencies at different radiation doses to the emulsion than the development with D-19 (ref. 29).

Throughout this recipe, the distilled water should have been previously boiled and cooled, after adding a few drops of bromine.

A 2% stock solution of gold chloride ($AuCl_3$ HCl $3H_2O$) is made up in the distilled water: this stock solution will keep for up to one month in a plastic bottle at 4°C. Immediately before use, 1 ml of this stock solution is diluted in 100 ml boiled distilled water, and the pH is adjusted to 7.0 by adding 0.5 N NaOH dropwise. Potassium thiocyanate, 0.25 g, and potassium bromide, 0.3 g, are then added and dissolved, and the solution made up to 500 ml with distilled water. This solution should only be used between 1 and 8 hours after making.

To make the developing solution, dissolve the following, in order, in 150 ml of the boiled distilled water: Elon, 0.45 g; ascorbic acid, 1.5 g; borax, 2.5 g; potassium bromide, 0.5 g; sodium sulphite, 7.5 g. Make up the final solution to 500 ml with boiled distilled water.

The developing process should be carried out with all solutions at 20°C, using the following routine:

Gold thiocyanate solution	5 minutes
Distilled water	rinse
Elon–ascorbic acid developer	4 minutes
Non-hardening fixer	1 minute
Distilled water, × 3	30 seconds

Salpeter and Szabo[48] found a considerably lower background with this developing process than with the D-19 routine which gave a comparable efficiency.

Without allowing the slide to dry after its final rinse, place it in distilled water for 15 minutes. Then, with a scalpel or razor-blade, scrape the emulsion and supporting membrane from around the edges of the slide, and strip them together gently from the slide on to the surface of distilled water, emulsion side uppermost. While the membrane is floating, place grids gently on the emulsion over the ribbons of sections. Then pick up the membrane, with emulsion and grids, from the surface of the water. This can be done quite simply using a perforated filter plate attached to an aspirator (Fig. 96), with a filter paper over the perforations. The membrane can be sucked gently against the filter paper and removed from the water. The aspirator is then turned off, and the filter paper, with membrane attached, taken off the filter plate and placed on a flat surface to dry, in a position protected from dust. Finally, the membrane is cut around the grids, which are then ready for viewing.

A number of authors, including Kopriwa[30], and Marshall and Faulkner[55], claim greater reproducibility of emulsion thickness from the use of a semi-automatic device for withdrawing the slide from the emulsion at constant speed.

Difficulties may occur with stripping the processed emulsion layer, section and support film off the slide. These are discussed in some detail by Salpeter and Bachmann[12]. First, it is clear that very careful cleaning of the slide, e.g. in chromic acid, before applying the collodion may etch the glass surface and make stripping very difficult indeed. It is enough that the slide should be "socially clean". The thinner the collodion layer, the better will be the contrast of the completed autoradiograph in the electron micros-cope. On the other hand, thicker layers may be easier to strip off. The collodion concentration may be increased to 0.7 or even 0.8% if difficulty is experienced with stripping. Alternatively, Crefeld[56] has suggested dipping the end of the slide a second time in 0.5% collodion to give a strengthened zone to start the separation of support film from glass.

If the membrane fails to float off the slide, return the slide to distilled water for a further 10 minutes and try again. This process can be repeated as long as one's patience lasts without damaging the specimen.

Cutting around the section with a scalpel may be best avoided, as it can stick the membrane to the slide along the line of the cut. The slide should never be allowed to dry out between processing and stripping, as this also makes stripping more difficult.

If all these items of folk lore fail and your temper is running out, score around the area carrying the sections with a sharp needle, and very carefully place a single drop of a 1 in 100 dilution of 40% hydrofluoric acid in distilled water on the score mark. Be careful to avoid spreading the acid over the upper surface of the slide towards the sections. After 2–3 minutes, rinse the acid off in several changes of distilled water. This procedure etches the glass from beneath the edges of the membrane, and nearly always succeeds in starting off the separation of collodion from glass.

(c) *Autoradiography with Eastman Kodak 129-01*[8]

The shelf-life of this emulsion appears to be around 2–3 months. The halide crystals are 300–500 Å in diameter. It is sensitive to low-energy particles only and, while useful for tritium and iodine-125, is unlikely to be sensitive enough for isotopes of higher energy. The appropriate safelight is Wratten "OA".

With the technique to be described, the predicted resolution has an HD value of 800 Å for a section 500 Å thick, and 1000 Å for a section 1200 Å thick. Exposure times should not exceed 12 weeks, as background may increase with longer exposures. Efficiencies should be about 14% with tritium and 38% with iodine-125.

Specimen preparation should follow the sequence outlined above for dipping in L4, except that the sections should have a grey or silver interference colour, equivalent to a thickness of 350–500 Å.

Take 1 g of 129-01 and add it to 2 ml of distilled water in a 10-ml measuring cylinder. Stand the cylinder in a waterbath at 50°C and stir gently. When the emulsion is uniformly molten and diluted, it is ready for use.

Take a test slide and, with a medicine dropper, place 3 drops of the molten diluted emulsion on it, pour them off back into the graduated cylinder immediately, and stand the slide upright to dry for a few minutes. The thickness of the emulsion layer can then be checked by taking the slide out of the darkroom and noting the interference colour in the area where sections are situated on the experimental slides. A monolayer should have a silver to pale-gold colour (Fig. 97). If the emulsion is too thick, it should be diluted empirically until the correct emulsion thickness is obtained. It is advisable to check the packing and thickness of silver halide crystals in the emulsion layer by direct observation in the electron microscope, until the procedure has become established.

When test slides appear to be satisfactory, as judged by their interference colour, cover the experimental slides in the same way, and leave them for 30 min in a vertical position to dry. Drying and exposure should follow the steps given on p. 407, except that exposure with a drying agent in inert gas is now essential.

Development should be in Dektol for 2 minutes at 24°C. Following development, the slides are rinsed, fixed and washed as described on p. 409. Stripping the emulsion and its support off the microscope slide in preparation for viewing follows the same procedure as for Ilford L4.

REFERENCES

1 J.J. Comer and S.J. Skipper, *Science*, 119 (1954) 141.
2 J. Liquier-Milward, *Nature*, 177 (1956) 619.
3 S.R. Pelc, J.D. Coombes and C.C. Budd, *Exptl. Cell Res.*, 24 (1961) 192.
4 L.G. Caro and R.P. van Tubergen, *J. Cell Biol.*, 15 (1962) 173.

412

5 J. Koehler, K.K. Muhlethaler and A. Frey Wyssling, *J. Cell Biol.*, 16 (1963) 73.

6 M.M. Salpeter and L. Bachmann, *J. Cell Biol.*, 22 (1964) 469.

7 P. Granboulan, *J. Roy. Microscop. Soc.*, 81 (1963) 165.

8 M.M. Salpeter and M. Szabo, *J. Histochem. Cytochem.*, 24 (1976) 1204.

9 L.G. Caro, *J. Cell Biol.*, 15 (1962) 189.

10 L. Bachmann and M.M. Salpeter, *Lab. Invest.*, 14 (1965) 1041.

11 M.M. Salpeter, L. Bachmann and E.E. Salpeter, *J. Cell Biol.*, 41 (1969) 1.

12 M.M. Salpeter and L. Bachmann, in M.A. Hayat (Ed.), *Principles and Techniques of Electron Microscopy, Vol. II*, Van Nostrand Reinhold, New York, 1972.

13 M.A. Williams, *Adv. Opt. Elect. Microscop.*, 3 (1969) 219.

14 N.M. Blackett and D.M. Parry, *J. Cell Biol.*, 57 (1973) 9.

15 N.M. Blackett and D.M. Parry, *J. Histochem. Cytochem.*, 25 (1977) 206.

16 M.M. Salpeter, F.A. McHenry and E.E. Salpeter, *J. Cell Biol.*, 76 (1978) 127.

17 D. Neumann, *Acta Histochem.*, 33 (1969) 217.

18 J. Jacob, *Int. Rev. Cytol.*, 30 (1971) 91.

19 M.M. Salpeter, H.C. Fertuck and E.E. Salpeter, *J. Cell Biol.*, 72 (1977) 161.

20 H.A. Fischer, H. Korr, H. Thiele and G. Werner, *Naturwissenschaften*, 58 (1971) 101.

21 L.-A. Buchel, *Proc. 5th Int. Congress Histochem. Cytochem.*, Bucharest, 1976, p. 397.

22 L.G. Caro, *J. Cell Biol.*, 41 (1969) 918.

23 J.N. Telford and F. Matsumura, *Stain Technol.*, 44 (1969) 259.

24 N.M. Maraldi, G. Biagini, P. Simoni and R. Laschi, in D.M. Prescott (Ed.), *Methods in Cell Physiology, Vol. V*, Academic Press, New York, 1972.

25 J.R. Williamson and H. van den Bosch, *J. Histochem. Cytochem.*, 19 (1971) 304.

26 R.W. Burry and R.S. Lasher, *J. Microscopy*, 104 (1975) 307.

27 H.A. Fischer and G. Werner, *Histochemie*, 29 (1972) 44.

28 A. Henrickson, *Brain Res.*, 85 (1975) 241.

29 M. Bouteille, *J. Microscop. Biol. Cell.*, 27 (1976) 121.

30 B.M. Kopriwa, *Histochemie*, 37 (1973) 1.

31 N.M. Maraldi, G. Biagini, P. Simoni and R. Laschi, *Histochemie*, 35 (1973) 67.

32 C.E. Stirling and W.B. Kinter, *J. Cell Biol.*, 35 (1967) 585.

33 P.M. Frederick and D. Klepper, *J. Microscopy*, 106 (1976) 209.

34 T.C. Appleton, *J. Microscopy*, 100 (1974) 49.

35 J.R.J. Baker and T.C. Appleton, *J. Microscopy*, 108 (1976) 307.

36 R.A. Haworth and J.A. Chapman, *J. Microscopy*, 106 (1976) 125.

37 G.M. Hodges and M.D. Muir, *Nature*, 247 (1974) 383.

38 G.M. Hodges and M.D. Muir, *J. Microscopy*, 104 (1975) 173.

39 K.A. Fisher and D. Branton, *J. Cell Biol.*, 70 (1976) 453.

40 L.D. Peachey, *J. Biophys. Biochem. Cytol.*, 4 (1958) 233.

41 L. Bachmann and P. Sitte, *Mikroskopie*, 13 (1958) 289.

42 M.A. Williams and G.A. Meek, *J. Roy. Microscop. Soc.*, 85 (1966) 337.

43 H.A. Fischer, *Brain Res.*, 85 (1975) 237.

44 J.P. Revel and E.D. Hay, *Exptl. Cell Res.*, 25 (1961) 474.

45 G.F.J.M. Vrensen, *J. Histochem. Cytochem.*, 18 (1970) 278.

46 W.C. Lewis and T.H. James, *Phot. Sci. Eng.*, 13 (1969) 54.

47 B.M. Kopriwa, *Histochemistry*, 44 (1975) 201.

48 M.M. Salpeter and M. Szabo, *J. Histochem. Cytochem.*, 20 (1972) 425.

49 H.A. Heremans, *Histochemie*, 25 (1971) 123.

50 G. Muller, *Exptl. Cell Res.*, 65 (1971) 386.

51 B.M. Kopriwa, *J. Histochem. Cytochem.*, 15 (1967) 501.

52 H. Weber, *Acta Biol. Med. Ger.*, 22 (1969) 159.

53 J.-I. Hiraoka, *Stain Technol.*, 47 (1972) 297.
54 J. Hubert and J. Bohatier, *Stain Technol.*, 50 (1975) 60.
55 J. Marshall and D. Faulkner, *J. Microscopy*, 109 (1977) 227.
56 W.H. Crefeld, *Histochemie*, 32 (1972) 281.
57 M.A. Williams in A.M. Glauert (Ed.), *Practical Methods in Electron Microscopy, Vol. 6*, North-Holland, Amsterdam, 1977.
58 L.A. Buchel, E. Delain and M. Bouteille, *J. Microscopy*, 112 (1978) 223.
59 K. Hayashi and R.L. Trelstad, *J. Histochem. Cytochem.*, 51 (1976) 68.

APPENDIX

Useful data

Avogadro's number = 6.025×10^{23}

1 curie = 3.7×10^{10} disintegrations per second

= 2.22×10^{12} disintegrations per minute

1 microcurie = 3.2×10^{9} disintegrations per day

1 day = 8.64×10^{4} second

$\pi = 3.1416$

$\varepsilon = 2.7183$

At a specific activity of 1 mCi/mmole, 1 disintegration per day is given by 1.88×10^{8} molecules.

At 1μCi/g, $1\,000\ \mu\text{m}^3$ of tissue will give approximately 3.2 disintegrations per day.

The relations between the mean track length (L) in μm, the initial energy of a β particle (E) in keV, and the mean number of grains in its track in G5 emulsion (G) are given by the following equations.

$\log L = 1.59 \qquad \log E - 1.51$

$\log G = 1.19 \qquad \log E - 0.74$

$\log G = 0.747 \qquad \log L + 0.385$

THE OPTIMAL ALLOCATION OF EFFORT IN GRAIN COUNTING BETWEEN THE LABELLED SOURCES AND BACKGROUND

To use these charts, first obtain a rough estimate of the ratio of counts over the labelled sources to be studied to counts over background: this ratio is the abscissa, ρ. The values of K indicate the number of similar labelled sources which will be examined: other values of K can be interpolated. The ordinates show the optimal number of grains to be counted for three stated values of the coefficient of variation (CV). Use graph (a) to determine the number of background grains to count, and graph (b) for the number of grains over the sources.

To use these charts for photometric estimations of grain density, or any other parameter of emulsion response, it is necessary first to convert the readings from arbitrary units into the corresponding number of grains. (From England and Miller, *J. Microscopy*, 92 (1970) 167)

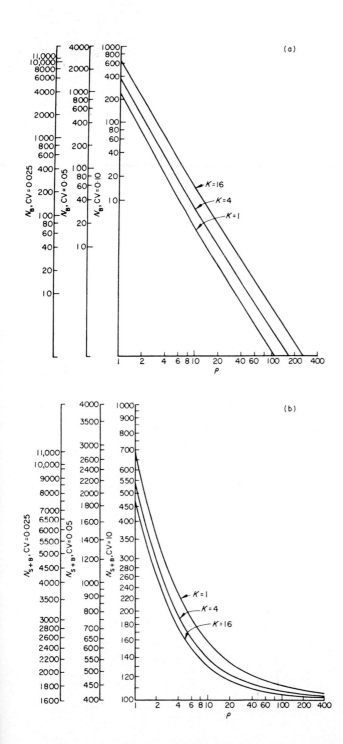

THE MEAN TRACK LENGTHS OF β PARTICLES IN ILFORD G5 EMULSION (DENSITY 3.815)

Initial energy (keV)	Track length (μm)	Intial energy (keV)	Track length (μm)
10	1.0	450	517
20	2.9	500	600
30	6.0	550	685
40	10.0	600	775
50	14.7	650	860
60	20.2	700	940
70	26.3	750	1035
80	33.2	800	1120
90	40.5	850	1210
100	48.5	900	1310
120	65.7	1000	1490
150	96.2	1200	1840
170	117	1400	2215
200	152	1600	2590
250	215	1800	2950
300	286	2000	3300
350	362	2200	3640
400	438	2400	4000

Values taken from range-energy curves of P. Demers, *Ionographie.*

PHYSICAL DATA ON ISOTOPES IN COMMON USE FOR AUTORADIOGRAPHY

Isotope	Half-life	Particle	Particle energy (keV)	γ-Rays
Calcium-45	165 days	β	250	None
Carbon-14	5,760 years	β	155	None
Chlorine-36	3.03×10^5 years	β	714	None
Chromium-51	27.8 days	E-N electron		Present
Cobalt-57	270 days	E-N electron		Present
Iodine-125	60 days	E-N electron		X-rays present
Iodine-131	8 days	β	250 (3%) 330 (9%) 610 (87%) 810 (1%)	Present
Iron-55	2.7 years	E-N electron		X-rays present
Iron-59	45 days	β	130 (1%) 270 (46%) 460 (53%) 1,560 (0.3%)	Present
Phosphorus-32	14.2 days	β	1,710	None
Sodium-22	2.6 years	Positron	540 (90%)	Present
Strontium-90	28 years	β	540	None
Yttrium-90	64.2 hours	β	2,250	None
Sulphur-35	87 days	β	167	None
Tritium (Hydrogen-3)	12.3 years	β	18.5	None

Note that those isotopes with extra-nuclear (E-N) electrons often have a complex decay scheme with a high percentage of electrons of very low energy. For instance, iodine-125 emits about 1.67 electrons per decay; electrons occur at the following energies, 2.77 keV (27.7%), 3.6 keV (48.8%), 22.5 keV (14.2%), 31.0 keV (6.7%) and 34.3 keV (1.2%)

EMULSIONS IN GENERAL USE IN AUTORADIOGRAPHY

Manufacturer	Crystal diameter (μm)	Sensitivity						Availability
		0	1	2	3	4	5	
Ilford	0.27						G5	a, b, d
Ilford	0.20	K0	K1	K2			K5	a, b, c
Ilford	0.14					L4		a, b, d
Eastman-Kodak	0.34						NTB-3	a
Eastman-Kodak	0.29			NTB				a
Eastman-Kodak	0.26				NTB-2			a
Eastman-Kodak	0.06			129–01				a
Kodak (U.K.)	0.20			AR-10				c

Availability: a, emulsion in gel form
               b, coated plates
               c, stripping-films
               d, emulsion layer on polyester film base

In addition, nuclear emulsions are available from the following suppliers for use within the country of origin only.

Japan:
1. Sakura NR-M2 and NR-H2 emulsions, from Konishiroku Photo Industry Co. Ltd., 3-1, Muromachi, Nihonbashi, Chnoko, Tokyo.
2. Fuji Et-2F; Fuji Film Co. Ltd., 2-26 Nishi-Azabu, Minatoku, Tokyo.

U.S.R.R.
1. A range of emulsions is available from NiiChim Photo Leningradsky Prospekt 47, Moscow. The characteristics of these emulsions is described in the following two sources.

   M.F. Merkulov and J.V. Kortukov, *J. Nauc. Prikl. Fotografii Kinematografii, 16* (1971) 326–335.

   O.I. Epifanova, *Radioautography*, Wysschaya Shkola, Moscow, 1977.

Index*

* EM = electron microscope; ARG = autoradiograph.

421

Development (continued)
–, chemical, 19–26, 32–35
–, colour, 38
–, effect of agitation, 25, 26, 32, 300
–, effect of concentration, 26, 32, 301
–, effect of temperature, 25, 26, 32, 33, 301, 378
–, effect on chemography, 172, 173
–, effect on reflectance, 219–221
–, fine grain techniques, 27, 405, 406, 411
–, kinetics of, 17–26, 33, 61, 62, 230, 385
–, latent image intensification and, 33–35, 79, 99–100, 104, 113, 173, 405, 406, 409
–, microscopy and, 21–25, 32, 33, 193, 300, 301
–, optimal conditions of, 21–25, 32, 33, 193, 300, 301, 314
–, physical, 27, 106, 405, 406
–, plateau, 25, 26, 385
–, radioactive grains, 38
–, reversal, 38
–, sensitivity and, 53, 94, 95, 103–106
–, temperature cycle, 378, 384
–, track ARGs, 379–381, 384, 385
Dichroic fog, 218
Direct deposition ARG, 37
Double hits, 96, 97, 202
Double isotope experiments, 305–309
–, double emulsion, 306–309
–, single emulsion, 305–306
Drying of emulsion, 17, 31, 356–358, 361–363
–, optimal conditions of, 266, 298, 325, 356–358, 361–363

Eastman-Kodak NTB emulsions, 295, 296, 315, 316, 364, 366–368, 420
–, dipping techniques, 353, 367, 368
–, drying of, 116, 297, 354, 361, 363, 367
–, exposure conditions, 362, 363, 367
–, latent image fading, 266, 361–363, 368
–, NTB-2, 92, 122, 297, 299, 307, 353, 354, 364, 366–368
–, NTB-3, 122, 297, 307, 353, 354, 364, 366–368, 376, 385
–, safelighting, 112, 113, 167, 297, 353, 367
–, shelf life, 373
–, tritium, efficiency with, 92, 364
Eastman-Kodak NTE emulsion, 79–83, 103–106, 294, 295, 389, 391, 392, 396–400, 403–405, 410, 411
–, efficiencies with, 103–106
–, latent image fading, 104, 105

–, resolution with, 79–83, 389
–, sensitivity of, 294, 389
–, variability of, 293, 389
Eastman-Kodak 129-01 emulsion, 103–106, 294, 295, 389, 395, 396, 397–400, 403–405, 410, 411
–, efficiencies with, 103–106
–, latent image fading, 104, 105
–, resolution with, 389
–, technique for, 401, 411
Efficiency, 10, 14, 57, 79, 87–110, 229, 244, 265
–, crystal size and, 94, 100, 101, 103, 185, 194, 315
–, definition of, 97, 100, 233, 234, 314, 373
–, development and, 98–100, 103–106, 230, 255, 405, 406
–, EM ARGs, 79, 103–106, 274–277, 390–392, 405, 406
–, emulsion sensitivity and, 94, 95, 100–106
–, emulsion thickness and, 93, 100–106, 109, 230–232, 255, 315, 352, 359–361
–, estimates of, 93, 100–106, 109, 317–322, 340–342, 363–366, 373, 374, 390–392
–, exposure time and, 96, 97, 100–103, 230
–, grain density ARGs, 101–103, 340–342, 363–366
–, increase by backscattering, 57, 315
–, latent image fading and, 95, 96, 107, 108, 265–269, 295
–, macroscopic ARGs, 88, 100, 101, 313–315, 317–322
–, particle energy and, 89–94, 100–103, 107
–, radiation dose and, 36, 97, 405
–, resolution and, 109, 294, 295, 387, 390–392, 395–397
–, source–emulsion geometry and, 89, 103–106, 109, 146, 233, 280, 315, 329, 330, 373, 374
–, source density and, 92, 93, 233, 390–392
–, source thickness and, 89–93, 101–103, 109, 163, 176, 177, 232, 233, 314, 318, 402
–, track ARGs, 88, 107–109, 373, 374
–, variations in source shape, 237, 244, 245, 373, 381
–, variations in source size, 233–238, 243, 244, 373, 381
Electron, 43–46, 50, 51
–, extra-nuclear, 57, 58
–, secondary, 47, 50, 52, 57, 58
Electron microscopy, 6, 76, 387
–, section thickness, 140, 141